T0211310

Communications in Computer and Information Science 615

Commenced Publication in 2007
Founding and Former Series Editors:
Alfredo Cuzzocrea, Dominik Ślęzak, and Xiaokang Yang

Editorial Board

More information about this series at http://www.springer.com/series/7899

Guntis Arnicans · Vineta Arnicane
Juris Borzovs · Laila Niedrite (Eds.)

Databases and Information Systems

12th International Baltic Conference, DB&IS 2016
Riga, Latvia, July 4–6, 2016
Proceedings

Springer

Editors

Guntis Arnicans
University of Latvia
Riga
Latvia

Vineta Arnicane
University of Latvia
Riga
Latvia

Juris Borzovs
University of Latvia
Riga
Latvia

Laila Niedrite
University of Latvia
Riga
Latvia

ISSN 1865-0929   ISSN 1865-0937 (electronic)
Communications in Computer and Information Science
ISBN 978-3-319-40179-9   ISBN 978-3-319-40180-5 (eBook)
DOI 10.1007/978-3-319-40180-5

Library of Congress Control Number: 2016940351

Printed on acid-free paper

This Springer imprint is published by Springer Nature
The registered company is Springer International Publishing AG Switzerland

Preface

This volume contains a selection of the papers presented at the 12th International Baltic Conference on Databases and Information Systems 2016 (DB&IS 2016). The conference was held during July 4–6, 2016, in Riga, Latvia. DB&IS 2016 continued the DB&IS series of biennial conferences, which have been held in Trakai (1994), Tallinn (1996, 2002, 2008, 2014), Riga (1998, 2004, 2010), and Vilnius (2000, 2006, 2012).

During this period, the DB&IS conference has become an international forum for researchers and developers in the field of databases, information systems, and related areas. The conference features original research and application papers on the theory, design, and implementation of today's information systems on a set of themes and issues.

The volume of *Communications in Computer and Information Science* series by Springer marks a new step for the conference. Having the proceedings published by Springer has raised the bar for our authors and will contribute to our on-going aim of improving the quality of the event.

DB&IS 2016 was organized by the Faculty of Computing, University of Latvia. The international Program Committee had 56 representatives from 21 countries all over the world. This year, 62 submissions from 16 countries were received. At least three reviewers evaluated each conference paper applying the single-blind type of peer review. As a result, 25 papers were accepted as full papers for publication in the present volume. The conference program was extended with several good talks, and additional papers were accepted for publication by the *Baltic Journal of Modern Computing* (BJMC). The conference was also accompanied by a doctoral consortium.

The selected papers span a wide spectrum of topics related to the development of information systems and data processing. The first significant topic covered is the development of ontology applications. The contributions address research on ontology visualization, ad hoc querying using controlled natural language, data mapping from one representation to another, as well as creating and exploiting concept maps in teaching and learning. The second topic is traditional for the DB&IS conferences — tools, technologies, and languages for model-driven development. There are papers that describe different kinds of models, metamodels, model transformations, tools for model-driven development, and business process modeling. The third group of papers addresses issues related to decision support systems and data mining, starting from specific novel techniques and finishing with domain-oriented systems that discover knowledge and improve human life. Natural language processing and building linguistic components for information systems are a rapidly developing research area promising to lead to excellent practical outcomes. This volume includes papers that deal with natural language processing and propose different usage of results. The other papers are related to advanced systems and technologies such as self-management of IS, the semantic-driven design of IS, human–computer interaction, improvement of IS performance, automatic usability evaluation, model-driven testing of real-time distributed systems, development of adaptive e-learning systems, and knowledge management performance measurement.

We would like to express our warmest thanks to all authors who contributed to the 12th International Baltic Conference on Databases and Information Systems 2016. Our special thanks to the invited speakers, Prof. Andris Ambainis, Prof. Gintautas Dzemyda, and Prof. Jaak Vilo, for sharing their knowledge. We are very grateful to the members of the international Program Committee and additional referees for their reviews and comments. We are grateful to the presenters, session chairs, and conference participants for their time and effort that made the DB&IS 2016 success. We also wish to express our thanks to the conference organizing team, the University of Latvia, Exigen Services Latvia, the IEEE, the IEEE Latvia Section, and other supporters for their contribution and making the event possible. Last, but not least, we would like to thank Springer for their excellent cooperation during the publication of this volume.

July 2016

Vineta Arnicane
Guntis Arnicans
Juris Borzovs
Laila Niedrite

Organization

The 12th International Baltic Conference on Databases and Information Systems (DB&IS 2016) took place during July 4–6, 2016 in Riga, Latvia, and was organized by the Faculty of Computing, University of Latvia.

General Chair

Juris Borzovs — University of Latvia, Latvia

Honorary Co-chairs

Janis A. Bubenko — Royal Institute of Technology and Stockholm University, Sweden

Arne Sølvberg — Norwegian University of Science and Technology, Norway

Program Co-chairs

Guntis Arnicans — University of Latvia, Latvia
Laila Niedrite — University of Latvia, Latvia

Organizing Chair

Vineta Arnicane — University of Latvia, Latvia

Finances Chair

Visvaldis Neimanis — University of Latvia, Latvia

Publicity Chair

Jeļena Poļakova — University of Latvia, Latvia

Information and Registration Chair

Darja Solodovņikova — University of Latvia, Latvia

Infrastructure Chair

Inga Medvedis — University of Latvia, Latvia

Steering Committee

Guntis Arnicans	University of Latvia, Latvia
Juris Borzovs	University of Latvia, Latvia
Albertas Čaplinskas	Vilnius University Institute of Mathematics and Informatics, Lithuania
Jānis Grundspeņķis	Riga Technical University, Latvia
Hele-Mai Haav	Institute of Cybernetics at Tallinn University of Technology, Estonia
Ahto Kalja	Tallinn University of Technology, Estonia
Mārīte Kirikova	Riga Technical University, Latvia
Audronė Lupeikienė	Vilnius University Institute of Mathematics and Informatics, Lithuania
Tarmo Robal	Tallinn University of Technology, Estonia
Olegas Vasilecas	Vilnius Gediminas Technical University, Lithuania

Program Committee

Guntis Arnicans	University of Latvia, Latvia (Co-chair)
Irina Astrova	Tallinn University of Technology, Estonia
Mikhail Auguston	Naval Postgratuate School, USA
Liz Bacon	CMS, Greenwich University, UK
Jānis Bārzdiņš	University of Latvia, Latvia
Andras Benczur	Eötvös Loránd University, Hungary
Jānis Bičevskis	University of Latvia, Latvia
Uldis Bojārs	University of Latvia, Latvia
Juris Borzovs	University of Latvia, Latvia
Christine Choppy	LIPN of the University of Paris 13, France
Albertas Čaplinskas	Vilnius University, Lithuania
Kārlis Čerāns	University of Latvia, Latvia
Valentina Dagienė	Vilnius University, Lithuania
Robertas Damaševičius	Kaunas University of Technology, Lithuania
Gintautas Dzemyda	Vilnius University, Lithuania
Dale Dzemydiene	Mykolas Romeris University, Lithuania
Vladislav V. Fomin	Vytautas Magnus University, Lithuania
Olga Fragou	Computer Technology Institute, Greece
Flavius Frasincar	Erasmus University Rotterdam, The Netherlands
Jānis Grundspeņķis	Riga Technical University, Latvia
Hele-Mai Haav	Tallinn University of Technology, Estonia
Andreas Harth	AIFB, Karlsruhe Institute of Technology, Germany
Mirjana Ivanovic	University of Novi Sad, Serbia
Hannu Jaakkola	Tampere University of Technology, Finland
Ahto Kalja	Tallinn University of Technology, Estonia
Audris Kalniņš	University of Latvia, Latvia
Mārīte Kirikova	Riga Technical University, Latvia

Dmitry Korzun	Petrozavodsk State University, Russia
Dalia Kriksciuniene	Vilnius University, Lithuania
Peep Küngas	University of Tartu, Estonia
Rein Kuusik	Tallinn University of Technology, Estonia
Mart Laanpere	Tallinn University, Estonia
Marion Lepmets	Regulated Software Research Centre, Dundalk Institute of Technology, Ireland
Audrone Lupeikiené	Vilnius University, Lithuania
Timo Mäkinen	Tampere University of Technology, Finland
Saulius Maskeliunas	Vilnius University, Lithuania
Raimundas Matulevičius	University of Tartu, Estonia
Jurijs Merkurjevs	Riga Technical University, Latvia
Emma Chávez Mora	The Catholic University of the Most Holy Conception, Chile
Nazmun Nahar	University of Jyvaskyla, Finland
Laila Niedrite	University of Latvia, Latvia (Co-chair)
Boris Novikov	St.-Petersburg University, Russia
Vladimir A. Oleshchuk	University of Agder, Norway
Algirdas Pakštas	London Metropolitan University, UK
Jaan Penjam	Institute of Cybernetics, Estonia
Ivan I. Piletski	Belarusian State University of Informatics and Radioelectronics, The Republic of Belarus
Jaroslav Pokorny	Charles University in Prague, Czech Republic
Tarmo Robal	Tallinn University of Technology, Estonia
Gunter Saake	University of Magdeburg, Germany
Jurģis Šķilters	University of Latvia, Latvia
Julius Štuller	Academy of Sciences of the Czech Republic, Czech Republic
Kuldar Taveter	Tallinn University of Technology, Estonia
Jaak Tepandi	Tallinn University of Technology, Estonia
Olegas Vasilecas	Vilnius Gediminas Technical University, Lithuania
Tatjana Welzer	University of Maribor, Slovenia
Robert Wrembel	Poznan Unviersity of Technology, Institute of Computing Science, Poland

Doctoral Consortium Co-chairs

Uldis Bojārs	University of Latvia, Latvia
Audronė Lupeikienė	Vilnius University, Lithuania
Raimundas Matulevičius	University of Tartu, Estonia

Additional Reviewers

Yury Chizhov	Riga Technical University, Latvia
Arnis Kiršners	Riga Technical University, Latvia
Arnis Lektauers	Riga Technical University, Latvia

Jevgeni Marenkov	Tallinn University of Technology, Estonia
Robertas Matusa	Vytautas Magnus University, Lithuania
Alex Norta	Tallinn University of Technology, Estonia
Enn Õunapuu	Tallinn University of Technology, Estonia
Petri Rantanen	Tampere University of Technology, Finland
Eeli Saarinen	Turku University, Finland
Jose Ignacio Abreu Salas	The Catholic University of the Most Holy Conception, Chile
Uldis Straujums	University of Latvia, Latvia
Ermo Täks	Tallinn Technical University, Estonia

Local Organizing Committee

Ella Arša	University of Latvia, Latvia
Mārtiņš Balodis	University of Latvia, Latvia
Ingrīda Cinkmane	University of Latvia, Latvia
Rudīte Ekmane	University of Latvia, Latvia
Anita Ermuša	University of Latvia, Latvia
Natālija Kozmina	University of Latvia, Latvia
Dace Mileika	University of Latvia, Latvia
Inga Medvedis	University of Latvia, Latvia
Rihards Rūmnieks	University of Latvia, Latvia
Darja Solodovņikova	University of Latvia, Latvia
Ārija Sproģe	University of Latvia, Latvia
Arnis Voitkāns	University of Latvia, Latvia

Sponsoring and Supporting Institutions

 University of Latvia

 Exigen Services Latvia

 IEEE

 Springer

Contents

Advanced Systems and Technologies

Business Process Modeling and Performance Measurement

Software Testing and Quality Assurance

Linguistic Components of IS

Information Technology in Teaching and Learning

Ontology, Conceptual Modeling and Databases

Towards Self-explanatory Ontology Visualization with Contextual Verbalization

Renārs Liepiņš[✉], Uldis Bojārs, Normunds Grūzītis,
Kārlis Čerāns, and Edgars Celms

Institute of Mathematics and Computer Science, University of Latvia,
Raina Bulvaris 29, Riga LV-1459, Latvia
{renars.liepins,uldis.bojars,normunds.gruzitis,
karlis.cerans,edgars.celms}@lumii.lv

Abstract. Ontologies are one of the core foundations of the Semantic Web. To participate in Semantic Web projects, domain experts need to be able to understand the ontologies involved. Visual notations can provide an overview of the ontology and help users to understand the connections among entities. However, the users first need to learn the visual notation before they can interpret it correctly. Controlled natural language representation would be readable right away and might be preferred in case of complex axioms, however, the structure of the ontology would remain less apparent. We propose to combine ontology visualizations with contextual ontology verbalizations of selected ontology (diagram) elements, displaying controlled natural language (CNL) explanations of OWL axioms corresponding to the selected visual notation elements. Thus, the domain experts will benefit from both the high-level overview provided by the graphical notation and the detailed textual explanations of particular elements in the diagram.

Keywords: Owl · Ontology visualization · Contextual verbalization

1 Introduction

Semantic Web technologies have been successfully applied in pilot projects and are now transitioning toward mainstream adoption in the industry. However, for this transition to go successfully, there are still hurdles that have to be overcome. One of them are the difficulties that domain experts have in understanding mathematical formalisms and their notations that are used in ontology engineering.

Visual notations have been proposed as a way to help domain experts to work with ontologies. Indeed, when domain experts collaborate with ontology experts in designing an ontology "they very quickly move to sketching 2D images to communicate their thoughts" [9]. The use of diagrams has also been supported

This work has been supported by the ESF project 2013/0005/1DP/1.1.1.2.0/ 13/APIA/VIAA/049 and the Latvian State Research program NexIT project No. 1 "Technologies of ontologies, semantic web and security".

© Springer International Publishing Switzerland 2016
G. Arnicans et al. (Eds.): DB&IS 2016, CCIS 615, pp. 3–17, 2016.
DOI: 10.1007/978-3-319-40180-5_1

by an empirical study done by Warren et al. where they reported that "one-third of [participants] commented on the value of drawing a diagram" to understand what is going on in the ontology [23].

Despite the apparent success of the graphical approaches, there is still a fundamental problem with them. When a novice user wants to understand a particular ontology, he or she cannot just look at the diagram and know what it means. The user first needs to learn the syntax and semantics of the notation – its mapping to the underlying formalism. In some diagrams an edge with a label P between nodes A and B might denote a property P that has domain A and range B, while in others it might mean that every A has at least one property P to something that is B. This limitation has long been noticed in software engineering [20] and, for this reason, formal models in software engineering are often translated into informal textual documentation by systems analysts, so that they can be validated by domain experts [5].

A similar idea of automatic conversion of ontologies into seemingly informal controlled natural language (CNL) texts and presenting the texts to domain experts has been investigated by multiple groups [12,16,22]. CNL is more understandable to domain experts and end-users than the alternative representations because the notation itself does not have to be learned, or the learning time is very short. Hoverer, the comparative studies of textual and graphical notations have shown that while domain experts that are new to graphical notations better understand the natural language text, they still prefer the graphical notations in the long run [15,19]. It leads to a dilemma of how to introduce domain experts to ontologies. The CNL representation shall be readable right away and might be preferred in case of complex axioms (restrictions) while the graphical notation makes the overall structure and the connections more comprehensible.

We present an approach that combines the benefits of both graphical notations and CNL verbalizations. The solution is to extend the graphical notation with contextual verbalizations of the axioms that are represented by the selected graphical element. The graphical representation gives the users an overview of the ontology while the contextual verbalizations explain what the particular graphical elements mean. Thus, domain experts that are novices in ontology engineering shall be able to learn and use the graphical notation rapidly and independently without special training.

In Sect. 2, we present the general principles of extending graphical ontology notations with contextual natural language verbalizations. In Sect. 3, we demonstrate the proposed approach in practice by extending a particular graphical ontology notation and editor, OWLGrEd, with contextual verbalizations in controlled English. In Sect. 4, we discuss the benefits and limitations of our approach, as well as sketch some future work. Related work is discussed in Sect. 5, and we conclude the article in Sect. 6.

2   Extending Graphical Notations with Contextual Verbalizations

This section describes the proposed approach for contextual verbalization of graphical elements in ontology diagrams, starting with a motivating example.

We are focusing particularly on OWL ontologies, assuming that they are already given and that the ontology symbols (names) are lexically motivated and consistent, i.e., we are not considering the authoring of ontologies in this article, although the contextual verbalizations might be helpful in the authoring process as well, and it would motivate to follow a lexical and consistent naming convention.

2.1 Motivating Example

In most diagrammatic OWL ontology notations, object property declarations are shown either as boxes (for example in VOWL [14]) or as labeled links connecting the property domain and range classes as in OWLGrEd [2]. Figure 1 illustrates a simplified ontology fragment that includes classes *Person* and *Thing*, an object property *likes* and a data property *hasAge*. This fragment is represented by using three alternative formal notations: Manchester OWL Syntax [8], VOWL and OWLGrEd. As can be seen, the visualizations are tiny and may already seem self-explanatory. Nevertheless, even in this simple case, the notation for domain experts may be far from obvious. For example, the Manchester OWL Syntax uses the terms *domain* and *range* when defining a property, and these terms may not be familiar to a domain expert. In the graphical notations, the situation is even worse because the user may not even suspect that the edges represent more than one assertion and that the assertions are far-reaching. In the case of *likes*, it means that everyone that likes something is *necessarily* a person, and vice versa.

We have encountered such problems in practice when introducing ontologies in the OWLGrEd notation to users familiar with the UML notation. Initially, it turned out that they are misunderstanding the meaning of the association edges. For example, they would interpret that the edge *likes* in Fig. 1 means "persons *may* like persons", which is true, however, they would also assume that other disjoint classes could also have this property, which is false in OWL because multiple domain/range axioms of the same property are combined to form an intersection. Thus, even having a very simple ontology, there is a potential for misunderstanding the meaning of both the formal textual notation (e.g., Manchester OWL Syntax) and the graphical notations.

The data property *hasAge* in Fig. 1 illustrates another kind of a problem. In some graphical notations (e.g., VOWL), data properties are represented by edges, and their value types – by nodes (using a style that is different from class nodes). In other notations (e.g., OWLGrEd), data properties are represented by labels inside the class node that corresponds to the data property's domain. While the representation and therefore the reading is similar to the object properties in VOWL, verbalization might help to novice OWLGrEd users.

2.2 Proposed Approach

We propose to extend graphical ontology diagrams with contextual on-demand verbalizations of OWL axioms related to the selected diagram elements, with

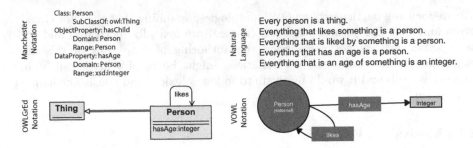

Fig. 1. A simplified ontology fragment alternatively represented by using Manchester OWL Syntax, VOWL and OWLGrEd, and an explanation in a controlled natural language

the goal to help users to better understand their ontologies and to learn the graphical notations based on their own and/or real-world examples.

The contextual verbalization of ontology diagrams relies on the assumption that every diagram element represents a set of ontology axioms, i.e., the ontology axioms are generally presented locally in the diagram, although possibly a single ontology axiom can be related to several elements of the diagram.

The same verbalization can be applied to all the different OWL visual notations, i.e., we do not have to design a new verbalization (explanation) grammar for each new visual notation, because they all are mapped to the same underlying OWL axioms. Thus, the OWL visualizers can reuse the same OWL verbalizers to provide contextual explanations of any graphical OWL notation.

By reusing ontology verbalizers, existing ontology visualization systems can be easily extended with a verbalization service. Figure 2 illustrates the proposed approach:

1. *Visualizer* is the existing visualization component that transforms an OWL ontology into its graphical representation.
2. The system is extended by a *User Selection* mechanism that allows users to select the graphical element that they want to verbalize.

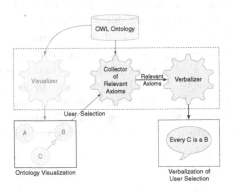

Fig. 2. Architecture of a contextual ontology verbalizer

3. *Collector* gathers a subset of the ontology axioms that correspond to the selected graphical element.
4. The relevant axioms are passed to *Verbalizer* that produces CNL statements – a textual explanation that is shown to the user.

By applying the proposed approach and by using natural language to interactively explain what the graphical notation means, developers of graphical OWL editors and viewers can enable users (domain experts in particular) to avoid misinterpretations of ontology elements and their underlying axioms, resulting in a better understand of both the ontology and the notation. For example, when domain experts encounter the ontology in Fig. 1, they would not have to guess what the elements of this graphical notation mean. Instead, they can just ask the system to explain the notation using the example of the ontology that they are exploring. When the user clicks on the edge *likes* in Fig. 1, the system shows the verbalization

> *Everything that likes something is a person. Everything that is liked by something is a person.*

which unambiguously explains the complete meaning of this graphical element.

The verbalization of ontology axioms has been shown to be helpful in teaching OWL to newcomers both in practical experience reports [18] as well as in statistical evaluations [12].

3   Case Study: Extending OWLGrEd with Contextual Verbalizations in ACE

The proposed approach is illustrated by a case study demonstrating the enhancement of extending OWLGrEd, a graphical notation and editor for OWL, with on-demand contextual verbalizations of the underlying OWL axioms using Attempto Controlled English (ACE) [6]. The CNL verbalization layer allows users to inspect a particular element of the presented ontology diagram and to receive a verbal explanation of the ontology axioms that are related to this ontology element.

A demonstration of our implementation is available online.[1]

3.1   Overview of the OWLGrEd Notation

The OWLGrEd notation [2] is a compact and complete UML-style notation for OWL 2 ontologies. It relies on Manchester OWL Syntax [8] for certain class expressions.

This notation is implemented in the OWLGrEd ontology editor[2] and its online ontology visualization tool[3] [13]. The approach proposed in this article is implemented as a custom version of the OWLGrEd editor and visualization tool.

---

[1] http://owlgred.lumii.lv/cnl-demo.
[2] http://owlgred.lumii.lv.
[3] http://owlgred.lumii.lv/online_visualization/.

In order to keep the visualizations compact and to give the ontology developers flexibility in visualizing their ontologies, the OWLGrEd notation provides several alternatives how a certain OWL axiom can be represented (e.g., either as a visual element or as an expression in Manchester OWL Syntax inside the class element). This makes the OWLGrEd notation a good case study for exploring the use of contextual verbalizations. In order to fully understand a visualization of an ontology, the domain experts would need to understand both the graphical elements and the expressions in Manchester OWL Syntax.

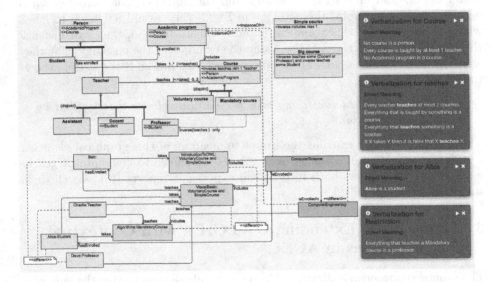

Fig. 3. An example mini-university ontology represented in the OWLGrEd notation

Figure 3 demonstrates the OWLGrEd notation through an example of a mini-university ontology. OWL classes (e.g., *Student, Person, Course* in Fig. 3) are represented by UML classes while OWL object properties are represented by roles on the associations between the relevant domain and range classes (e.g., *teaches, takes, hasEnrolled*). OWL datatype properties are represented by attributes of the property domain classes. OWL individuals are represented by UML objects (e.g., *Alice, Bob, ComputerScience*).

Simple cardinality constraints can be described along with the object or datatype properties (e.g., *every student is enrolled in exactly 1 academic program*), and *inverse-of* relations can be encoded as inverse roles of the same association.

Subclass assertions can be visualized in the form of UML generalizations that can be grouped together using generalization sets (the "fork" elements). Disjointness or completeness assertions on subclasses can be represented using UML generalization set constraints (e.g., classes *Assistant, Docent* and *Professor* all are subclasses of *Teacher* and are pairwise disjoint). OWLGrEd also introduces a graphical notation for property-based class restrictions – a red arrow between the nodes. For example, the red arrow between classes *MandatoryCourse* and *Professor* corresponds to the following restriction in the Manchester notation:

```
MandatoryCourse SubClassOf inverse teaches only Professor
```

Class elements can have text fields with OWL class expressions in Manchester OWL Syntax. While OWLGrEd allows to specify class expressions in a graphical form, more compact visualizations can be achieved by using the textual Manchester notation (e.g., in the descriptions of *Course*, *SimpleCourse* and *BigCourse*). The '<' textual notation is used for sub-class and sub-property relations, '=' for equivalent classes/properties and '<>' for disjoint classes/properties (e.g., *Person* <> *Course*).

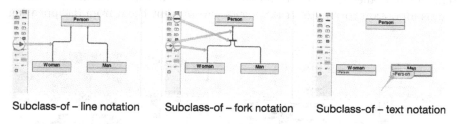

Subclass-of – line notation   Subclass-of – fork notation   Subclass-of – text notation

Fig. 4. Options for representing the class hierarchy in OWLGrEd

Figure 4 illustrates the multiple ways how OWL axioms can be represented in diagrams using the OWLGrEd notation. It shows how the generalization (a *subclass-of* relation) can be represented using a line notation, a more compact "fork" notation and a text notation that may be preferable in some cases (e.g., for referring to a superclass that is defined using an OWL class expression and is not referenced anywhere else).

On the assertion or individual level (ABox), there are two options for stating class assertions (instances): by using the *instanceOf* arrow to the corresponding class element or by stating the class name or expression in the box element denoting the individual.

We refer to [1,2] for a more detailed explanation of the OWLGrEd notation and editor, as well as the principles of its visual extensions [4].

3.2 Adding Verbalizations to OWLGrEd

In order to help domain experts to understand the visualized ontology diagram, they are presented with explanations – textual representation (verbalization) of all OWL axioms corresponding to a given element of the ontology diagram.

The verbalization can help users even in relatively simple cases, such as object property declarations where user's intuitive understanding of the domain and range of the property might not match what is asserted in the ontology. The verbalization of OWL axioms makes this information explicit while not requiring users to be ontology experts. The value of contextual ontology verbalization is even more apparent for elements whose semantics might be somewhat tricky even for more experienced users (e.g., *some*, *only* and cardinality constraints on properties, or generalization forks with *disjoint* and *complete* constraints).

The CNL verbalization layer for which an experimental support has been added to the OWLGrEd editor enhances the ontology diagrams with an interactive means for viewing textual explanations of the axioms associated with a particular graphical element.

By clicking a mouse pointer on an element, a pop-up widget is thrown, containing a CNL verbalization of the corresponding axioms in Attempto Controlled English. By default, the OWLGrEd visualizer minimizes the number of verbalization widgets shown simultaneously by hiding them after a certain timeout. For users to simultaneously see the verbalizations for multiple graphical elements, there is an option to "freeze" the widgets and prevent them from disappearing.

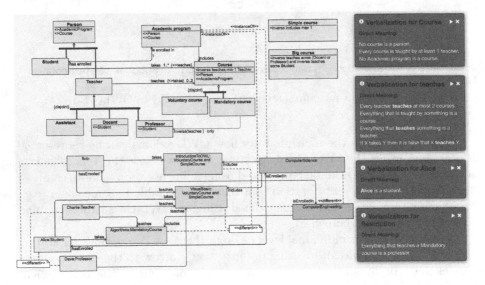

Fig. 5. The example ontology in the OWLGrEd notation (see Fig. 3) with CNL verbalizations (explanations) of the selected diagram elements.

Figure 5 shows an example of multiple verbalizations displayed on the diagram introduced in Fig. 3. They describe the ontology elements that represent the class *Course*, the object property *teaches*, the individual *Alice* and the restriction on the class *MandatoryCourse*. Verbalizations are implicitly linked to the corresponding elements using the element labels. While it might be less convenient to identify the implicit links in a static image, the interactive nature of the combined ontology visualization and verbalization tool makes it easier for users to keep the track. Visual cues (e.g., a line between a verbalization and the diagram element) could be added to make the linking more noticeable. However, to keep the visualization simple, such cues are not currently employed.

The object property *teaches*, represented in the diagram by an edge connecting the class *Teacher* to the class *Course*, has the following verbalization in ACE (see Fig. 5):

Every teacher teaches at most 2 courses.

Everything that is taught by something is a course.

Everything that teaches something is a teacher.

If X takes Y then it is false that X teaches Y.

Note that the specific OWL terms, like *disjoint*, *subclass* and *inverse*, are not used in the ACE statements. The same meaning is expressed implicitly – via paraphrasing – using more general common sense constructions and terms.

In this case, the edge represents not only the domain and range axioms of the property but also the cardinality of the range and the restriction that *teaches* is disjoint with *takes* (expressed by the if-then statement).

The property restriction on the class *MandatoryCourse*, shown in the diagram as a red line connecting the class *MandatoryCourse* to the class *Professor*, is another case when a CNL explanation is essential. Its meaning is expressed in ACE by the following statement:

Everything that teaches a mandatory course is a professor.

In this case, similarly as in the case of the *disjoint* restrictions, the ACE verbalizer has rewritten the axiom in a more general but semantically equivalent form avoiding the use of the determiner *only* (*nothing but* in ACE) [10]. At the first glance, it might seem confusing for an expert, however, such a semantic paraphrase can be helpful to better understand the consequences of the direct reading of the axiom:

Every mandatory course is taught by nothing but professors.

The steps involved in the verbalization of an ontology diagram element (as implemented in OWLGrEd) are:

1. Every diagram element represents a set of OWL axioms.
2. The set of axioms corresponding to this element is ordered by axiom type and sent as a list to the verbalization component.
3. The verbalization component returns a corresponding list of CNL statements.
4. The resulting CNL statements (the textual explanation of the diagram element) are displayed to the user.

The translation from OWL to ACE is done by reusing the readily available verbalizer from the ACE toolkit [11].[4]

---

[4] http://attempto.ifi.uzh.ch/site/resources/.

In order to acquire lexically and grammatically well-formed sentences (from the natural language user's point of view), additional lexical information may need to be provided, e.g., that the property *teaches* is verbalized using the past participle form "taught" in the passive voice (*inverse-of*) constructions or that the class *MandatoryCourse* is verbalized as a multi-word unit "mandatory course". This information is passed to the OWL-to-ACE verbalizer as an ontology-specific lexicon.

In the case of controlled English, the necessary lexical information can be largely restored automatically from the entity names (ontology symbols), provided that English is used as a meta-language and that the entity names are lexically motivated and consistently formed.

If aiming for multilingual verbalizations, domain-specific translation equivalents would have to be specified additionally, which, in general, would be a semi-automatic task.

An appropriate and convenient means for implementing a multilingual OWL verbalization grammar is Grammatical Framework (GF) [17] which provides a reusable resource grammar library for about 30 languages.[5] Moreover, an ACE grammar library based on the GF general-purpose resource grammar library is already available for about 10 languages [3]. This allows for using English-based entity names and the OWL subset of ACE as an interlingua, following the two-level OWL-to-CNL approach suggested in [7].

In fact, we have applied the GF-based approach to provide an optional support for lexicalization and verbalization in OWLGrEd in both English and Latvian, a highly inflected Baltic language.

4   Discussion

This section discusses the use of contextual ontology verbalization, focusing on its applicability to various graphical notations, extending the scope of axioms to include in verbalizations and the potential limitations of the approach.

4.1   Applicability to Other Notations

The proposed approach is applicable to any ontology visualization where graphical elements represent one or more OWL axioms. The value of using verbalization functionality is higher for more complex notations (e.g., OWLGrEd) where graphical elements may represent multiple axioms but even in simple graphical notations, where each graphical element corresponds to one axiom, users will need to know how to read the notation. Contextual ontology verbalization addresses this need by providing textual explanations of diagram elements and the underlying OWL axioms.

A more challenging case is notations where some OWL axioms are represented as spatial relations between the elements and are not directly represented

---

[5] http://www.grammaticalframework.org/.

by any graphical elements (e.g., Concept Diagrams represent *subclass-of* relations as shapes that are included in one another [21]). In order to represent these axioms in ontology verbalization they need to be "attached" to one or more graphical elements that these axioms refer to. As a result, they will be included in verbalizations of relevant graphical elements. In the case of Concept Diagrams, the *subclass-of* relation, which is represented by shape inclusion, would be verbalized as part of the subclass shape.

4.2 Extending the Scope of Verbalization

The scope of OWL axioms that are included in CNL explanations of diagram elements can be adjusted by modifying the *Collector* component (see Fig. 2) that selects OWL axioms related to a particular element. In our primary use case, the choice of OWL axioms to verbalize is straightforward – for each element only the axioms directly associated with this element (i.e. the axioms that this element represents) are used for generating CNL verbalizations. Depending on the graphical notation the scope of axioms for verbalization may need to be expanded, as we pointed out in Sect. 4.1.

Enlarging the scope of axioms to verbalize may also be useful for generating contextual documentation of a selected element. In contrast to the primary use case, where the verbalization was used to explain the notation, in the case of contextual documentation, we want to show the user all the axioms that are related to the selected element (e.g., when the user selects a class node, the system would also show verbalizations of domain and range axioms where this class is mentioned). Such approach is widely used in textual ontology documentation (e.g., in FOAF vocabulary specification[6]).

Another use case for enhanced verbalizations is running inference on the ontology and including inferred OWL axioms in verbalizations. Since it is not practical to show all inferred axioms in the graphical representation, contextual verbalizations are a suitable place for displaying this information. Verbalization of inferred axioms may also be useful for ontology debugging. For instance, in the ontology in Fig. 3, it might be derived that every individual belonging to the class *BigCourse* also belongs to the class *Course*, however the same would not necessarily be true for every individual belonging to the class *SimpleCourse* (one can term this a design error present in the ontology).

To support these use cases, the scope of verbalization axioms can be increased to include additional axioms. It is important to note that by doing this (i.e. by adding axioms not directly represented by the element) we would lose the benefit of having equivalent visual and verbal representations of the ontology and the resulting verbalizations might not be as useful for users in learning the graphical notation. This limitation can be partially alleviated by visually distinguishing between CNL sentences that represent direct axioms (i.e. axioms that directly correspond to a graphical element) and other axioms included in the verbalization.

---

[6] http://xmlns.com/foaf/spec/#sec-crossref.

4.3   Limitations

Due to the interactive nature of the approach it might not work well for documenting ontologies when diagrams are printed out or saved as screenshots. While the example in Fig. 5 shows how multiple verbalizations are displayed simultaneously and that they can still be useful when saved as screenshots, the image would become too cluttered if a larger amount of verbalizations were displayed simultaneously therefore a naive approach of showing all verbalizations on a diagram at once would not work for documentation.

The combined ontology visualization and verbalization approach can be adapted to documenting ontologies by exporting fragments of the ontology diagram showing a particular graphical element along with verbalized statements corresponding to this element. The resulting documentation would have some redundancy because one CNL statement may be relevant to multiple concepts. However, it has been shown that such "dictionaries" are perceived to be more usable than the alternative, where all axioms verbalizations are displayed just once, without grouping [24].

Verbalization techniques that are a part of the proposed approach have the same limitations as ontology verbalization in general. In particular, verbalization may require additional lexical information to generate grammatically well-formed sentences. To some degree, by employing good ontology design practices and naming conventions as well as by annotating ontology entities with lexical labels, this limitation can be overcome. Another issue is specific kinds of axioms that are difficult or verbose to express in natural language without using terms of the underlying formalism.

As it was mentioned, the contextual verbalizations could be generated in multiple languages, provided that the translation equivalents have been provided while authoring the ontology. This would allow domain experts to explore ontologies in their native language and would be even more important to the regular end-users exploring ontology-based applications.

5   Related Work

To the best of our knowledge, there are no publications proposing combining OWL ontology visualizations with contextual CNL verbalizations but there has been a movement towards cooperation between both fields. In ontology visualizations, notations have been adding explicit labels to each graphical element that describes what kind of axiom it represents. For example, in VOWL a line representing a *subclass-of* relation is explicitly labeled with the text "Subclass of". This practice makes the notation more understandable to users as reported in the VOWL user study where a user stated that "there was no need to use the printed [notation reference] table as the VOWL visualization was very self-explanatory" [14]. However, such labeling of graphical elements is only useful in notations where each graphical element represents one axiom. In more complex visualizations where one graphical element represents multiple axioms there would be no place for all the labels corresponding to these axioms. For example, in the OWLGrEd notation, class boxes can represent not just class definitions but also *subclass-of* and *disjoint classes* assertions. In such cases, verbalizations provide understandable explanations. Moreover, in some notations (e.g., Concept Diagrams [21]) there might be no graphical elements at all for certain kinds of axioms, as it was mentioned in Sect. 4.1.

In the field of textual ontology verbalizations there has been some exploration of how to make verbalizations more convenient for users. One approach that has been tried is grouping verbalizations by entities. It produces a kind of a dictionary, where records are entities (class, property, individual), and every record contains verbalizations of axioms that refer to this entity. The resulting document is significantly larger than a plain, non-grouped verbalization because many axioms may refer to multiple entities and thus will be repeated in each entity. Nevertheless, the grouped presentation was preferred by users [24]. Our approach can be considered a generalization of this approach, where a dictionary is replaced by an ontology visualization that serves as a map of the ontology.

An ad-hoc combination of verbalization and visualization approaches could be achieved using existing ontology tools such as Protégé by using separate visualization and verbalization plugins (e.g., ProtégéVOWL[7] for visualization and ACEView[8] for verbalization). However, this would not help in understanding the graphical notation because the two views are independent, and thus a user cannot know which verbalizations correspond to which graphical elements. Our approach employs closer integration of the two ontology representations and provides contextual verbalization of axioms that directly correspond to the selected graphical element, helping users in understanding the ontology and learning the graphical notation used.

6   Conclusions

Mathematical formalisms used in ontology engineering are hard to understand for domain experts. Usually, graphical notations are suggested as a solution to this problem. However, the graphical notations, while preferred by domain experts, still have to be learned to be genuinely helpful in understanding. Until now the only way to learn these notations was by reading the documentation.

In this article, we proposed to use the CNL verbalizations to solve the learning problem. Using our approach the domain expert can interactively select a graphical element and receive the explanation of what the element means. The explanation is generated by passing the corresponding axioms of the element through one of the existing verbalization services. The service returns natural language sentences explaining the OWL axioms that correspond to the selected element and thus explaining what it means.

We demonstrated the proposed approach in a case study where we extended an existing ontology editor with the contextual CNL explanations. We also pre-

---

[7] http://vowl.visualdataweb.org/protegevowl.html.
[8] http://attempto.ifi.uzh.ch/aceview/.

sented the architecture of the extension that is general enough to apply to wide range of other ontology notations and tools.

In conclusion, we have shown how to extend graphical notations with contextual CNL verbalizations that explain the selected ontology element. The explanations help domain experts to rapidly and independently learn and use the notation from the beginning without a special training, thus making it easier for domain experts to participate in ontology engineering without extended training, which solves one of the problems that hinder the adoption of Semantic Web technologies in the mainstream industry.

References

1. Bārzdiņš, J., Bārzdiņš, G., Čerāns, K., Liepiņš, R., Sproģis, A.: OWLGrEd: A UML-style graphical notation and editor for OWL 2. In: Proceedings of the 7th International Workshop on OWL: Experience and Directions (OWLED) (2010)
2. Bārzdiņš, J., Bārzdiņš, G., Čerāns, K., Liepiņš, R., Sproģis, A.: UML style graphical notation and editor for OWL 2. In: Günther, H., Forbrig, P. (eds.) BIR 2010. LNBIP, vol. 64, pp. 102–114. Springer, Heidelberg (2010)
3. Camilleri, J., Fuchs, N., Kaljurand, K.: ACE grammar library. Technical report MOLTO Project Deliverable D11.1 (2012)
4. Cerans, K., Ovcinnikova, J., Liepins, R., Sprogis, A.: Advanced OWL 2.0 ontology visualization in OWLGrEd. In: 10th International Baltic Conference on Databases and Information Systems VII-Selected Papers (DB&IS), pp. 41–54 (2012)
5. Frederiks, P.J., Van der Weide, T.P.: Information modeling: the process and the required competencies of its participants. Data & Knowl. Eng. **58**(1), 4–20 (2006)
6. Fuchs, N.E., Kaljurand, K., Kuhn, T.: Attempto controlled English for knowledge representation. In: Baroglio, C., Bonatti, P.A., Małuszyński, J., Marchiori, M., Polleres, A., Schaffert, S. (eds.) Reasoning Web. LNCS, vol. 5224, pp. 104–124. Springer, Heidelberg (2008)
7. Gruzitis, N., Barzdins, G.: Towards a more natural multilingual controlled language interface to OWL. In: Proceedings of the 9th International Conference on Computational Semantics, pp. 1006–1013 (2011)
8. Horridge, M., Patel-Schneider, P.F.: OWL 2 Web Ontology Language Manchester Syntax. W3C Working Group Note (2009)
9. Howse, J., Stapleton, G., Taylor, K., Chapman, P.: Visualizing ontologies: a case study. In: Aroyo, L., Welty, C., Alani, H., Taylor, J., Bernstein, A., Kagal, L., Noy, N., Blomqvist, E. (eds.) ISWC 2011, Part I. LNCS, vol. 7031, pp. 257–272. Springer, Heidelberg (2011)
10. Kaljurand, K.:Attempto controlled English as a semantic web language. Ph.D. thesis, University of Tartu (2007)
11. Kaljurand, K., Fuchs, N.E.: Verbalizing OWL in attempto controlled English. In: Proceedings of OWL: Experiences and Directions (OWLED) (2007)
12. Kuhn, T.: The understandability of OWL statements in controlled English. Semant. Web **4**(1), 101–115 (2013)
13. Liepins, R., Grasmanis, M., Bojars, U.: OWLGrEd ontology visualizer. In: ISWC Developers Workshop 2014. CEUR (2015)
14. Lohmann, S., Negru, S., Haag, F., Ertl, T.: VOWL 2: user-oriented visualization of ontologies. In: Janowicz, K., Schlobach, S., Lambrix, P., Hyvönen, E. (eds.) EKAW 2014. LNCS, vol. 8876, pp. 266–281. Springer, Heidelberg (2014)

15. Ottensooser, A., Fekete, A., Reijers, H.A., Mendling, J., Menictas, C.: Making sense of business process descriptions: an experimental comparison of graphical and textual notations. J. Syst. Softw. **85**(3), 596–606 (2012)
16. Power, R., Third, A.: Expressing OWL axioms by English sentences: dubious in theory, feasible in practice. In: Proceedings of the 23rd International Conference on Computational Linguistics: Posters. pp. 1006–1013. Association for Computational Linguistics (2010)
17. Ranta, A.: Grammatical Framework, a type-theoretical grammar formalism. J. Funct. Program. **14**(2), 145–189 (2004)
18. Rector, A., Drummond, N., Horridge, M., Rogers, J., Knublauch, H., Stevens, R., Wang, H., Wroe, C.: OWL pizzas: practical experience of teaching OWL-DL: common errors and common patterns. In: Motta, E., Shadbolt, N.R., Stutt, A., Gibbins, A. (eds.) EKAW 2004. LNCS, pp. 63–81. Springer, Heidelberg (2004)
19. Sharafi, Z., Marchetto, A., Susi, A., Antoniol, G., Guéhéneuc, Y.G.: An empirical study on the efficiency of graphical vs. textual representations in requirements comprehension. In: 21st International Conference on Program Comprehension, pp. 33–42. IEEE (2013)
20. Siau, K.: Informational and computational equivalence in comparing information modeling methods. J. Database Manag. **15**(1), 73–86 (2004)
21. Stapleton, G., Howse, J., Bonnington, A., Burton, J.: A vision for diagrammatic ontology engineering. In: Visualizations and User Interfaces for Knowledge Engineering and Linked Data Analytics. CEUR (2014)
22. Stevens, R., Malone, J., Williams, S., Power, R., Third, A.: Automating generation of textual class definitions from OWL to English. J. Biomed. Semant. **2**(S–2), S5 (2011)
23. Warren, P., Mulholland, P., Collins, T., Motta, E.: The usability of description logics. In: Presutti, V., d'Amato, C., Gandon, F., d'Aquin, M., Staab, S., Tordai, A. (eds.) ESWC 2014. LNCS, vol. 8465, pp. 550–564. Springer, Heidelberg (2014)
24. Williams, S., Third, A., Power, R.: Levels of organisation in ontology verbalisation. In: Proceedings of the 13th European Workshop on Natural Language Generation, pp. 158–163. Association for Computational Linguistics (2011)

Self-service Ad-hoc Querying Using Controlled Natural Language

Janis Barzdins[1], Mikus Grasmanis[1], Edgars Rencis[1(✉)],
Agris Sostaks[1], and Juris Barzdins[2]

[1] Institute of Mathematics and Computer Science,
University of Latvia, Riga, Latvia
{Janis.Barzdins,Mikus.Grasmanis,Edgars.Rencis,
Agris.Sostaks}@lumii.lv
[2] Faculty of Medicine, University of Latvia, Riga, Latvia
Juris.Barzdins@lu.lv

Abstract. The ad-hoc querying process is slow and error prone due to inability of business experts of accessing data directly without involving IT experts. The problem lies in complexity of means used to query data. We propose a new natural language- and *semistar* ontology-based ad-hoc querying approach which lowers the steep learning curve required to be able to query data. The proposed approach would significantly shorten the time needed to master the ad-hoc querying and to gain the direct access to data by business experts, thus facilitating the decision making process in enterprises, government institutions and other organizations.

Keywords: Ad-hoc querying · Star ontologies · Controlled natural language · Hierarchical data

1   Introduction and Problem Statement

The amount of data collected by enterprises, government institutions and other organizations grows significantly every year. Data alone do not guarantee a success – data should be transformed into information, and it should be used accordingly in order to succeed. This process has often been referred to as Business Intelligence (BI). Typically, BI tools offer wide possibilities of data analysis, however they require a significant amount of investment and IT expertise. Needless to say that BI processes involve IT experts whose task is to translate business requirements and queries into a language which is understandable by computer. For example, the Children's Clinical University Hospital (Riga, Latvia) collects sensitive data of clinical processes. Data are stored in the relational database and maintained by the team of local IT experts. The IT experts take part in the clinical processes. They translate hospital managers, practitioners and researchers questions into SQL queries. Although there are predefined reports, the business requirements are changing very often and, consequently, IT experts are overloaded. Therefore business decision processes are very slow and error-prone because of miscommunication and hurry. Direct access to data by domain

G. Arnicans et al. (Eds.): DB&IS 2016, CCIS 615, pp. 18–34, 2016.
DOI: 10.1007/978-3-319-40180-5_2

experts would be a solution. However the problem is that domain experts do not possess the required skills to formulate queries by themselves, because of the complexity of query languages used to retrieve answers from data stores. In our previous work [1] we have defined so called "3How" problem which consists of three main problems related to this context:

(1) how to describe data to be easily perceived by domain experts;
(2) how to query data simply enough for domain experts;
(3) how to perform query efficiently enough in order to get an answer to a sufficiently wide class of queries in reasonable time.

Actually, the problem has been relevant for more than 40 years. The SQL (SEQUEL) language [2] is the *de facto* standard of querying relational databases where most of data are being stored at the moment. However, it turned out that the way data are stored and retrieved in the relational databases was too complicated for domain experts. There are similar languages to SQL (e.g. SPARQL for RDF ontologies [3]) which require a very precise formulation of the textual query (both syntax and semantics) and deep knowledge of underlying technology, thus making them too sophisticated to learn and use. Therefore there have been attempts to make wrappers for these languages – e.g. graphical query builders like Graphical Query Designer for SQL Server [4], ViziQuer [5] and Ontology Based Data Access (OBDA) approach [6], particularly, the OptiqueVQS [7] for SPARQL and RDF databases, or form-based tools using wizards and standard GUI elements (e.g. tables and lists) like SAP Quick Viewer SQVI [8]. There are also other proposals which provide the means for direct data access. One of the most well-known approaches is Self-Service Business Intelligence (SSBI) which was proposed by Microsoft [9]. It provides a rich set of tools (Power BI) allowing the end-user to build sophisticated data visualizations and make data analysis mainly through spreadsheet applications.

Yet, there is a significant drawback of the mentioned approaches, namely, a steep learning curve which is required to learn a new query language and to understand the way data are stored. In order to make the learning of ad-hoc querying easier we propose a new query language which is based on the controlled natural language. We rely on simple *semistar* data ontologies which resemble the structure of documents which are backing up the business processes in the organization. Health records are good examples of such documents in a hospital. *Semistar* ontology defines a vocabulary (terms) which are allowed to use while querying the data. This allows us to control the language in the way which is familiar to the domain expert. Therefore, the complexity of precise understanding of the rich natural language can be reduced in our implementation of natural language query interface. It should.be noted that our query system's ability to explain to the user how his query has been understood plays an important role for ad-hoc querying.

Experiments have shown that the proposed query language can be taught in a short time to medical students. Even after a short (2 h) lecture for a group of medicine students almost every participant could understand the given examples of queries written in the proposed language. Most of the students were also able to formulate queries for the proposed questions about the clinical processes of the hospital. This paper is organized as follows. Section 2 sums up the related work done by other authors. In Sect. 3 we introduce the concept of *semistar data ontology* which is heavily exploited in designing the proposed query language. The language itself is described in Sect. 4 where we give its syntax and semantics together with its typical examples. In Sect. 5 we briefly outline the basics of the implementation of the language, and in Sect. 6 we describe the practical experiments we have performed to test our language from different points of view. Section 7 concludes the paper.

2　Related Work

A viable option to query data simply enough for domain experts (2nd problem of "3How") is a natural language. Therefore in this section we will discuss the existing natural language interfaces to databases (NLIDB-s). A lot of work has been done in this area [10–14]. "Natural language is an effective method of interaction for casual users with a good knowledge of the database, who perform question-answering tasks, in a restricted domain" [15]. It should be noted that the formulation of the precise query itself is a hard problem for users without mathematical background, e.g. Ogden et al. showed that users would not be able to specify the meaning of "and" clearly enough for unambiguous understanding of a query [16]. Related research [17] in the knowledge base area has reported that there are still lots of problems in the understanding of complex queries.

In order to make an NLIDB system usable by domain experts it is necessary to solve the problem of *linguistic coverage*. People do not know what the computer "knows", i.e. when a natural language is used there is no common context in the conversation between the user and the database [18]. It is very important to explain to users what the database "knows" and what can be asked. Database schemas used by IT experts (like ER models) are too complex and contain too many technical details to be useful for the explanation of the underlying data. Computers, on the other hand, cannot properly understand what users mean by their queries because of richness and ambiguity of the natural language. In order to achieve a *consensus* between the user and the computer an intermediate representation of data schema is needed.

Many NLIDB solutions rely on the user's domain knowledge and meaningful names used in the schema's elements. These solutions search for similarities between the names and the terms written by user [10, 12]. There are NLIDB solutions that use ontologies (*lexicons, vocabularies*) to represent the data schema [10, 19, 20]. Ontologies define concepts, their properties and relationships which can be used by user. Ontology is automatically obtained from the relational database schema. Ontology hides some technical details, but still requires the understanding of basic entity-relationship model principles. Thus, the traditional NLIDB approaches have not reached wide usage, at least not for deep querying with nontrivial calculations. Therefore other approaches have to be studied in order to solve the "3How" problem. It is the main goal of this paper.

3 Semistar Data Ontologies

Analysis of current situation suggests that it is hopeless to try to develop an easily perceptible query language that can be used on arbitrary ontologies, because such language has not yet appeared after 30 years since the invention of relational databases. Therefore we introduce an important subset of data ontologies called the *Semistar ontologies* (see also [1, 21, 22]). A typical example of a semistar ontology is depicted in Fig. 1. This is a simplified version of data ontology used in Riga Children's Clinical University Hospital (the actually used ontology consists of 25 classes and 142 attributes). Semistar ontologies are data ontologies whose graphical representation corresponds to a star-like structure (having no loops) with a restriction on multiplicities, such that all associations must have the multiplicity equal to 1 at the end of the association that is nearer to the star's center. Semistar ontologies have only one type of associations between basic classes – the "has" relation (e.g. Patient has HospitalEpisodes, HospitalEpisode has TreatmentWards, etc.). Besides basic classes, a semistar ontology can also contain other classes called the classifiers (depicted with white background in Fig. 1). Associations between basic classes and classifier classes are coded as attributes (e.g. familyDoctor: CPhysician).

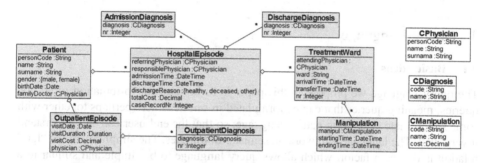

Fig. 1. Simplified semistar ontology used in Riga Children's Clinical University Hospital.

Semistar ontology is a practically important and expressive subset of all data ontologies, and practical use-cases often exploit exactly this type of ontologies. As can be seen in Fig. 1, hospital ontology viewed from patients' and physicians' point of view comes out to be a semistar ontology. Our experience shows that even in more general cases, when some ontology is not a semistar ontology, one can usually find an important subset of it to comply to principles of semistar ontology. We can always think of a semistar ontology as a subject-oriented ontology where the role of the subject can be performed by a patient (in case of medical domain), a customer (in case of some service domain), etc.

We allow attributes of basic classes to have two kinds of data types – the primitive types and the classifiers. We use the following predefined data types and operations:

- Integer (e.g. 75, −75), Real (e.g. 0.75, −75.0), operations: +, −, *, /;
- Boolean (true, false), operations: and, or, not;

- String (e.g. "abc"), operations: substring (e.g. "abcde".substring(2,3)="bc");
- Date (e.g. 2015.06.17), unary operations: year(), month(), day(), dayOfWeek(), binary operation: - (e.g. 2015.06.17−2015.05.12 = 1M5D);
- DateTime (e.g. 2015.06.17T10:45), unary operations: year(), month(), day(), hour(), minute(), second(), date(), binary operation: − (subtraction);
- Duration (e.g. 3Y4M5DT6H7M30.25S), unary operations: years(), months(), days(), hours(), minutes(), seconds().

If some attribute a has a classifier class as data type and this classifier class has some attribute k, then also $a.k$ denotes a valid *attribute expression* and its data type will be that of attribute k. If x is an instance of some class, for which attribute a is defined, then also $x.a$ (or $x.a.k$, if type of a is a classifier class) denotes a valid attribute expression. We can build more complex attribute expressions from simpler ones using the abovementioned operations allowed for the given data types. Some examples of attribute expressions: *personCode, x.personCode, x.familyDoctor.surname, x.admissionTime.month(), (dischargeTime-admissionTime).days()*, etc.

We can now compare two attribute expressions (or constants) to obtain *attribute conditions*, e.g. *personCode=250285-10507, x.personCode=250285-10507, person Code.substring(1,4)=2502, dischargeTime-admissionTime>25d* (meaning – 25 days), *x.birthDate.year()>=1985, familyDoctor<>nil* (a family doctor exists), etc.

4 Query Language

4.1 Basic Ideas

The query language we propose in this paper is to be used for formulation of ad-hoc queries, meaning queries that can be formulated in one sentence (perhaps together with some subordinate clauses) in natural language, so that the end-user can still understand it very well. The language will work on semistar ontologies. The simplicity of the "has" relation is the main factor, which allows query language to be simple and similar to a natural language. It is therefore convenient to build queries in a controlled natural language. This feature allows the language to be easily perceptible by non-IT specialists.

Let us introduce an example query that will be exploited further in this section – *count Patients, who have at least one HospitalEpisode, which has Manipulation with manipul.code=02078*. This natural language sentence is understandable by the domain expert. Let us now inspect a bit more complicated query: *count Patients, who have at least one HospitalEpisode, which has at least one TreatmentWard, which has at least one Manipulation with manipul.code=02078*. This sentence may cause a certain ambiguity as it is not clear whether the asked *Manipulation* refers to *HospitalEpisode* or to *TreatmentWard*. It could be used in both meanings. In other words, relative pronouns such as "who" and "which" not always give us accurate understanding of what we relate to. To cope with such situations we introduce a concept of so called short name in our controlled natural language. Formally, the short name is a variable over instances of the given class – *count Patients p, where exists p.HospitalEpisode e, where exists e.TreatmentWard t, where exists t.Manipulation m, where m.manipul.code=02078*. Now we are able to specify also the second of abovementioned meanings – *count Patients p, where exists p.Hospi-talEpisode e, where exists e.TreatmentWard t, where exists e.Manipulation m, where m.manipul.code=02078*. We have also unified other components of the natural language, e.g. we use the keyword "where" instead of "who", "which" and "with", and the keyword "exists" instead of "have/has at least one". The dot notation after the short name must be perceived as the "of" relation – *count Patient p, where exists HospitalEpisode e of Patient p, where exists TreatmentWard t of HospitalEpisode e, where exists Manipulation m of TreatmentWard t, where manipul.code of Manipulation m equals 02078*.

Formally speaking, the short name must be used before every attribute name to get rid of ambiguities. However, in cases when it is clear to which class the particular attribute refers the short name can be omitted. We also allow omitting other features of the language that are not critical for understanding of queries (e.g. one can omit the empty parentheses after the unary Date and DateTime operations like *year()* or *minute()*).

Let us now introduce some basic notations that we will use describe the query language. We will use the terms *parent class* and *child class* to refer to classes that are higher or lower in the "have" hierarchy. For example, the class "TreatmentWard" has two parent classes – "HospitalEpisode" (direct parent) and "Patient" (further ancestor) and one child class "Manipulation". If x is an instance of the class "TreatmentWard", then its parent instances will be denoted as $x.HospitalEpisode$ and $x.Patient$. In both cases they denote exactly one instance, i.e. that of the class "HospitalEpisode" and of the class "Patient", respectively. We use the same dot notation also for accessing instances of child classes, but in this case we obtain a set of instances. For example, $x.Manipulation$ would be a set of manipulations reachable from the given treatment ward x.

In more complicated cases another concept of *brother class* is important. If x is an instance of the class "HospitalEpisode", then by $x.HospitalEpisode$ we understand $y.HospitalEpisode$, where $y=x.Patient$ (i.e. y is the closest parent of x, which is also parent class of the given class "HospitalEpisode"). Similarly, if x is an instance of the class "TreatmentWard", then $x.OutpatientEpisode=y.OutpatientEpisode$, where $y=x.Patient$.

If *AClass* is an arbitrary class of the ontology, we will use the term *AClass attribute expression* to denote attribute expressions of both *AClass* itself and all of its parent classes (we assume here that parents and children share no common attribute names). We cannot use attribute expressions of child classes here, because there can be many children instances for the given *AClass* instance. We will be able to access these instances by introducing quantors *exists/notexists* later.

We can also perceive our query language as an analogue to some many-sorted predicate language with a difference that it is written in such syntax that is more user-friendly. There has been an attempt to create such a language [23], though it has not led to a practical implementation.

4.2 Syntax and Semantics of the Query Language

Queries are to be written in a controlled natural language and are based on seven sentence templates. The main part of the templates is the so called *selection condition*, which is a selection condition over instances of the given class. We assume that selection conditions are to be written in a natural language. We describe the used language constructs more formally at the end of this section. However, the sentence templates described in this section can be understood without knowing the precise syntax of selection conditions.

T1. COUNT AClass [x] WHERE <selection condition>

Semantics: counts instances of *AClass*, which satisfy the selection condition. Examples:

- *COUNT Patients, WHERE EXISTS HospitalEpisode, WHERE referringPhysician = familyDoctor* (count of patients who have been referred to hospital by their family doctors);
- *COUNT HospitalEpisodes, WHERE dischargeTime-admissionTime>15d* (how many episodes have lasted longer than 15 days);
- *COUNT HospitalEpisodes e1, WHERE EXISTS HospitalEpisode e2, WHERE e1<>e2 AND e2.admissionTime>e1.dischargeTime AND e2.admissionTime-e1. dischargeTime<30d* (how many there have been such episodes, after which the patient has returned to hospital in less than 30 days).

T2. {SUM/MAX/MIN/AVG/MOST} <attribute expression> FROM AClass [x] WHERE <selection condition>

Semantics: selects instances of *AClass*, which satisfy the selection condition, calculates the attribute expression for each of these instances obtaining a list to which the specified aggregate function is then applied. Examples:

- *SUM totalCost FROM HospitalEpisodes, WHERE dischargeReason=healthy AND birthDate.year()=2012* (how much successful treatments of patients born in 2012 have cost);
- *MOST diagnosis.code FROM DischargeDiagnoses, WHERE nr=1 AND dischargeReason=deceased* (get the most frequent main (nr=1) death diagnosis).

T3. SELECT FROM AClass [x] WHERE <selection condition> ATTRIBUTE <attribute expression> ALL DISTINCT VALUES

Semantics is obvious. Examples:

- *SELECT FROM HospitalEpisodes, WHERE dischargeReason=deceased, ATTRIBUTE responsiblePhysician.surname ALL DISTINCT VALUES*;
- *SELECT FROM DischargeDiagnoses, WHERE nr=1 AND dischargeReason= deceased, ATTRIBUTE diagnosis.code ALL DISTINCT VALUES*.

T4. SHOW [n/ALL] AClass WHERE <selection condition>

Semantics: shows n or all instances of *AClass* which satisfy the selection condition.

T5. FULLSHOW [n/all] AClass WHERE <selection condition>

Semantics: the same as "show", but shows also the child class instances attached to the selected *AClass* instances.

T6. SELECT AClass x WHERE <selection condition>, DEFINE TABLE <x-expr'1> [(COLUMN C₁], …, <x-expr'n> [(COLUMN Cₙ)] [, KEEP ROWS WHERE <C₁ selection condition>] [, SORT [ASCENDING/DESCENDING] BY COLUMN Cᵢ] [, LEAVE [FIRST/LAST] n ROWS]

Semantics: selects all instances of *AClass*, which satisfy the selection condition, then makes a table with columns C_1 to C_n, which for every selected *AClass* instance *x* contains an individual row, which in column C_1 contains the value of the <x-expr'1>, …, in column C_n contains the value of the <x-expr'n>. Then it is possible to perform some basic operations with the table like filtering out unnecessary rows, sorting the rows by values of some column and then taking just the first or the last n rows from the table. Examples:

– *SELECT HospitalEpisodes x, WHERE dischargeReason=deceased, DEFINE TABLE x.surname (COLUMN Surname), x.dischargeTime.date() (COLUMN Dying_ date), (COUNT x.Manipulation, WHERE manipul.code=02078) (COLLUMN Count_02078), (SUM manipul.cost FROM x.Manipulation, WHERE manipul. code=02078) (COLUMN cost_02078);*
– *SELECT CPhysicians k, WHERE name=Gatis AND EXISTS HospitalEpisode, WHERE responsiblePhysician=k, DEFINE TABLE surname (COLUMN Physician_ surname), (COUNT HospitalEpisodes, WHERE responsiblePhysician=k) (COLUMN Episode_count), (MOST diagnosis.code FROM AdmissionDiagnoses, WHERE nr=1 AND responsiblePhysician=k) (COLUMN Most_frequent_main_ diagnosis), KEEP ROWS WHERE Episode_count>5, SORT DESCCENDING BY COLUMN Episode_count, LEAVE FIRST 10 ROWS.*

Let us now talk a bit more precisely about the means for defining columns. The expression <*x-expr'i*> defines the value of column C_i in the row that corresponds to the *AClass* instance *x*. This expression can be defined in one of four ways:

<x-expr'i> ::= <x-dependent attribute expression> | <x-dependent count expression> | <x-dependent {SUM/MAX/MIN/MOST] expression> | <x-dependent child attribute selector expression>

<x-dependent attribute expression> examples: *x.surname*, x.discharge Time.date(). *Prefix "x." can be used before attributes of both* x *and its parents. Semantics is obvious.*

<x-dependent count expression> examples: *(COUNT x.Manipulations, WHERE manipul.code=02078), (COUNT HospitalEpisodes, WHERE responsiblePhysician=x).* In the first example we use the prefix *x* in "x.Manipulations" to denote that we do not select from the whole set of manipulations, but only from those that are reachable from *x*.

<x-dependent {SUM/MAX/MIN/MOST} expression> examples: *(SUM manipul.cost FROM x.Manipulations, WHERE manipul.code=02078), (MOST diagnosis.code FROM AdmissionDiagnoses, WHERE nr=1 AND responsiblePhysician=x).*

<x-dependent child attribute selector expression> examples: *(x. DischargeDiagnosis, WHERE nr=1).diagnosis.code, (x.TreatmentWard, WHERE nr=*).ward* (by * we denote the number of the last instance of TreatmentWard connected to the given HospitalEpisode *x*). This is a new kind of construction whose general form is as follows: *(x.<name of x children class A>, WHERE <selection condition>).<name of attribute a of class A>.* Its value is defined in the following way – we start by taking all instances of class A that are reachable from *x*, then select of them those instances that satisfy the selection condition and then create a list of values of the attribute *a* of the selected instances. The most important case here is the one where this list contains only one instance, e.g. in the following table definition example:

SELECT HospitalEpisodes x, WHERE dischargeReason=deceased, DEFINE TABLE x.surname (COLUMN Surname), x.dischargeTime.date() (COLUMN Dying_date), (x. DischargeDiagnosis, WHERE nr=1).diagnosis.code (COLUMN main_diagnosis), (x. TreatmentWard, WHERE nr=).ward (COLUMN last_ward).*

T7. There are two more cases in the definition of the table, where table rows come from other source, not being instances of some class. Being very similar these two cases form two subtemplates of the last template:

a) **SELECT FROM AClass [a] WHERE <selection condition> ATTRIBUTE <attribute expression> ALL DISTINCT VALUES x, DEFINE TABLE…**

b) **SELECT FROM INTERVAL (start-end) ALL VALUES x, DEFINE TABLE…**

Semantics of both cases is obvious. Examples:

– *SELECT FROM TreatmentWards ATTRIBUTE ward ALL DISTINCT VALUES x, DEFINE TABLE x (COLLUMN Ward), (SUM manipul.cost FROM Manipulations, WHERE ward=x) (COLUMN Cost);*

– *SELECT FROM INTERVAL (1-12) ALL DISTINCT VALUES x, DEFINE TABLE x (COLUMN Month), (COUNT HospitalEpisodes, WHERE admissionTime.month() =x) (COLUMN Episode_count) (MOST diagnosis.code FROM AdmissionDiagnoses, WHERE nr=1 AND admissionTime.month()=x) (COLUMN Most_frequent_main_diagnosis).*

Let us conclude this section by defining more formally the constructs of a controlled natural language allowed in the selection conditions. They can, of course, be guessed from the examples given above.

```
<AClass selection expression> ::= AClass [<short name>]
[WHERE <selection condition>]
   <selection   condition>   ::=   <attribute   condition>   |
<quantor  condition>  |  (<selection  condition>  {AND|OR}
<attribute  condition>)  |  (<selection  condition>  {AND|OR}
<quantor condition>)
   <quantor  condition>  ::=  {[NOT]EXISTS  |  FORALL}  [short
```

Short name provides a name for the given object and can be any string different from the class and attribute names. The abovementioned grammar provides a formal language (for formulating selection expressions) that is close to a natural language and therefore easily perceptible. From a natural language's point of view selection expressions are sentences in a controlled natural language that exploit both words of a natural language (like *AND, OR, WHERE, EXISTS, NOTEXISTS*) and "foreign" words – attribute expressions whose syntax and semantics were described above. The grammar is only needed as a guide how to build the selection expressions. We do not use it in teaching the language to domain experts. We, instead, use the same principle exploited when a child learns to speak a natural language, i.e. learning from examples. It was therefore necessary to first see the sentence templates together with some examples and only afterwards see the formal grammar underlying parts of these templates.

5 Implementation of the Query Language

In this section we will lay out the basics of the system underlying the query implementation. The very basic component here is the system metamodel which describes the classes, associations and attributes used in the particular hospital metamodel on which the query is to be executed. A simplified version of the system metamodel can be seen in Fig. 2. This metamodel is coded in Java, where each logical class of the metamodel corresponds to one Java class. There are only two types of multiplicities used for roles in the system metamodel – 1 and *. These are coded as Java attributes of types *A* and *List<A>* respectively (where *A* is the name of the class to which the respective association end is attached).

The overall architecture of the system is depicted in Fig. 3. The system metamodel described above forms the basis of the architecture and serves as an input for three types of generators seen in Fig. 3.

The first generator is used for generating information about abstract data types for the querying system. It takes as an input information about classes from the abovementioned system metamodel and gives as an output a set of Java classes together with their respective attributes (including associations). These classes are the ones

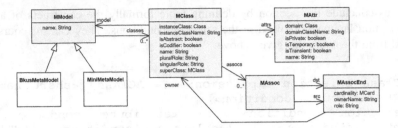

Fig. 2. The system metamodel (a simplified version).

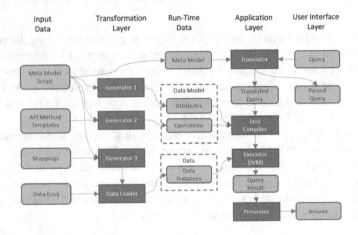

Fig. 3. The overall system architecture.

describing the underlying hospital information system. Let us call them hospital classes. The query will be formed and executed on hospital classes.

The second generator generates operations for the hospital classes generated by the first generator. It takes as an input the system metamodel and the operation templates and generates necessary operations for hospital classes. These operations lay a foundation to which the natural language queries will be translated. The structure of operation templates and their consisting parts are outside the scope of this paper.

The third generator generates the Java code (called the Loader) that loads particular instances of hospital classes from .csv files into the hospital classes themselves. The generator takes the system metamodel and a mapping from .csv to Java classes as inputs and gives the fully functioning Loader code as an output. The Loader can then read the .csv files obtained from the database of the hospital information system and generate from them respective instances of the hospital classes (each .csv file more or less corresponds to one hospital class). The Loader must be able to cope with incomplete or even incorrect data, which are typical situations, as our experience has shown. For example, the loading process must work well, when some attribute is not set in the .csv file, when the value of some attribute does not comply with the expected data type, when the foreign key points to a non-existent instance of another table etc.

The mapping from .csv to Java classes is currently very simple, it only links text fields to attributes. The development of this mapping is a broad topic itself, and we have several ideas about how it can be improved – it can take into account only things needed for the execution of a particular query, it can be bi-directional etc. This is, however, not in the scope of this paper.

All three generators are started one after another, when the system is started. It can take some time, depending on the volume of data to be read from .csv files. When the process has ended all the necessary data are located in RAM, and the system is ready to execute user queries.

Query execution consists of a Read-Evaluate-Print loop – user enters a query in a controlled natural language, it is then translated to Java syntax (largely exploiting the operations generated for hospital classes by the second generator), executed in Java, and finally the obtained result is depicted back to the user. There is one side branch to this general schema as can be seen in Fig. 3 – evaluation of the natural language query is performed by the query translator during the query forming phase, and the parsed query is shown back to the user immediately so that he can alter the query accordingly if necessary. To be able to understand and parse the query, the translator also exploits the system MM to obtain information about class names, attributes and associations.

Large part of the system functionality is composed of the operations generated for hospital classes during the system start-up. Let us now take a bit deeper insight into how these class methods look like. These are functions written in the functional programming style:

- $F<T_1,T_2,...,T_n,R>$ – function of n arguments (with types T_1, T_2, ..., T_n) whose return type is R;
- $pred<T>$ – predicate of one argument with type T (the same as $F<T,bool>$).

There are two types of functions, namely, global functions over all instances of some class and more local functions over only those instances of some class that can be reached from some given instance. For example, there is a method $countA(pred<A>)$ generated for every class A. It returns the total count of instances of class A, for which the given predicate returns true. Similarly, there is a method $countAB(b,pred<A>)$, which only inspects those instances of class A that are reachable through links from the given instance b of class B. A list of the main methods can be seen in Table 1.

In the process of designing the operation structure and syntax two contradicting things were to be taken into account – logical clearness and potential performance. One of the first goals for this system was its ability to respond to queries rapidly (in no more than 1–2 s on one year data of Riga Children's Clinical University Hospital) and to do that on a regular computer at hospital, not on a supercomputer. Thus, the implementation efficiency was the main factor that dictated the design of the class methods to which queries are translated. It is therefore not the best solution from the logical clearness point of view.

Concluding the description of the query execution engine, it can be noted that we have developed our own No-SQL embedded in-memory database with a functional query language and with a good API in the form of the generated operations. We have eliminated the need for interpreter and reflection, which has given us an opportunity to improve the performance of query execution. Of course, some improvements have been

Table 1. List of the main methods of hospital classes.

Prefix of method name	Global method	Method with context
count	+	+
sum	+	+
min/max	+	+
avg	+	+
most	+	−
countDistinct	+	−
findOne/findAll	+	+
any/all	+	+
map	+	+
table	+	+

made possible thanks to the fact that we exploit the data only in read-only mode. We have explored other in-memory databases before and found no solution suitable for our specific needs. We will not analyze deeper other approaches in this paper.

Let us conclude with an example showing the translated result of a query.

Query in our language: *COUNT Patient p, WHERE EXISTS HospitalEpisode e, WHERE EXISTS Manipulation m, WHERE manipul.code=02078*

Translated query in Java: *countPatient(p->anyHospitalEpisodePatient(p,e->anyManipulationHospitalEpisode(e,m->m.manipul.code.equals("02078"))));*

6   Proof of Concepts

We showed in Sect. 3 that semistar ontologies cover quite a wide spectrum of practically important data ontologies including hospital data schemas from patients' and physicians' point of view. In this section we will briefly inspect the other aspects of the "3How" problem mentioned in Introduction, i.e. (1) Is the offered query language expressive enough for practical use-cases and simple enough for understanding by domain experts; and (2) Does the query language have sufficiently efficient implementation?

The first aspect consists of two parts. The expressiveness of the query language was demonstrated by turning it into a working language for Riga Children's Clinical University Hospital when annual reports had to be generated. It turned out to be expressive enough for this task. During the two year period, when it was used for report generation, the language underwent a continuous improvement process. It was important to achieve such a level that managers of wards are able to formulate themselves all the necessary queries without going to a programmer with every 5th or 10th query to write the desired query in SQL. Results of such queries were either single numbers or data fields, or tables of data fields. In case of tables, we do not undertake all the necessary calculations and table operations provided by other applications such as MS Excel. Our aim is to generate a table containing all the necessary data that can then be exported to a spreadsheet or an R tool (a tool for statistical analysis).

The second part of the first abovementioned aspect regards the possibility for domain experts to learn the language. To test this aspect we performed both individual experiments with potential end-users and group tests. General situation from the language teaching point of view was best demonstrated in an experiment with experienced nurses who study to obtain Master Degree at the Medical Faculty, University of Latvia. We presented a two hour long lecture about the language and the tool for querying the data. One third of that time was devoted to explanation of the underlying data ontology (to explain to non-programmers what is a class, an attribute, etc.). Afterwards the language was explained on examples, and homework was given to test the level of understanding. The homework consisted of two parts. Firstly, students had to understand sentences written in our controlled natural language and to write them in a good really natural language. Secondly, they had to work in the opposite direction turning natural language sentences into our formal language. Results obtained from this experiment can be seen in Table 2. The main conclusion here is that another two hours long lecture after the completion of homework would be beneficial for a better understanding of the proposed query language.

Table 2. The results of the experiment

Task execution level (%)	Number of students succeeded (n = 15)				
	≥ 90	$\geq 75 < 90$	$\geq 50 < 75$	$\geq 25 < 50$	< 25
Understanding of queries	9	2	2	1	1
Writing of queries	3	5	2	3	2

A very important factor related to the teaching process is the fact that the underlying data ontology was anonymized (from patients', physicians' and wards' point of view), but real. Our experience shows that students being domain experts of this ontology rapidly got very interested in the querying process and started to perceive this as a game. This fact had a beneficial impact on the learning process.

Finally, let us talk a bit about the ability to implement the language efficiently. The main principles of such implementation were already given in Sect. 4. To test the performance of our implementation we gathered 120 typical query examples from real annual report analysis of intensive care ward and from discussions with managers of other wards. The complexity of these queries is similar to those demonstrated in Sect. 4. The volume of data over the period of year 2015 was the following – there were about 35'000 hospital episodes and 70'000 outpatient episodes in Riga Children's Clinical University Hospital (in total the data took up less than 2 GB RAM). The performance on such queries and data volume can be seen in Fig. 4, where queries are sorted in an increasing order by their execution time.

We can see that the vast majority of these 120 queries executes in less than 0.3 s. According to statistics there are about 350'000 hospital episodes together in all hospitals in Latvia per year (about ten times more than in Riga Children's Clinical University Hospital). It means that all these data would take up less than 20 GB RAM. Currently a quad-core computer with 32 GB RAM costs about 1'000 euro. Since the

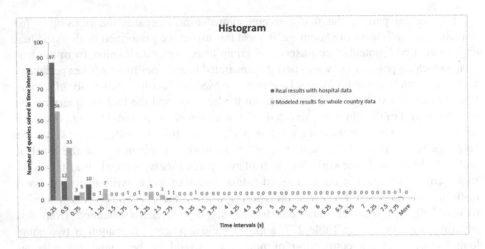

Fig. 4. Performance of the query execution.

semistar data ontology is granular [21, 22], the query execution can be done in parallel on all four cores thus improving the execution time four times (our experiment seen in Fig. 4 was performed on only one core). We can conclude that the performance of the query execution over data of all the hospitals in Latvia would only be 2.5–3 times slower than it is now providing the ability to answer a vast majority of queries in less than one second.

We are, of course, not limited by only one computer with four cores. We can also use several computers connected via high throughput Ethernet thus reducing the waiting time even more (e.g. one second on ten year data of all Latvian hospitals). Of course, sufficient performance on very large data volumes is another research topic that requires more studies.

By working on the proof of concepts we can conclude that practical testing of our approach has demonstrated that the "3How" problem can be successfully solved at least for the scope of the health system in Latvia.

7   Conclusions and Future Work

Practical experiments with our query language have shown that there are yet at least two important features that must be added to increase its usability – subset definition feature (*DEFINE InfectiousDisease = SELECT ...*) and attribute definition feature (*DEFINE HospitalEpisode.duration = dischargeTime-admissionTime*). These and similar features are currently under development and require some technical work to be implemented. Another useful feature would be to obtain event distribution in time, which could further be analyzed in MS Excel using its time axis component.

In this paper we have sketched a formal natural language for formulating queries that is still quite a bit controlled. Our future goals include reducing the level of control, so that the language would become more and more usable for domain experts being non-programmers. Queries would be formulated very inaccurately (perhaps providing only some basic keywords), and the system could try to understand the query the user has wanted to formulate and offer the resulting query (or more than one potential queries) back to the user for affirmation. This would be the next step towards a really user-friendly query language.

Acknowledgments. This work is supported by the Latvian National research program SOPHIS under grant agreement Nr.10-4/VPP-4/11.
Authors are also very thankful to Lolita Zeltkalne for language consulting.

References

1. Barzdins, J., Rencis, E., Sostaks, A.: Data ontologies and ad hoc queries: a case study. In: Haav, H.M., Kalja, A., Robal, T. (eds.) Proceedings of the 11th International Baltic Conference, Baltic DB&IS, pp. 55–66. TUT Press (2014)
2. Chamberlin, D.D., Boyce, R.F.: SEQUEL: a structured English query language. In: Proceedings of ACM SIGFIDET Workshop, Ann Arbor, Mich., pp. 249–264, May 1974
3. Prud'hommeaux, E., Seaborne, A.: SPARQL Query Language for RDF. W3C Recommendation, 15 January 2008. http://www.w3.org/TR/rdfsparql-query
4. Microsoft. Graphical Query Designer User Interface. https://msdn.microsoft.com/en-us/library/ms365414.aspx. Accessed 3 Feb 2016
5. Zviedris, M., Barzdins, G.: ViziQuer: a tool to explore and query SPARQL endpoints. In: Antoniou, G., Grobelnik, M., Simperl, E., Parsia, B., Plexousakis, D., Leenheer, P., Pan, J. (eds.) ESWC 2011, Part II. LNCS, vol. 6644, pp. 441–445. Springer, Heidelberg (2011)
6. Kogalovsky, M.R.: Ontology-based data access systems. Program. Comput. Softw. **38**(4), 167–182 (2012)
7. Soylu, A., Kharlamov, E., Zheleznyakov, D., Jimenez-Ruiz, E., Giese, M., Horrocks, I.: Ontology-based visual query formulation: an industry experience. In: Bebis, G., et al. (eds.) ISVC 2015. LNCS, vol. 9474, pp. 842–854. Springer, Heidelberg (2015). doi:10.1007/978-3-319-27857-5_75
8. SAP. QuickViewer. http://help.sap.com
9. Aspin, A.: Self-service business intelligence. In: High Impact Data Visualization with Power View, Power Map, and Power BI, pp. 1–18. Apress (2014)
10. Li, F., Jagadish, H.V.: Constructing an interactive natural language interface for relational databases. PVLDB **8**(1), 73–84 (2014)
11. Llopis, M., Ferrández, A.: How to make a natural language interface to query databases accessible to everyone: an example. Comput. Stand. Interfaces **35**(5), 470–481 (2013)
12. Papadakis, N., Kefalas, P., Stilianakakis, M.: A tool for access to relational databases in natural language. Expert Syst. Appl. **38**, 7894–7900 (2011)
13. Androutsopoulos, I., Ritchie, G.D., Thanisch, P.: Natural language interfaces to databases – an introduction. Nat. Lang. Eng. **1**(1), 29–81 (1995)
14. Popescu, A.M., Armanasu, A., Etzioni, O., Ko, D., Yates, A.: Modern natural language interfaces to databases: composing statistical parsing with semantic tractability. In: COLING (2004)
15. Capindale, R.A., Crawford, R.G.: Using a natural language interface with casual users. Int. J. Man-Mach. Stud. **32**, 341–361 (1990)

16. Ogden, W.C., Kaplan, C.: The use of AND and OR in a natural language computer interface. In: Proceedings of the Human Factors Society 30th Annual Meeting, pp. 829–833. The Human Factors Society, Santa Monica (1986)

17. Yin, P., et al.: Answering questions with complex semantic constraints on open knowledge bases. In: Proceedings of the 24th ACM International on Conference on Information and Knowledge Management. ACM (2015)

18. Chin, D.: An analysis of scripts generated in writing between users and computer consultants. In: National Computer Conference, pp. 637–642 (1984)

19. Santoso, H., et al.: Ontology extraction from relational database: concept hierarchy as background knowledge. Knowl.-Based Syst. **24**(3), 457–464 (2011)

20. Bartolini, R., Caracciolo, C., Giovanetti, et al.: Creation and use of lexicons and ontologies for NL interfaces to databases. In: Proceedings of the International Conference on Language Resources and Evaluation, vol. 1, pp. 219–224 (2006)

21. Barzdins, J., Rencis, E., Sostaks, A.: Granular ontologies and graphical in-place querying. In: Short Paper Proceedings of the PoEM 2013. CEUR-WS, vol. 1023, pp. 136–145 (2013)

22. Barzdins, J., Rencis, E., Sostaks, A.: Fast ad hoc queries based on data ontologies. In: Haav, H.M., Kalja, A., Robal, T. (eds.) Frontiers of AI and Applications. Databases and Information Systems VIII, vol. 270, pp. 43–56. IOS Press (2014)

23. Yang, J.S.H., Chin, Y.H., Chung, C.G.: Many-sorted first-order logic database language. Comput. J. **35**(2), 129–137 (1992)

Database to Ontology Mapping Patterns in RDB2OWL Lite

Kārlis Čerāns[(✉)] and Guntars Būmans

Institute of Mathematics and Computer Science, University of Latvia, Raina
Blvd. 29, Riga LV-1459, Latvia
{karlis.cerans,guntars.bumans}@lumii.lv

Abstract. We describe the RDB2OWL Lite language for relational database to
RDF/OWL mapping specification and discuss the architectural and content
specification patterns arising in mapping definition. RDB2OWL Lite is a simplification of original RDB2OWL with aggregation possibilities and order-based
filters removed, while providing in-mapping SQL view definition possibilities.
The mapping constructs and their usage patterns are illustrated on mapping
examples from medical domain: medicine registries and hospital information
system. The RDB2OWL Lite mapping implementation is offered both via
translation into D2RQ and into standard R2RML mapping notations.

Keywords: Database to ontology mapping · Ontologies · RDF · Mapping
patterns

1 Introduction

Exposing the contents of relational databases (RDB) to semantic web standard RDF [1]
and OWL [2] formats enables the integration of the RDB contents into the Linked Data
[3] and Semantic web [4] information landscape. An important benefit of
RDB-to-RDF/OWL mapping is also the possibility of creating a conceptual model of
the relational database data on the level of RDF Schema/OWL and further on accessing
the RDB contents from the created semantic/conceptual model perspective. Since the
RDB-to-RDF/OWL mapping connects the technical data structure (the existing RDB
schema) with the conceptual one, it would typically have not a one-to-one data
structure reproduction, but would rather involve some data transformation means.

The task of mapping relational databases to RDF/OWL format is nowadays well
understood and widely studied, just to mention D2RQ [5], Virtuoso RDF Graphs [6],
Ultrawrap [7] and Spyder [8] among different RDB-to-RDF/OWL mapping languages
and tools, together with W3C standard R2RML [9]. Most of RDB-to-RDF/OWL
mapping languages and approaches, however concentrate most on clear mapping
structure and efficient implementation with less attention paid to the concise mapping
writing possibility suitable both for manual mapping information creation and using of
the mapping information as semantic-level documentation of the relational database
structure. There are higher-level mapping definition means in Spyder tool [8], as well
as ontop [10], where the mappings are described as separate artifacts to be considered

© Springer International Publishing Switzerland 2016
G. Arnicans et al. (Eds.): DB&IS 2016, CCIS 615, pp. 35–49, 2016.
DOI: 10.1007/978-3-319-40180-5_3

besides both the database and ontology structures. An earlier work of the authors [11, 12] on RDB2OWL language has shown a principal possibility of reusing both the target ontology and source database structures in the mapping definition via placing the mapping information in a compact textual form into the annotations of the target ontology entities. A distinguishing RDB2OWL language feature is the possibility to introduce user defined functions and a number of meta-level features such as multi-class conceptualization this way obtaining compact and more reader-oriented mapping specifications.

RDB2OWL Lite is a simplification of original RDB2OWL by removing in-mapping aggregation possibilities and order-based filters. We show here by analysis of database-to-ontology mapping architectures that the aggregation operations can be performed already on the level of SQL views within the databases; limited forms of aggregation can be performed also on the end-user query level either directly in SPARQL 1.1 language [13] that supports aggregate query notation, or in visual query language ViziQuer [14].

RDB2OWL Lite provides also new practically motivated mapping features such as in-mapping SQL view definition. This feature can be efficiently exploited in translation of RDB2OWL mappings into R2RML and it can turn the database pre-processing phase unnecessary. New notations for linked view referencing in data property maps and class instance URI specification are introduced, as well.

The RDB2OWL Lite implementation is available both via earlier existing mapping translation into D2RQ (D2RQ mapping running is supported by D2R server [5]), and in a novel way via translation into the standard R2RML notation [9], supported by a number of tools, including ontop [10] and R2RML Parser [15].

The RDB2OWL Lite mapping definition is the integrated with basic ontology inferencing facilities. The RDB2OWL implementation suite is able to build in the ontology subclass, sub-property and inverse properties inference into the D2RQ or R2RML mapping generation phase so allowing an inference-agnostic runtime mapping frameworks to serve the RDF data from the mapping that include also the inferred knowledge.

A particular usage scenario for RDB2OWL mappings is a semantic re-engineering of existing relational databases (cf. e.g. [16, 17]) aiming at existing data access from the data ontology conceptual model perspective. This scenario has motivated most of the mapping constructions built into RDB2OWL. The tool has been used to re-work the Latvian medicine registries example [16] with the ontology containing 173 classes, 219 object properties and 816 data properties into a maintainable mapping solution. The tool has been tested on mapping the hospital information system, analyzed initially in [18], onto the corresponding data ontology, as well.

In what follows, Sect. 2 describes architectural patterns of database-to-ontology mapping solutions, Sect. 3 introduces RDB2OWL Lite basic mapping patterns, followed by advanced mapping patterns in Sect. 4, tool implementation notes in Sect. 5 and application cases in Sect. 6. Section 7 concludes the paper.

2 On Database to Ontology Mapping Architecture

One of the goals of the database-to-ontology mappings is to serve as a bridge between the conceptual ontology level and technical database level in answering queries, formulated on the conceptual level over the data residing in the database (the other goal of the mappings can be to provide the conceptual-level documentation of the source database structure).

Figure 1 outlines the basic database-to-ontology mapping architecture schema, involving both the mapping application and query processing steps in accordance to the principal mapping application goal considered in this section.

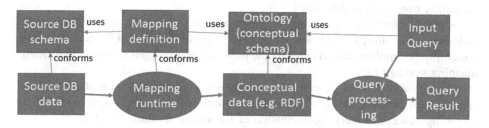

Fig. 1. The basic database-to-ontology mapping architecture schema

There are two variants of the considered mapping architecture, namely (i) "batch processing" option, where the conceptual data corresponding to the source data base are computed and stored in an appropriate repository (e.g., a RDF data store), and (ii) "on the fly" option, where the mappings are executed on-demand, as needed to answer the queries asked.

The conceptual query infrastructure, as depicted in Fig. 1, is usually thought of as an addendum to the production user interface that is well suited for everyday data manipulations as well as the standard, predefined queries over the data.

The Fig. 1 query infrastructure can be, in principle, connected directly to the production database (we call this option **Scenario A**), with the clear benefit of avoiding necessity to make a copy of the source database, what would be important e.g. in case of very large data sets. The **Scenario B** option, however, foresees using a copied and possibly transformed database as the source for the mappings and queries. In the medical domain such a transformation may typically involve data anonymization or de-personalization, as what has been done both in examples from [16] and [18] considered in this paper. Such anonymization or de-personalization, although losing some aspects of the data, may substantially extend the user base of the query mechanism, as well as to simplify the user access controls needed for the query mechanism. Not giving any explicit preference to any of the two scenarios, note that within Scenario A, in addition to the user and permission management issues, also the eventual extra workload of the custom queries over the production database is to be taken into account.

We note also that the Scenario B with attaching the mapping and query infrastructure to the (transformed) copy of the production database, allows for an easy possibility to introduce a SQL-level pre-processing step of the source database. Such a pre-processing step is essential for both use cases considered here, and it generally may involve:

- adding new static information in form of new data tables and/or views (e.g. to introduce classifier information that is missing in the original database)
- adding new dynamic information in form of data-dependent views, including sophisticated value computation and/or aggregate operations (in the case of full pre-processing step, the data dependent views can even be materialized).

Given the appropriate administrative and technical arrangements, the pre-processing step can be introduced to some extent also within Scenario A, however, since this introduction involves modification of the production database, its practical implementation may run into difficulties.

A partial solution to the need of the pre-processing step in the form of SQL tables and/or views is offered by the new RDB2OWL Lite in-mapping SQL view definition possibility (its implementation relies on the presence of analogous construct in R2RML mapping standard [9]).

The use of SQL-level pre-processing, however, is best to be kept to the minimum, since it runs contrary to the other database-to-ontology mapping usage goal that is providing conceptual-level documentation of the source database structure.

3 RDB2OWL Basic Mapping Patterns

The correspondence between a relational database and an OWL ontology/RDF schema is defined by means of an RDB2OWL Lite mapping (in what follows we shall use simply RDB2OWL to mean RDB2OWL Lite, if not explicitly said otherwise). The RDB2OWL mapping, as described in [11, 12] consists of elementary mappings, or maps, describing the connections to the database for ontology/RDF schema classes (class maps), object properties (object property maps) and data properties (data property maps), ascribed by annotation assertions to the respective ontology/schema entities[1] (i.e. the classes, object properties (intuitively, the roles linking two classes) and data properties (intuitively, the attributes ascribing literal values to class instances).

The general pattern of class map as well as object and data property map definition follow the structural principle of mapping ontology classes to database tables, data properties to table columns and object properties to links between tables.

More precisely, a class map in RDB2OWL is characterized by a table expression, possibly involving a join of several tables and a filter, and a pattern for instance resource URI construction from a table expression row. A class map can be attached to a single ontology/schema class, meaning that it describes instances for this class.

An object property map contains references to its subject and object class maps that are included within the property map's (extended) table expression structure; such a reference brings the class map's table expression as a component into the property

---

[1] We use `rdb2owl:DBExpr` as the annotation property, where `rdb2owl: = <http://rdb2owl.lumii.lv/2012/1.0/rdb2owl#>`.

map's table expression and provides the pattern of the property triple subject resp. object URI generation from the property map's table expression row. An object property map is always attached to a single object property within the ontology.

Similarly, a data property map is characterized by a corresponding (extended) table expression that contains a reference to the property map's subject class map, and a value expression describing the computation of the property literal value that corresponds to the URI, as computed by the subject class map's URI pattern. A data property map is always attached to a single data property within the ontology.

The textual form of RDB2OWL maps relies on textual presentations of table expressions, instance URI generation patterns and expression values; it is optimized to take into account the available ontology and RDB schema context information.

We illustrate the basic RDB2OWL mapping patterns first on a simplistic mini-University example. Figure 2 contains a mini-University OWL ontology in OWLGrEd[2] ontology editor notation [19] (showing OWL classes as UML classes, OWL object and data properties as UML association roles and attributes)[3]. The ontology in Fig. 2 contains both its structure definition (the classes and properties) and further OWL constructs (e.g. class expressions and datatype facet restrictions), as well as it is enriched by a custom *{DB: <annotation_value>}* notation for the database connection annotations (cf. [20]). The corresponding RDB schema is outlined in Fig. 3.

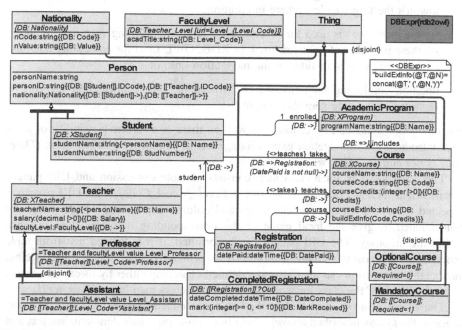

Fig. 2. Mini-university ontology with RDB2OWL mapping annotations The '{DB: …}' notation is used for annotation assertions with rdb2owl:DBExpr property

---

[2] The ontology editor can be downloaded from http://owlgred.lumii.lv/.

[3] The ontology with annotations in a standard OWL RDF/XML format is available at http://rdb2owl. lumii.lv/demo/UnivExample.owl.

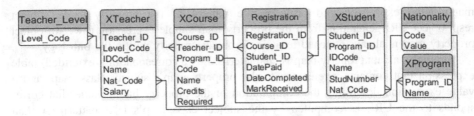

Fig. 3. A mini-University RDB schema

A *table expression* in RDB2OWL in the simplest and most common case is just a table name, such as *XTeacher*, or *Teacher_Level*; a filter expression can be added to a table expression, using a semicolon, such as '*XCourse;Required = 1*'. The table expressions involving several tables can be formed either in

(a) item list notation with comma-separated, possibly alias-labeled table expressions, such as '*XTeacher T, XCourse C; T.Teacher_ID = C.Teacher_ID*', or

(b) navigation list notation, such as '*XTeacher[Teacher_ID]-> [Teacher_ID] XCourse*' that can be shortened to '*XTeacher-> XCourse*' by omitting navigation columns that are (i) target table PKs and (ii) their matching sole source table FKs towards the target tables; where in addition:

   – the mark '=>' is used for reverse, i.e. PK-to-FK mapping specification and
   – there is option of '=' mark (new in RDB2OWL Lite) for coinciding database table PK's on both sides of the navigation link, or

(c) notation combining both (a) and (b), where the whole navigation list can be regarded as a single item in the item list.

The navigation links can be chained, as well as there is a further possibility of local filter introduction in navigation expressions, such as *XStudent =>Registration:(DatePaid is not null)-> XCourse*.

The textual syntax for a *class map* consists of table expression and URI pattern specification, such as '*Teacher_Level [uri = Level_{Level_Code}]*'; the URI pattern is expected to have a list of constant strings and column expressions whose values are concatenated to give the local name of the corresponding ontology class instance within the ontology. The URI pattern can be omitted, if it is formed just from table expression's leftmost table name followed by its primary key column(s) value. A class map's table expression can refer also to a defined class map either by its explicit name, or by the name of the class for which it is the sole class map (e.g. *[[Teacher]]*).

The *object property map* is described as an extended table expression (in accordance to the syntax described above) where two sub-expressions denote its subject and object class maps respectively. These sub-expressions can be either:

(a) explicitly marked by an alias <s> or <t>, respectively, in some place within the table expression's structure,

(b) defined by the mark <s> or <t> as a table expression item itself; in this case the sole class maps defined explicitly for the property domain or range class are considered the subject and object class maps for the property map, or

(c) in the case of missing explicit <s> and/or <t> marks these marks are assumed <s> for the leftmost and < t > for rightmost item within the table expression.

These conventions on object property map as well as table expression syntax allow denoting e.g. the object property map simply by '->', if the property corresponds to the sole FK-to-PK mapping between the tables serving as table expressions in the sole class maps of property domain and range classes,

The *data property map* can be described by a column name/column expression that is to be evaluated in the context of the sole class map attached to the property domain class. Alternatively, *<table_expression>.<value_expression>* notation can be used for data property map description, where the table expression is required to contain either explicit or implicit reference *<s>* to its source class map that is either specified explicitly or is the sole class map ascribed to the property map's subject class.

4 Advanced Mapping Patterns in RDB2OWL Lite

Although the basic RDB2OWL Lite mapping patterns allow defining a wide range of database and ontology correspondences that already are not one-to-one, the practical use cases require advanced mapping pattern usage, either for end user convenience, or for bringing in new mapping definition facilities. We explain the advanced mapping patterns on the basis of the introduced use cases from the medical domain.

4.1 User Defined Functions and Multi-class Conceptualization

We review first two of the most important advanced mapping definition patterns present in the initial RDB2OWL, namely, user defined functions and multi-class conceptualization.

The user defined functions allow to move part of the code for data property map out of the annotations ascribed to the data properties themselves. This can be useful in the case of complex correspondence definition, as well as it is essential in the case, if several parts of the mapping are based on dependencies of the same form.

The example in Fig. 2 contains also an illustration of a function definition using an ontology level *rdb2owl:DBExpr* annotation, the defined function *buildExtInfo* is then used in a *courseExtInfo* property value expression for the *Course* class. The RDB2OWL functions can be used for repeating mapping fragments, as well as for increasing mapping readability in the case of complex mapping constructions. The function body can introduce also additional table expressions into the context of evaluating its value expression; such table expressions are joined to the table expression defining the context of evaluation of the function-calling value expression.

The user-defined functions, used extensively in the Latvian Medicine Registries example are the following:

$boolT(@X) = (CASE(@X)$ *WHEN 1 THEN 'true' WHEN 0 THEN 'false' END)* $^{\wedge\wedge}xsd{:}Boolean$, that, given a 1 or 0 value in a database table field, returns a Boolean value, correctly formatted for xml documents;

$split4(@X) = [Numbers;len(@X) > n*4].(substring(@X,n*4 + 1,4))$, that, given the table or view *Numbers* in the database with a column N containing integer values from 0 up to some limit (e.g., 20), allows the application *split4(FieldX)* to split character string in the *FieldX* column into set of its consecutive substrings of length 4;

$xToInt(@X) = [;@X <> ''].(@X)^{\wedge\wedge}xsd{:}integer$, that for a database field with varchar type containing integer values eliminates value generation in the case of database field value being an empty string.

The *multiclass conceptualization* is a mapping pattern where one database table T is mapped to several ontology classes $C1, C2,..., Cn$ each one reflecting some subset of T columns as the class' properties. The RDB2OWL mapping decoration ?Out in a class map placed in the annotation of a class', say Ci, implies the inclusion into Ci only for those instances that have at least one property defined, whose domain is within Ci. In the introductory example of Fig. 2 the decoration *?Out* in the database annotation for *CompletedRegistration* class means relating to this class only those rows of the described class map table expression that contain a value in at least one property whose domain is this class.

The multiclass conceptualization feature is extensively used for the subclasses of an Anamnesis class in the Medicine registries example (8 classes with total 220 data properties).

4.2   Support for Views on Class Map Table

The data property maps for data properties can, by default, reference the table expression context ascribed to the class map of data property domain class (say, C). A typical situation would be that the class C is mapped onto some database table (say, T), but the data property could correspond to a column that is available in some database view that is based on the table T.

To handle this practically important situation, a novel annotation form *[= view].* *viewCol* is introduced into RDB2OWL Lite, where a row in view is linked to the row in T on the basis of the equality of the values in the T primary key column(s). This feature finds an extensive usage in the hospital information system application case, as described further in Sect. 6.2 and illustrated in Fig. 5 therein.

4.3   Sub-class and Sub-property Inference

The direct entity-by-entity translation of RDB2OWL mappings into D2RQ or R2RML mappings would create a situation with data mappings available only for the classes directly mapped towards the database. For instance, if an ontology class, say, *Student*, is said to correspond to a database table *XStudent*, and there is a super-class *Person* of the *Student* class, there would be no data mapping from the *XStudent* table into the *Person* class. If the resulting D2RQ or R2RML mappings are used in the "RDF dump" mode for generating the RDF triples corresponding via the mapping to the source database contents, and the generated RDF triples are stored in a RDF triple store such as Virtuoso, the triple store can perform the inference corresponding to the sub-class relation in the ontology. The RDF triple stores would also typically support the sub-property and inverse properties inference.

If, however, the R2RML or D2RQ mapping is used for on-the-fly RDB data access via the target ontology structure, the mapping support tools would typically not support similar inference capabilities. For this situation there is the "inference" option available in the RDB2OWL mapping tool that implements the inference step during the target mapping generation phase. So, the RDB2OWL tool would together with the target D2RQ or R2RML class map for the *Student* class, generate also a similar class map for the *Person* class. The RDB2OWL tool for D2RQ or R2RML mapping generation currently supports subclass, sub-property and inverse properties inference. It would not be difficult to add mappings also for e.g. property chain conditions in the target ontology to the RDB2OWL mapping generation (an inference mode not supported by most RDF triple stores) via automated "chaining" of the mapping expressions corresponding to the chained properties. For instance, one could add a property chain requirement *takes > inv(student) o course* to the ontology of Fig. 2 and obtain the chained mapping =>Registration-> for the *takes* property between the *Student* and *Course* classes (this mapping is then to be translated into the target mapping language); however, the usage of RDB2OWL constructs directly in mapping description allow for a finer-grained mapping specification, e.g. with adding conditions such as =>Registration:(DatePaid is not null)->, not expressible in OWL.

The sub-class and sub-property inference mechanisms are fundamental to the Latvian Medicine Registries ontology, the use of these facilities is outlined further in Sect. 6.1.

4.4 In-Mapping SQL Views

RDB2OWL Lite allows introducing virtual database views within the database reference annotations using the keyword 'View' followed by its name and its definition in SQL syntax, e.g.

DBRef(View[courseView as SELECT c.* FROM "COURSE" c],
    View[courseView2 as SELECT
    COURSE_ID, 'course_name = ' + name AS full_name FROM COURSE]).

The views introduced in this way into the mapping description can be referred to in the same way as views contained in the database itself within the table expressions used in class map and property map definitions.

Using the RDB2OWL Lite tool for the R2RML mapping generation, the in-mapping SQL views are translated into R2RML queries, this way allowing the view application without actually storing the view definition in the database. This option would be important while implementing the Scenario A of database-to-ontology mapping organization (cf. Sect. 2), where the mapping and query structure is connected directly to the production database.

Unfortunately the D2RQ mapping language does not support a similar construction of entering an SQL query as the class map definition, therefore in the case of using

D2RQ mappings the additional pre-processing step creating the defined views in the source database is to be applied and, therefore, the source database for the ontology-level data access cannot be kept intact.

5 Mapping Implementation in RDB2OWL Tool

The RDB2OWL mapping tool[4] takes as the input annotated OWL ontology as well as a connection to the source database (in order to read the source database schema) and it produces a D2RQ or R2RML mapping for accessing the source database. The D2RQ mappings are supported via D2RQ platform[5] SPARQL endpoint or RDF dump tool. There are also tools supporting the R2RML mappings available. The database connection information (JDBC driver name and connection information) can be defined as an annotation in the data ontology, or it can be entered directly in the RDB2OWL tool.

The RDB2OWL mapping processing in the tool follows a number of steps:

- Load ontology: the ontology with the attached RDB2OWL annotations is loaded into the internal RDB2OWL mapping model (cf. [11, 21]); the parsing of the annotations into the RDB2OWL model items is performed.
- Load source database schemas: the database schema information is loaded into the model.
- Finalize abstract mapping: the advanced mapping constructs (e.g. named class maps, shorthands, defaults and user defined functions) are resolved into basic mapping constructions bringing the mapping into the "semantic" RDB2OWL model (a variant of RDB2OWL Raw metamodel of [11]); the obtained model is then used as the basis for D2RQ/R2RML mapping code generation, as well as it can be used for generating direct SQL scripts (work in progress) or providing yet other implementations
- D2RQ or R2RML Mapping generation on the basis of the created semantic RDB2OWL model with the inference option turned either on or off.

These steps can be performed either one-by-one, or automatically by a single command "Create mapping from ontology".

The RDB2OWL tool is implemented in JAVA programming language, however, its current implementation uses libraries for the internal RDB2OWL model handling that are working only in Microsoft Windows environment (there is work in progress to remove this restriction).

The internal structure of the models used in the RDB2OWL tool implementation is described in [21].

The RDB2OWL mapping processing tool contains the basic mapping validity checks regarding the referenced table or column presence in the database. The practical use cases suggest, however, that more extensive mapping validity checks would be worthwhile especially in the case of large-sized mappings.

---

[4] http://rdb2owl.lumii.lv/.

[5] http://d2rq.org/.

6   RDB2OWL Lite Application Cases

In this section the experience with two application cases – the Latvian Medicine registries, reported originally in [16] and a hospital information system, reported originally in [18] – is reviewed.

6.1   Latvian Medicine Registries

The RDB2OWL language has been used successfully in re-working of the Latvian Medicine Registries example [16]. The language features used in the mapping fall into the RDB2OWL Lite language subset considered here, using however the database pre-processing step adding new database tables or views (either of the mechanisms can be used equally well) in the following situations:

- classifiers existing semantically but not being introduced explicitly in the database schema
- the auxiliary *Numbers* table/view for string splitting
- a few views for auxiliary value calculations on the basis of the existing table data (a situation where the calculation description on the SQL level is simpler than the corresponding mapping description on RDB2OWL level).

The database pre-processing step is a solution that can be termed as acceptable for the considered use case since the data offered for semantic re-engineering were not the operational data of any working information system but were extracted from the real data base and then anonymized.

Figure 4 shows a small excerpt of the Medicine Registries example with class names translated into English and database table and column names changed, still conveying the general principle of overall ontology-to-database mapping creation.

We note the main class of illness case that consists of subclasses (Trauma, Cancer and Diabetes in the example; there are three more in the practical example); each of the subclasses has its own link to the patient and the diagnosis (these links are retrieved directly from the database, as defined by '->' DBExpr annotations), yet the semantic

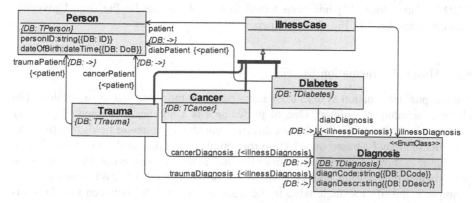

Fig. 4. A fragment of medicine registries ontology and database mapping

analysis of the data can be performed on the level of the person, illness case and diagnosis, as, for instance, in the following ViziQuer query (Fig. 5).

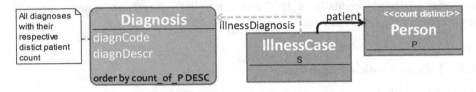

Fig. 5. A ViziQuer query example

The practical working solution of the Medicine Registries example have been achieved via generated D2RQ mapping with the source data residing on the Microsoft SQL Server. It has been possible to explore the generated SPARQL endpoint both in the ad hoc mode by the ViziQuer tool [14, 22] and in browsing mode by the OBIS tool [23] within the relational database semantic re-engineering framework described in [17]. The usage of D2RQ tool both in on-the-fly SPARQL endpoint creation and in the RDF dump mode with the RDF data loaded further on into the Virtuoso server have been considered with the better runtime performance achieved with the data stored in Virtuoso server. The tool running experience on a year 2015 laptop with 16 GB RAM has shown that D2RQ mapping creation from RDB2OWL mapping (only schema level involved) takes about half minute; the 8 GB RDF dump with 43 million triples is created on the D2R server in about 12 min and can be loaded into OpenLink Virtuoso Server 7.2.2 in less than 5 min. The Virtuoso server is able to answer most queries with up to three classes and aggregation, e.g. the one in Fig. 5, within few seconds.

As explained in Sect. 4.4, he RDB2OWL to D2RQ translation approach, does not support using the new mapping-level database view definition facility in a non-touch source database mode. There are initial experiments for R2RML mapping generation for the Latvian Medical Registries, however, it is still as issue to find the match between the R2RML mapping code generated by the RDB2OWL tool and suitable R2RML mapping processor that is able to handle the generated code. The R2RML-Parser tool [15] has been tested to be able to handle the mini-University example considered in Fig. 2 in this paper.

6.2   Hospital Information System

The hospital information system use case is based on moderate-size data ontology (20 classes, including classifiers). One of its features is a number of computed property inclusion in the ontology that are not directly available as database fields; the property computation may be based on selection row numbers and date and time manipulation.

Figure 6 shows a short excerpt from the information system ontology showing the patients and their curing episodes within the hospital. Each of OWL classes in the example is linked to a single table in the database, the link 'e' between the classes is based on a primary-to-foreign key relation between the corresponding database tables.

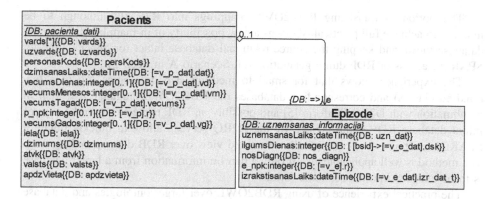

Fig. 6. A hospital information system fragment

The data property maps are described either as a single column in the table referenced by the respective class-to-table map, or they are defined using the new in RDB2OWL Lite table-to-view extension mechanism (cf. Sect. 4.2) allowing identification of table and view rows on the basis of table primary key column values (these columns are to be present also in the view).

The main need for SQL-level view constructions in the mapping implementation arises due to need of date and time calculations that cannot be easily performed on the semantic level, as well as for the row-number property computations. Given the presence of the additional value-computing views in the database, it would have been possible to describe the mappings also in the earlier RDB2OWL versions, however, with much less convenience in mapping writing and reading ability.

Since the initial use case of hospital information system data analysis involved an anonymized copy of the real production data, it is acceptable to extend the database with the extra value-computing views necessary for the mapping definition, as required by the more stable D2RQ-oriented application. There is, however, a challenge to create a fully working mapping towards the hospital information system in an environment where the source database is to be kept intact.

The issue of missing classifiers that has been an important one in the case of the Latvian Medicine Registries, shows up also in the Hospital Information System use case in the phase of "natural" query formulations over the semantic data store, so this issue is also to be considered during the database enriching process.

7 Conclusions

The RDB2OWL Lite mapping language and tool allows easy creation of wide range of database-to-ontology mappings involving basic one-to-one correspondence between ontology classes, data and object properties and database tables, columns and relations respectively, as well as more advanced constructs (e.g. adding filters, sub-class relations, value processing, etc.).

The option of translating RDB2OWL mappings into R2RML, although to be matured to achieve full practicality, opens a new possibility of in-mapping defining the database views and keeping the source relational database intact for either on-the-fly SPARQL access or RDF dump generation (cf. Scenario A in Sect. 2).

The experience shows that for small to medium size ontologies (up to a few hundred classes) and corresponding databases the RDB2OWL language and tool, in combination with D2RQ Platform [5] and possibly an RDF triple store (in the case of initial database RDF dump created by the D2RQ server), allow to quickly obtain a SPARQL endpoint over conceptually structured view over RDB data up and running. The method is well applicable also if only a part on information from a larger database is to be exposed into the semantic form.

The practical experience of using RDB2OWL over larger ontologies and database schemas, as in Latvian medicine registries example [16] shows that the tool is well usable, however, some further ontology and mapping engineering means for cross-checking the validity and completeness of the created mappings would be helpful (these can be used both on the RDB2OWL annotation level, as well as on the level of the generated executable D2RQ/R2RML mapping code).

The direct RDB2OWL mapping text size both in the considered mini-University and in the Latvian medicine registries [16] examples is about 8-9 times smaller, if compared with the generated D2RQ mapping text size, thus suggesting RDB2OWL as an option for manual database-to-ontology correspondence specification.

Regarding the OWL constructions allowed in the ontologies that are RDB2OWL mapping targets, we note that in fact, any OWL features also beyond the RDF Schema notation can be permitted. Figure 2 contains a number of such features, for instance equivalent class assertions and datatype facet restrictions. We can note that the equivalent class assertions for the subclasses of the *Teacher* class are "enforced" by the RDB2OWL mapping definition; while the facet restrictions in the current mapping version are left to be checked as constraints over the obtained RDF triple set.

Although the RDB2OWL tool has a well-defined task of creating a conceptual-level SPARQL endpoint for a relational database, permitting its standalone use, a particular usage context of the RDB2OWL database-to-ontology mapping tool is within a larger semantic database platform involving also the OWLGrEd ontology editor [19], the visual custom SPARQL query generation tool ViziQuer [14, 22] and the SPARQL endpoint browser OBIS [23].

Acknowledgements. This research has been supported by Latvian State Research program (2014-2017) NexIT project No. 1 'Technologies of ontologies, semantic web and security'.

References

1. Resource Description Framework (RDF). http://www.w3.org/RDF/
2. Motik, B., Patel-Schneider P.F., Parsia B. (eds.): OWL 2 Web Ontology Language Structural Specification and Functional-Style Syntax (2009). https://www.w3.org/TR/2009/REC-owl2-syntax-20091027/

3. Linked Data. http://linkeddata.org
4. Berners-Lee, T., Hendler, J., Lassila, O.: The semantic Web. Sci. Am. **5**, 29–37 (2001)
5. D2RQ. Accessing Relational Databases as Virtual RDF Graphs. http://d2rq.org/
6. Blakeley, C.: RDF views of SQL data (declarative SQL schema to RDF mapping). OpenLink Software (2007)
7. Sequeda, J.F., Cunningham, C., Depena, R., Miranker, D.P.: Ultrawrap: using SQL views for RDB2RDF. In: Poster Proceedings of the 8th International Semantic Web Conference (ISWC2009), Chantilly, VA, USA (2009)
8. Revelytix Spyder Tool. http://www.revelytix.com/content/spyder
9. R2RML: RDB to RDF Mapping Language. http://www.w3.org/TR/r2rml/
10. Calvanese, D., Cogrel, B., Komla-Ebri, S., Lanti, D., Rezk, M., Xiao, G.: How to stay ontop of your data. In: Databases, Ontologies and More. ESWC (Satellite Events), pp. 20–25 (2015)
11. Čerāns, K., Būmans, G.: RDB2OWL: a RDB-to-RDF/OWL mapping specification language. In: Barzdins, J., Kirikova, M. (eds.), Databases and Information Systems VI, IOS Press, pp. 139–152 (2011)
12. Būmans, G., Čerāns, K.: Advanced RDB-to-RDF/OWL mapping facilities in RDB2OWL. In: Grabis, J., Kirikova, M. (eds.) BIR 2011. LNBIP, vol. 90, pp. 142–157. Springer, Heidelberg (2011)
13. SPARQL 1.1 Overview. W3C Recommendation, 21 March 2013. http://www.w3.org/TR/sparql11-overview/
14. Čerāns, K.; Ovčiņņikova, J., Zviedris, M.: SPARQL aggregate queries made easy with diagrammatic query language ViziQuer. In: Proceedings of ISWC 2015 PD. http://ceur-ws.org/Vol-1486/paper_68.pdf
15. R2RML Parser. https://github.com/nkons/r2rml-parser
16. Barzdins, G., Liepins, E., Veilande, M., Zviedris, M.: Semantic latvia approach in the medical domain. In: Proceedings of 8th International Baltic Conference on Databases and Information Systems. Haav, H.M., Kalja, A. (eds.), pp. 89–102. TUT Press (2008)
17. Čerāns, K., Barzdins, G., Būmans, G., Ovcinnikova, J., Rikacovs, S., Romane, A., Zviedris, M.: A relational database semantic re-engineering technology and tools. Baltic J. Mod. Comput. (BJMC) **3**(3), 183–198 (2014)
18. Barzdins, J., Rencis, E., Sostaks, A.: Fast ad hoc queries based on data ontologies. In: Frontiers in Artificial Intelligence and Applications, Databases and Information Systems VIII, vol. 270, pp. 43–56. IOS Press (2014)
19. Barzdins, J., Barzdins, G., Cerans, K., Liepins, R., Sprogis, A.: OWLGrEd: a UML style graphical notation and editor for OWL 2. In: Proceedings of OWLED 2010 (2010)
20. Čerāns, K., Ovčiņņikova, J., Liepiņš, R., Sproģis, A.: Advanced OWL 2.0 ontology visualization in OWLGrEd. In: Caplinskas, A., Dzemyda, G., Lupeikiene, A., Vasilecas, O. (eds.) Databases and Information Systems VII, Frontiers in Artificial Intelligence and Applications, vol. 249, pp. 41–54. IOS Press (2013)
21. Čerāns, K., Būmans, G.: RDB2OWL: a language and tool for database to ontology mapping. In: Proceedings of CAiSE FORUM (2015). http://ceur-ws.org/Vol-1367/paper-11.pdf
22. Zviedris, M., Barzdins, G.: ViziQuer: a tool to explore and query SPARQL endpoints. In: Antoniou, G., Grobelnik, M., Simperl, E., Parsia, B., Plexousakis, D., De Leenheer, P., Pan, J. (eds.) ESWC 2011, Part II. LNCS, vol. 6644, pp. 441–445. Springer, Heidelberg (2011)
23. Zviedris, M., Romane, A., Barzdins, G., Cerans, K.: Ontology-based information system. In: Kim, W., Ding, Y., Kim, H.-G. (eds.) JIST 2013. LNCS, vol. 8388, pp. 33–47. Springer, Heidelberg (2014)

Tools, Technologies and Languages for Model-Driven Development

Models and Model Transformations
Within Web Applications

Sergejs Kozlovics[✉]

Institute of Mathematics and Computer Science,
University of Latvia (Riga, Latvia), Raina blvd. 29, Riga 1459, Latvia
sergejs.kozlovics@lumii.lv

Abstract. Unlike traditional single-user desktop applications, web applications have separated memory and computational resources (the client and the server side) and have to deal with multiple user accounts. This complicates the development process. Is there some approach of creating web applications without thinking about web-specific aspects, as if we are developing stand-alone desktop applications? We say, "yes", and that is where models and model transformations come in handy. The proposed model-driven approach simplifies the development of web applications and makes it possible to use a single code base for deploying both desktop and web-based versions of the software.

Keywords: Models · Model transformations · Web applications

1 Introduction

In 2001, Model-Driven Architecture (MDA) was considered a promising approach for software development [30]. Indeed, models are a universal tool for system and data modeling; they can be used at different levels of abstraction — from domain-specific languages familiar to domain experts to platform-specific aspects and neat implementation details. Automated transformations between such models could replace traditional compilers. Although MDA and its further developments like Model-Driven Engineering (MDE), Model-Driven Software Development (MDSD), and other MD* have known success stories, some experts consider that the model-driven approach (in general) has "missed the boat" [13,14,18,29]. Although models are still in honor within academic researchers, most practitioners continue to use traditional technologies (relational databases and popular programming languages such as Java, C#, etc.; not models and transformation languages) due to lack of stable, production-ready model-driven infrastructure.

Nevertheless, another "boat", by which models could go, appears on the horizon — web-based software development. Currently, the majority of web application is developed using well-known server-side technologies (PHP, SQL and no-SQL databases, etc.), web protocols, and the client-side HTML+CSS+JavaScript stack. While creating a web-based application, the developer has to think about

© Springer International Publishing Switzerland 2016
G. Arnicans et al. (Eds.): DB&IS 2016, CCIS 615, pp. 53–67, 2016.
DOI: 10.1007/978-3-319-40180-5_4

server-side code, the client-side code, the communication issues, and the user-specific aspects (authentication and access control). This complicates the development of web applications, since network-specific issues have to be considered *in addition* to the primary functionality of the software. Moreover, it may be hard to choose where the particular computation-intensive code has to be executed — at the server side or at the client side. For instance, we faced this dilemma when considering layout computation for graph-like diagrams.

In this paper we show that if we "resurrect" models and take them on board, web-based applications can be developed much easier. In our approach, models are used as a memory (Random Access Memory, RAM) analog, which is automatically synchronized between the client and the server, thus, making network communication transparent. We also show how models help to manage the resources (processor and memory) automatically and transparently between multiple users, who can use the application simultaneously. Thus, the developers can just assume a single user PC as a target.

The paper is structured as follows. First, we describe our previous approach of using models and model transformations in classical desktop applications (Sect. 2). Then we adapt it to meet the requirements of the web (Sect. 3). We continue by providing solutions for certain issues arising when moving models to the web. The "Related Work" section lists several alternative approaches for developing web applications. It is followed by the conclusion, which presents our experimental results and points to further research directions.

2   Traditional Approach

In 2008, we have proposed a domain-specific desktop tool building approach called the Transformation-Driven Architecture, TDA [7,22]. TDA has been successfully used to implement several domain-specific tools such as ProMod, OWLGrEd, and VisiQuer [4–6]. The main concepts of TDA are model transformations, engines, and interface metamodels (Fig. 1). Model transformations are

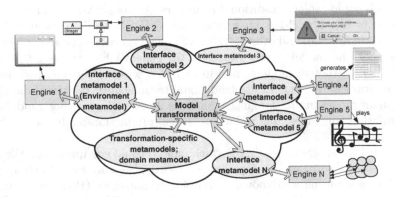

Fig. 1. The outline view on the Transformation-Driven Architecture, TDA.

used to implement business logic. Unlike MDA, TDA uses model transformations at runtime. Engines are pluggable modules that provide certain auxiliary functionality (such as services and graphical presentations) for transformations. While transformations are usually written in a platform-independent way using model transformation languages or traditional programming languages, engines are usually implemented using platform-specific libraries and technologies. There may be multiple variations of the same engine for different platforms (different operating systems).

Model transformations communicate with engines via instances of interface metamodels. There are special classes called events and commands (subclasses of the Event class and the Command class). When a transformation needs to call some engine, it creates a corresponding command instance, sets its attribute values and creates necessary links to specify the command arguments and the context. Then the command instance is linked to the submitter object in the model, which is treated as a request to execute that command. Engines, in their turn, are able to emit events, when certain actions (e.g., user clicks) occur. In order to catch an event, some event-handling transformation has to be registered as an event listener. Events are emitted in the same way as commands (e.g., an event object is created and linked to the submitter). On the one hand, such event/command mechanism keeps transformations away from technical issues of calling different engines, which may be written in different programming languages. On the other hand, engines can be written in traditional languages, without the need to know how to call particular model transformation, which may be written in some specific transformation language or in some ordinary high-level language. All calls between the engines and transformations are performed automatically when corresponding links to the submitter are created. TDA has different adapters for different programming and transformation languages.

Models are stored in a model repository (in-memory repository, in most cases). Engines and transformations access models via a common API (we call it Repository Access API, RAAPI[1]) implemented for various programming languages and platforms. Certain RAAPI wrappers have been developed to provide query-based repository access (e.g., the lQuery language [24]). There are also some code generators that produce C++/Java classes that can be used to access objects stored in the model repository as if they were C++/Java objects. Thus, RAAPI (or one of its wrappers) is the only API a particular transformation or engine has to be aware of to be able to work within TDA.

3   Bringing Models to the Web

Now we show how the above-mentioned architecture for desktop tools can be scaled for web-based applications. The first approximation is as follows:

---

[1] http://tda.lumii.lv/raapi.html.

- Transformations, which implement business logic, remain intact. They are implemented using traditional high-level or transformation languages and executed at the server side.
- Engines, which mainly implement graphical presentations, are executed in the client browser. Engines must be re-written in JavaScript (or other language that translates to JavaScript), utilizing HTML and CSS, to provide a neat user experience without the need to install support for non-JavaScript languages. We may think of web engines as engine variations for the web platform. The interface metamodels of web engines remain intact, thus, we do not need to re-write transformations (but a TDA adapter for web-based engines is needed).
- Communication between engines and transformations now has to be implemented using network technologies.

Rewriting engines in the JavaScript+HTML+CSS stack as well as introducing a web-server to deliver the code of web engines to the client browser is a technical straight-forward process (once it is done, we need to maintain just the web-based engine, since it can be used also for desktop tools). But ensuring the communication between engines and transformations over the network involves more complex issues, including:

- bi-directional communication (we have to communicate in both directions by means of commands and events);
- asynchronous execution of commands and events;
- accessing the model repository from the server and from the client.

The next approximation is to get rid off the traditions and allow transformations to be executed right in the client browser and allow non-interactive engines to run at the server side. Thus, certain transformations and engines can be launched without the round-trip delay between the client and the server.

To complete the picture, we introduce multiple users. This includes user authentication and sharing server resources (processor and memory) between multiple users, with the potential scaling in mind.

The issues we have just mentioned are addressed in the next section.

Notice that among these issues, every web application has to consider potential security risks. We assume that best practice recommendations for preventing typical attacks are always kept in mind (e.g., escaping of HTML strings, using secure HTTPS/WebSocket connections, session checks, etc.) [20]. Security issues are mostly technical and are not considered in this paper.

4   Dealing with the Issues

4.1   Bi-Directional Communication Issues

The traditional HTTP protocol, designed in 1992, was developed to be a client-server stateless protocol: a client initiates a request and waits for the server to respond; each next request is treated as independent, since no state is stored at the protocol level. To use HTTP for bi-directional communication between transformations and engines, where multiple users may be working with different models, we need some means for the server to initiate the communication (in order transformations running on the server side could send commands to engines running at the client side). A session identifier is also required for each authenticated user. These are well-known issues. Widely used solutions for bi-directional HTTP include long polling, when the client asks the server for commands at certain intervals, and COMET, where the client initiates a request, but the server delays the response until some command has to be sent to the client [15]. The traditional way to identify the session is to use the JSESSIONID cookie along with the list of active sessions at the server. Certain libraries, such as DWR[2], are able to factor out these technical issues. Still, in order to pass events and commands over the network, we need to serialize and deserialize them.

The WebSockets protocol, standardized in 2011 (drafts from 2010), is intended for high-speed bi-directional communication [16,19]. The protocol does not perform a handshake each time a message is sent. Moreover, message encoding overhead is minimal (compared to heavyweight HTTP headers). Currently, all recent versions of popular web-browsers support web sockets. Still, if we go for web sockets, we need some TDA event/command serialization or synchronization solution. We discuss it in Sect. 4.3.

To ensure the correct execution of commands and event handling, we need two additional TDA components, which are present regardless of the particular technique used for network communication. They are the web engine adapter and the client-side command manager[3].

Web engine adapter. When a transformation creates a command and stores it in the model repository, TDA calls the corresponding engine via a specific adapter. There are different adapters for different types of engines, usually, depending on the programming language or calling conventions (DLL, Java class, .NET assembly, etc.). We can assume that web engines have to be called via a special TDA "JavaScript" adapter. However, unlike traditional engine adapters, which work locally, this adapter starts the web-server (if it has not been done before), serializes the command, and sends it to the client browser.

Client-side command manager. Each TDA-based tool has Environment Engine, which is responsible for creating the main application window and attaching/detaching child windows. In case of web-based TDA, Environment Engine occupies one browser window (or tab), while other windows are attached as embedded frames (iframes) by means of some windowing library such as jQueryUI[4] or Dojo[5]. Each TDA engine has some function for processing commands. For desktop-based TDA tools, TDA takes care of calling the appropriate function for the given command, since the engines are attached locally. However, for web-based applications, engines are at the client side. Thus, we need

---

[2] http://directwebremoting.org/.

[3] we do not need a manager for events, see Sect. 4.3.

[4] http://jqueryui.com/.

[5] http://dojotoolkit.org/.

some client-side command manager, which determines the correct engine and its iframe, and passes the command to that frame. For web-based TDA, the manager can be a part of Environment Engine. It listens for command messages from the server. Then, given a command object, the manager searches for a corresponding *Frame* object in the model repository (each presentation engine must have created such object). Since Environment Engine already maintains a map that associates *Frame* objects with iframes, the command manager can use this map to get the correct iframe and forward the command message to it.

4.2 Asynchronous Issues

The major arguments for introducing asynchronous calls between the client and the server are as follows:

- When some engine (running in the browser) emits an event, some event-handling transformation is called at the server side. In order not to freeze the browser (taking into a consideration the network latency and the event handling transformation execution time), event handling should be asynchronous.
- When some transformation at the server side creates a command for some presentation engine, the engine usually needs to repaint some GUI elements. There is only one JavaScript thread in the browser; the thread is common to all engines, thus, command processing should not block other engines. Moreover, a separate GUI thread (which is the JavaScript browser thread is our case) is a de facto best practice standard (otherwise, deadlocks are inevitable) [10]. Since there is only one GUI thread, all GUI operations must be enqueued (as it is in the case of JavaScript operations), and thus, they cannot be synchronous[6].
- When some transformation at the server side creates a command for some engine, it must not block the server. Then other users can use the server resources, while the asynchronous command is being executed at the first user's browser.

Based on this, we require all events and commands of web-based engines to be asynchronous[7]. Thus, when a command or an event is being submitted, the caller thread is not blocked. If a callback is needed, an engine can emit an event, when it finishes processing the command, and a transformation can issue a command after the event has been handled. In the latter case, if the engine needs to repaint its presentation, it can use some optimistic prediction technique to visualize the expected state before the transformation finishes (the state can be adjusted later, if needed).

     To support asynchronous communication, we introduce the *AsyncCommand* class in the metamodel (all events have been already asynchronous in TDA).

---

[6] Similar approach is used in traditional GUI libraries, such as Java Swing (the function SwingUtilities.invokeLater), JavaFX (the function Platform.runLater), Borland/Embarcadero VCL (the Synchronize function), etc.

[7] Any bi-directional communication technique mentioned in Sect. 4.1 can be used asynchronously.

Since all GUI commands of existing presentation engines are asynchronous, we can just make them subclasses of *AsyncCommand*. The TDA event/command mechanism now checks whether the given command is asynchronous. If yes, the command is forwarded to the corresponding engine. For desktop-based tools, TDA supports synchronous command calls as well (thus, when the transformation emits a command, the control is returned only when the command execution has finished).

4.3   Accessing the Model Repository from the Server and from the Client

Since model transformations use the model repository intensively, it is reasonable to run the repository at the server side. Before sending commands (with their context) to engines, commands are serialized. We may expect that the serialization should contain all the necessary information for the engine. However, engines may need to access objects that are not directly linked to the commands (not in the context). The full context serialization (or the full repository serialization in the worst case) is unreasonable, if the engine has to visualize just some of the objects. Also, when an event occurs, the engine needs to store it in the repository. Thus, some means to access the repository from the client side (engine side) is needed. There are two approaches:

- Provide some client-side query mechanism, while keeping the repository at the server side.
- Synchronize the repository between the client and the server

The first approach requires some query language. Our first approximation is to provide functions such as *findObjects*, *loadObjects*, *storeObjects*, and *deleteObjects*. The arguments and the result are in the JSON syntax (see Fig. 2).

The second (synchronization) approach requires some means to synchronize the server-side repository with some client-side data structure, containing the same model. The synchronization can be done in several ways:

- By means of bi-directional HTTP implementations (e.g., COMET). Since HTTP connection has to be established on each message (and this involves certain delay), it is reasonable to synchronize models in batch mode. For instance, repository write operations can be recorded at the server side while a transformation is being executed. When a command is being issued, the collected write operations are serialized and sent to the client browser. Likewise, while an engine is performing some operations, all model changes are recorded and then sent back to the server on events.
- By means of web sockets. The benefits of web sockets are:
  - the connection has to be established only once;
  - keeping the connection alive involves almost no overhead;
  - the connection is asynchronous, but the order of messages is preserved;
  - data do not need to be serialized, since binary communication is possible.

```
                                    {
                                      reference: 1001,
                                      class: "Person",
        {                             name: "John",
          isKindOf: "Person",         children: [ {
          name: "John"                  reference: 1002
        }                             }, {
                                        reference: 1003
                                      }]
                                    }
              (a)                             (b)
```

Fig. 2. (a) A JSON object representing a query for *findObjects* for finding a Person with the given name. (b) A possible result of that query. Links are encoded as JSON arrays. Two special attributes, *reference* and *class*, specify the object identifier in the model repository and the class name, respectively. The *loadObjects* function can be used then to get attribute values for objects 1002 and 1003.

Thus, all repository write operations can be sent to a web socket right away, without introducing a special buffer for batch processing.

- By means of existing infrastructures, which provide automatic data synchronization. For instance, we can use the Meteor[8] infrastructure for that. Meteor stores data on the server side in a MongoDB and implements a common query language for both the server and the client, while keeping data synchronization transparent. While MongoDB is optimized for efficient queries, it has slow write operations.

Unfortunately, the client-side query mechanism as well as the HTTP synchronization requires data serialization/deserialization to/from JSON syntax. It proved to be slow in our experiments, where it may take around 2 seconds to serialize/deserialize graph diagrams of moderate size (around 100 elements), including network delay. The Meteor/MongoDB approach is optimized for efficient queries, but it has slow write operations (around 10000 write operations per second on a 3.4 GHz i7 processor, including optimizations). Our experiments show that existing transformations use write operations quite intensively (see Sect. 4.5), thus, only a small number of users can be connected to a web application without exceeding the processor power limits. The only feasible solution (from the above) is to use web sockets. In fact, web sockets allow data synchronization to be performed in parallel with computation. Still, we need some means to represent data at the client side. Binary JavaScript objects stored according to the syntax from Fig. 2 can be used for that. If we use binary web sockets, expensive JSON serialization can be replaced with lightweight object creations or attribute updates. While such JavaScript objects can be "touched" directly for read access, write access requires special functions ("setters"), since we need to listen to the changes to be able to synchronize them back to the server. In the example from Fig. 2, the object would be augmented with the functions *setName* and *setChildren*.

---

[8] https://www.meteor.com.

The server-side repository automatically takes care of launching engines and transformations, when command and event objects are put into the repository and connected to the submitter object. The client-side repository replica, however, needs some adjustments. In case of Meteor, we can introduce a client-side listener, which listens to new command objects and passes them to the client-side command manager. In case of HTTP/web sockets, when a message is received at the client, the message is analyzed. If the message contains a repository write operation for connecting a command object to the submitter, the client forwards the command to the client-side command manager.

With events, the process is much simpler. The client just creates an event and links it to the submitter at the client side. When the event reaches the server (during synchronization), the server-side repository will process it as usual.

We have already mentioned that code generators can be used to generate wrappers for repository classes in different programming languages (C++, Java, etc.) in the traditional TDA. They can still be used for server-side code, but synchronized client-side JavaScript objects are already native JavaScript objects. Thus, regardless of the programming language (and the server or client side), developers of transformations and engines may treat repository objects as native OOP-objects in RAM. Moreover, since the synchronization between the client and the server is automatic, the developers can assume they are writing applications for a single PC. We believe that this is an important benefit that models can bring. Models are like lens, beyond which network aspects can be hidden. In addition, function calls can also be implemented in a way native to the particular programming language by providing glue code for commands and events in the repository.

4.4 Server-Side Engines and Client-Side Transformations

Most TDA engines are graphical and, in case of web applications, are executed in the browser. Still, some engines can perform certain computation without the need to visualize anything, but requiring server resources (e.g., intensive computations are inefficient, if running as browser scripts). On the other hand, it may be reasonable to develop certain model transformations in JavaScript to be executed directly in the browser (e.g., small GUI event handlers, which may need just to adjust the presentation). Thus, server-side engines and client-side transformations are needed.

Technically, if an engine does not have any own graphical presentation, it can be executed at the server side by means of existing TDA event/command mechanism. When a command for a not-web-based engine is created, it is passed to some adapter at the server side, which executes the command.

For client-side event handlers the process is not that easy. We need a client-side event manager, which monitors write operations that are sent from the client to the server. If an event is being sent, the client-side event manager gets the associated event listener (its name is stored in the repository). If the corresponding event listener is written in JavaScript and has to be run at the client side, the event manager executes that listener right away. The event may still be posted to the server, where the corresponding client-side JavaScript adapter will be searched. Since no such adapter exists (or, we may create a fictive adapter, which ignores all its events), the event will not be executed at the server side. Another option is just not to post the event to the server, but then the client and the server repositories would not be identical until the event object is deleted at the client side.

If all transformations are client-side transformations (called as event handlers), then the whole application can be run within a web browser. To persist the model, either a server-side repository, or a third-party cloud storage can be used. For instance, DropBox[9] and OneDrive[10] files can be used to store the serialized models (public APIs are available). If no server-side transformations and engines are present, then no model synchronization is needed. Models can be loaded from the cloud, when the web application is loaded, and saved on exit (or on regular basis).

4.5 Multi-user Issues

For authentication we can use a traditional login/password approach, or delegate the authentication to third parties (such as Google or Facebook). Web-based Environment Engine implement the client-side authentication, while the server process performs necessary checks and marks the given user (associated with the given HTTP session) as logged in. For Meteor-based variant, we can use Meteor authentication with plugins.

For traditional desktop-based TDA we used our proprietary repositories (MIL_REP/OUR and JR) as well as ECore files to store models [8,31,33]. Since there may be multiple users accessing their models at the same time, for web-based TDA we can use a traditional or no-SQL database as well. Databases implement all necessary services, such as disk cache, optimized search, support for multiple threads/processes, etc., which are essential in a multi-user environment.

It is reasonable that one server process is dedicated to the web-server. Another one may be dedicated for the database. Depending on the number of processor cores at the server, we can create N worker processes, which can be used to execute server-side transformations and engines. Ideally, N would be equal to the number of processor cores minus 2, since 2 processes are occupied by the web server and the database. Each server-side transformation or engine call is enqueued and then processed by one of the N "workers".

Another solution is to create N workers, where each worker has its own in-memory database (instead of a common single database). Thus, we do not need a dedicated database process, and if the number of users does not exceed N, they can use the server in parallel. When the $N+1$-th user comes in, the in-memory database of the user, who was idle for the longest time, is flushed to the disk, and the repository of the new user is loaded in that place.

---

[9] https://www.dropbox.com.
[10] https://onedrive.live.com.

In case of some runtime exception in a server-side transformation, the corresponding worker process can be terminated (and a new "healthy" process can be launched for further transformations). In case of a common database, the previous repository state (before the error) has to be restored. In case of in-memory database, no actions are required (since repository flushing to disk is performed only after successful execution of server-side transformations).

For a server having 8 GB RAM, we can assume 4 GB are free. Taking into a consideration our experience with desktop-based TDA, the 100 MB upper bound for each in-memory repository seems to be reasonable. Thus, a single server can serve up to 40 users without swap. Based on our existing experience, we can assume that each transformation performs 1000 model operations on average[11]. Thus, we have 40000 model operations per 40 users, where 10000, could be write operations and 30000 read operations. Our in-memory repositories can deal with 40000 operations in a few milliseconds. Since repository actions are synchronized asynchronously at once, we just need to add the network delay (usually, 100–200 ms), thus the total time is below one second, which is considered adequate [26,27]. MongoDB (used by Meteor), in its turn, has slow write operations (10000 write operations could be executed in 1000 ms on a 3.40 GHz processor, if we use the batch mode). Thus, we can assume 10000 write operations take one second, while other 30000 read operations take another second, resulting in 2 s, which is less efficient and is at the bound of the "seamless" user experience [28].

Notice that the calculations above are at the full load of 40 users, who emit events each 1 or 2 s. In practice, transformations are called occasionally, thus, more users can be connected and using server resources without interference (still, for the in-memory solution, repository flushing/loading may be required). As a result, the number of simultaneously connected users may reach several hundreds. For thousands of users, we need to configure load-balancing between multiple servers. While the number of 10000 users is considered appropriate for existing operating systems (the C10K problem), serving millions of users (the C10M problem) requires bypassing the OS by using sophisticated techniques (and currently this is not our goal).

5   Related Work

In 2007, de Castro et al. presented an MDA-based approach for developing service-oriented web applications [17]. This approach has been applied using the AMMA tools and the ATL transformation language for modeling web applications [1,3,12]. A different, but also MDA-based approach was used in Visual-Wade[12] (currently obsolete). It was intended as a out-of-box product for generating PHP code for web applications from source models, which could be defined graphically with a few mouse clicks. While models were used at the development stage, traditional databases such as MySQL, PostreSQL, and ORACLE were

---

[11] These are transformations with non-intensive computation, as transformations in existing TDA-based tools.

[12] http://visualwade.software.informer.com/.

used at runtime. Other similar tools include WebRatio[13] (using Web Modeling Language, WebML[14]) and OpenUWE (using the ArgoUWE, an ArgoUML-base tool) followed by MagicUWE [11,21]. The WebSA approach also uses MDA, but explicitly focuses on the functional requirements of web applications [9]. WebTA is a transformation engine specifically designed to bring transformations into WebSA [25]. An extensive survey on different MDA-based approaches for web applications can be found in the article by Schwinger and Koch [32]. Still, all these approaches use models and transformations at development time. This differs from our approach, where models and transformations are executed at runtime. Currently, the MDA/MDE field is in the state of stagnation (especially, after the Bezivin "Why did MDE miss the boat?" talk in 2011 [13]). We believe that our web-based approach can give a new breath to models, thus new results in the field can appear.

While we use either a model repository or a database to store models, one can use third-party cloud storage for that (reasonable, when all transformations and engines are running at the client side). While we can use any file format for that, using spreadsheets (like Google spreadsheet[15] or Microsoft Excel online[16], which have JavaScript APIs) and the appropriate encoding (e.g., sheets are classes, rows are objects, and columns are attribute values), we also get a free tabular repository browser for debug purposes as a by-product.

The EASA Spreadsheet Deployment platform is an interesting approach to creating web application from Excel files [2]. The Excel file is treated as a source model, from which an out-of-box web application is obtained.

Google Apps Script[17] is a platform to developing web applications intended to be run in a web browser. Google provides a graphical form designer, where JavaScript code can be attached to user events (clicks). Since applications are being executed in a web browser, additional services are required to persist data (e.g., to store user projects). Since Google Apps Script integrates with Google services, Google Drive can be used as a storage device. Still, if we need certain server-side computation or access to some proprietary database, the integration does not work; we need to create a web-service or some API for that. Thus, the system becomes heterogeneous. In contrast, our TDA-based approach provides persistency automatically, since all the models are saved in a repository. Both the server and the client use the same model (and they do not need to be aware of model synchronization). The TDA event/command mechanism is a unified way to call transformations (functions) or web-services regardless of their location and particular protocols used.

RollApp[18] is an interesting solution for bringing existing desktop applications to the web. It builds a Linux-based virtual environment, where the content of the

---

[13] http://www.webratio.com/.
[14] http://www.webml.org/.
[15] https://apps.google.com/products/sheets/.
[16] https://office.live.com/start/Excel.aspx.
[17] https://www.google.com/script/start/.
[18] https://www.rollapp.com/.

application window is sent to the browser (technically, this can be implemented easily, since X Window System already implements that feature). File dialogs are redirected to the cloud storage. Since not all programs are supported and mainly they are open source, we can assume that certain minor modifications of code are required. RollApp is a paid subscription. A a free plan, where the changes could not be saved, is also available. Although this solution provides a universal way of bringing desktop applications to the web, it requires much more server resources (processor and memory for creating a virtual environment) than creating web applications by means of traditional web technologies, where resources can be shared among multiple users more efficiently.

The m-Power[19] platform branch can be traced back to 1983. The goal is to build a web-interface for legacy applications. The platform uses traditional databases and Java for the resulting applications, and the process is not model-driven. We believe that models help the developers think at a higher level of abstraction, which is proposed by our approach. However, in case of legacy applications, which usually are not model-based, m-Power could be a better solution.

6 Conclusion

The paper provided a sketch of a TDA-based solution for developing web applications using models and model transformations at runtime. We have implemented a prototype, using the ECore repository for model storage. The prototype includes:

- the web-based Environment Engine (utilizing the Dojo toolkit for attaching child windows);
- a simple client-side query language (utilizing the JSON syntax mentioned in the paper) for accessing server-side repository from the client;
- some web-based engines, which have been developed or re-written in JavaScript.

A recent TDA-based DataGalaxy tool can be considered as approbation of the main ideas of the approach [23]. Since this tool has only web-based engines, they can be used in both desktop and web modes without change. Java transformations can also be used either as ordinary TDA transformations, or server-side transformations. Thus, the same code base can produce both desktop-based and web-based versions of DataGalaxy. This is the main strength of the approach.

We are working on developing web versions of engines used in our desktop-based ontology editor OWLGrEd, thus, OWLGrEd (as well as some other tools) can be launched in the web in the near future. The main benefit of our approach is that web based tools can be treated by developers as desktop-based tools. Another benefit comes from models. Models provide a universal platform-independent encoding for data as well as for operations. In the future this can lead to a high-level network-transparent RAM analog. We believe that the potential power of models will eventually reveal itself, if we start using models in the web.

---

[19] http://www.mrc-productivity.com.

Acknowledgments. The work has been supported by Latvian State Research programme (2014–2017) NexIT project No.1 'Technologies of ontologies, semantic web and security'.

References

1. ATL use case - modeling web applications. http://www.eclipse.org/atl/usecases/webapp.modeling/
2. EASA   Spreadsheet   Deployment.   http://www.easasoftware.com/solutions/spreadsheet-deployment/
3. Allilaire, F., Idrissi, T.: ADT: Eclipse development tools for ATL. In: Proceedings of Second European Workshop on Model Driven Architecture (2004)
4. Barzdins, G., Liepins, E., Veilande, M., Zviedris, M.: Ontology enabled graphical database query tool for end-users. Frontiers in Artificial Intelligence and Applications, vol. 187, pp. 105–116. IOS Press (2008)
5. Barzdins, J., Barzdins, G., Cerans, K., Liepins, R., Sprogis, A.: OWLGrEd: a UML style graphical notation and editor for OWL 2. In: Proceedings of OWLED 2010 (2010)
6. Barzdins, J., Cerans, K., Kalnins, A., Grasmanis, M., Kozlovics, S., Lace, L., Liepins, R., Rencis, E., Sprogis, A., Zarins, A.: Domain specific languages for business process management: a case study. In: Proceedings of DSM 2009 Workshop of OOPSLA 2009, pp. 34–40, Florida, USA (2009)
7. Barzdins, J., Kozlovics, S., Rencis, E.: The Transformation-Driven Architecture. In: Proceedings of DSM 2008 Workshop of OOPSLA 2008, pp. 60–63, Nashville, Tennessee, USA (2008)
8. Barzdins, J., Barzdins, G., Balodis, R., Cerans, K., Kalnins, A., Opmanis, M., Podnieks, K.: Towards semantic Latvia. In: Vasileckas, O., Eder, J., Caplinskas, A. (eds.) Proceedings of Seventh International Baltic Conference on Databases and Information Systems, Communications, Lithuania, Vilnius pp. 203–218 (2006)
9. Beigbeder, S.M., Castro, C.C.: An MDA approach for the development of web applications. In: Koch, N., Fraternali, P., Wirsing, M. (eds.) ICWE 2004. LNCS, vol. 3140, pp. 300–305. Springer, Heidelberg (2004)
10. kgh blog: Multithreaded toolkits: A failed dream? Originally. https://community.oracle.com/people/kgh/blog/2004/10/19/multithreaded-toolkits-failed-dream, http://tecnologia.revistacocktel.com/multithreaded-toolkits-a-failed-dream/. Accessed 5 May 2016
11. Busch, M., Koch, N.: MagicUWE – a CASE tool plugin for modeling web applications. In: Gaedke, M., Grossniklaus, M., Díaz, O. (eds.) ICWE 2009. LNCS, vol. 5648, pp. 505–508. Springer, Heidelberg (2009)
12. Bzivin, J., Jouault, F., Rosenthal, P., Valduriez, P.: The AMMA platform support for modeling in the large and modeling in the small. Technical report, LINA, Universite de Nantes (2005)
13. Bzivin, J.: Why did MDE miss the boat? In: First International Workshop on Combined Object-Oriented Modeling and Programming (COOMP 2011) (2011)
14. Cabot, J.: Clarifying concepts: MBE vs MDE vs MDD vs MDA. Post at MOdeling LAnguages. http://modeling-languages.com/clarifying-concepts-mbe-vs-mde-vs-mdd-vs-mda/
15. Carbou, M.: Reverse Ajax, Part 1: Introduction to Comet. http://www.ibm.com/developerworks/library/wa-reverseajax1/index.html. Accessed 5 May 2016

16. Carbou, M.: Reverse Ajax, Part 2: WebSockets. http://www.ibm.com/developerworks/library/wa-reverseajax2/index.html. Accessed 5 May 2016
17. de Castro, V., Vara, J., Marcos, E.: Model transformation for service-oriented web applications development. In: Workshop Proceedings of 7th International Conference on Web Engineering, pp. 284–198 (2007)
18. Dubray, J.J.: Why did MDE miss the boat? (A summary). InfoQ News, 27 Oct 2011. http://www.infoq.com/news/2011/10/mde-missed-the-boat
19. IETF: The WebSocket protocol. RFC 6455. https://tools.ietf.org/html/rfc6455
20. Kern, C.: Securing the tangled web. Commun. ACM **57**(9), 38–47 (2014). http://dx.org/10.1145/2643134
21. Knapp, A., Koch, N., Moser, F., Zhang, G.: ArgoUWE: a CASE tool for web applications. In: Proceedings of the 1st International Workshop on Engineering Methods to Support Information Systems Evolution (EMSISE 2003) (2003)
22. Kozlovics, S., Barzdins, J.: The transformation-driven architecture for interactive systems. Autom. Control Comput. Sci. **47**(1/2013), 28–37 (2013). Allerton Press Inc
23. Kozlovics, S., Rucevskis, P.: Manipulating and visualizing data by means of data galaxies. Frontiers in Artificial Intelligence and Applications, vol. 270, pp. 85–98. IOS Press (2014)
24. Liepiņš, R.: Library for model querying: IQuery. In: Proceedings of the 12th Workshop on OCL and Textual Modelling, OCL 2012, pp. 31–36. ACM, New York (2012)
25. Meliá, S., Gómez, J., Serrano, J.L.: WebTE: MDA transformation engine for web applications. In: Baresi, L., Fraternali, P., Houben, G.-J. (eds.) ICWE 2007. LNCS, vol. 4607, pp. 491–495. Springer, Heidelberg (2007)
26. Miller, R.: Response time in man-computer conversational transactions. In: Proceedings of AFIPS Fall Joint Computer Conference, vol. 33, 267–277 (1968)
27. Nielsen, J.: Usability Engineering. Morgan Kaufmann, San Francisco (1993)
28. Nielsen, J.: Website response times (2010). https://www.nngroup.com/articles/website-response-times/
29. Object Management Group: MDA Success Stories. http://www.omg.org/mda/products_success.htm
30. Object Management Group: Model Driven Architecture. http://www.omg.org/mda/
31. Opmanis, M., Čerāns, K.: Multilevel data repository for ontological and meta-modeling. In: Databases and Information Systems VI - Selected Papers from the Ninth International Baltic Conference, DB&IS 2010 (2011)
32. Schwinger, W., Koch, N.: Web engineering: the discipline of systematic development of web applications (chap.) In: Modeling Web Applications, pp. 39–64. Wiley, Hoboken (2006)
33. Steinberg, D., Budinsky, F., Paternostro, M., Merks, E.: EMF Eclipse Modeling Framework, 2nd edn. Addison-Wesley, Reading (2008)

Metamodel Specialization for DSL Tool Building

Audris Kalnins[✉] and Janis Barzdins

Institute of Mathematics and Computer Science,
University of Latvia, Riga, Latvia
{audris.kalnins,janis.barzdins}@lumii.lv

Abstract. Most of domain-specific tool building and especially diagram editor building nowadays involves some usage of metamodels. However normally the metamodel alone is not sufficient to define an editor. Frequently the metamodel just defines the abstract syntax of the domain, mappings or transformations are required to define the editor. Another approach [8] is based on a fixed type metamodel, there an editor definition consists of an instance of this metamodel to be executed by an engine. However there typically a number of functionality extensions in a transformation language is required. The paper offers a new approach based on metamodel specialization. First the metamodel specialization based on UML class diagrams and OCL is defined. A universal metamodel and an associated universal engine is described, then it is shown how a specific editor definition can be obtained by specializing this metamodel. Examples of a flowchart editor and UML class diagram editor are given.

Keywords: Metamodeling · Metamodel specialization · DSL tools · Diagram editors

1 Introduction

Metamodeling typically is the basis for most DSL tool definition platforms nowadays, but the principles how metamodels are used vary significantly. Many graphical tool building platforms are related to Eclipse EMF and GMF frameworks [1, 2]. They all are oriented towards "classical" diagram building tasks where at first a domain metamodel describing the abstract syntax must be defined in EMF, only then the graphical concrete syntax (presentation metamodel) is described as a GEF metamodel, with a mapping metamodel between the both added (in GMF). There are some improvements of the basic Eclipse approach such as ObeoDesigner [3] where the presentation metamodel can be defined as a viewpoint of the domain metamodel, or Eugenia [4] where the presentation and mapping metamodels are defined as annotations to the domain metamodel and then generated using a transformation language – thus there the basic Eclipse pattern – start with the domain metamodel is preserved. A completely different platform – Microsoft DSL [5] uses a similar pattern by starting with a domain metamodel and then adding the presentation and mapping metamodels and ending up in code generation, only metamodels are created in a "dialect" of UML. A completely domain specific metamodeling language GOPPRR is used in the MetaEdit [6] platform

© Springer International Publishing Switzerland 2016
G. Arnicans et al. (Eds.): DB&IS 2016, CCIS 615, pp. 68–82, 2016.
DOI: 10.1007/978-3-319-40180-5_5

where the graphical syntax metamodel can be defined directly but with limited functionality and still involving some code generation. A common feature to all these platforms and some similar is that for each new DSL a new metamodel must be created in some metamodeling language.

The platform devoted most directly to graphical DSL tool (graphical modeling language editor) definition is the platform developed by IMCS UL – TDA [7, 8] (the platform initially was named GrTP [9]). There a fixed Tool definition metamodel is proposed which defines type classes for all DSL elements – GraphDiagram, Node, Edge, Compartment, Palette etc., in addition style classes for all these elements are also present in this metamodel. Then a concrete DSL and an editor for it is defined as an instance set of this metamodel. However the complete metamodel for an editor contains also the runtime elements – the classes GraphDiagram, Node, Edge etc., thus instances of completely different nature – Node, NodeType and NodeStyle (and so on) coexist in a runtime model corresponding to this metamodel. In addition, these instances of semantically different layers must be properly linked. The approach is quite usable for simple DSLs where the type instance set defining the language can be created by an auxiliary tool – the Configurator [10], without deep knowledge of the metamodel. However even for slightly more complicated languages and editors the mechanism of extension points related to events of relatively low abstraction level and associated transformations has to be used. To create such a transformation (in a special Lua/lQuery language [11]) the developer has to have a deep knowledge of the metamodel. The editor runtime is based on a Universal Interpreter – a type metamodel interpreter which interacts with the created custom transformations and the support engines for managing diagram layout and user dialogs (Presentation Engine and Dialog Engine).

During the TDA development attempts have been made to use also alternative styles of metamodel usage, closer to the topic of this paper – metamodel specialization. One such attempt [8, 12] was to use an extended UML stereotype mechanism and a stereotype specialization. Another one [13] was the extension of UML class specialization by non-standard concepts borrowed from OWL and forcing a class to be a subclass of another in a dynamic way. However none of these ideas are based on a clean UML usage and none of them were completed and implemented.

This paper proposes a completely new approach to editor definition for Domain Specific Modeling Languages (DSML) on the basis of their graphical syntax. The approach is based on a consistent use of metamodel specialization. In Sect. 2 the metamodel specialization based solely on standard UML class diagram elements and OCL is precisely defined. It should be noted that though the concept of subclass is widely used in metamodel building the whole metamodel specialization to a new more detailed metamodel is a new idea in metamodeling. The only reference where the term "metamodel specialization" has been explicitly used is [14] where the concept has been used for DSML extension, but within a significantly different context. Section 3 introduces the concepts of Universal metamodel (UMM) for the graphical diagram domain and a Universal Engine (UE) related to this metamodel. The concepts are explained on a very simple flowchart editor example. Section 4 describes a complete UMM for DSML definition and the main functionality of UE providing a realistic editor behaviour. Section 5 gives a complete specialization of UMM for a realistic flowchart editor. Section 6 presents some most interesting fragments of a class diagram editor defined by a specialization, there the proposed approach for defining the internal structure of texts in a diagram is illustrated as well. Both these examples confirm that DSML definition by specialization is the cleanest and most understandable way. Finally some basic principles of an implementation of the approach are given in Sect. 7.

2 Metamodel Specialization

Class specialization is a well-known concept in UML. In a sense, it is a cornerstone in building understandable class diagrams. It is also a widely used approach in building metamodels in MOF. However, there is a variation of specialization which can provide a completely new idea in building class models. It is the specialization of a whole metamodel.

A widely used UML concept is an abstract class, which cannot have instances directly. To be more precise, here we need two concepts of an abstract class. The first one is a *fully abstract* class which is being defined as a union of instance sets of all its subclasses. This kind of abstract classes will be used here for purely technical purposes to pull up common properties for several subclasses.

The other kind of abstract classes will be used to assign an intuitive semantics to a set of classes and their properties. We will call such classes *semiabstract*. Certainly, this definition itself is completely intuitive as well.

Now let us give an example of semiabstract classes in Fig. 1.

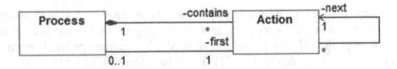

Fig. 1. Example of semiabstract classes

This very simple class diagram containing two semiabstract classes still gives some guidelines of its intended meaning – it represents a simple kind of Process metamodel containing a sequence of actions. Now let us create a specialization of this metamodel by creating subclasses of classes in Fig. 1. This specialization presents a simplest kind of Business Trip process – see Fig. 2.

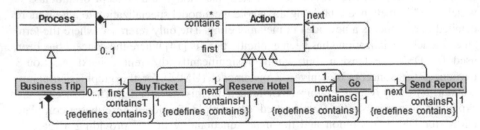

Fig. 2. Simple process metamodel specialization – Business Trip (Color figure online)

The specialization still represents a metamodel for Business Trip. But classes of the specialization are no more semiabstract – they can have instances, e.g., TripToBerlin, ReserveLufthansaFlight etc. The specialization diagram relies un UML *redefines* feature for subclasses – inherited association ends having the same name as for the superclass are redefined automatically, but renamed ends use the explicit *redefines* modifier. Our intuitive semantics of the specialized metamodel completely complies with the semantics assumed for semiabstract classes of the generic Process metamodel. We will call a generic metamodel which is being specialized a Universal Metamodel (or UMM for short) – see more details on this in the next section. Classes of a UMM will be shown here with a white background, but the specialized classes – with a coloured one.

In order to make the metamodel examples more readable and compact in this paper we use a custom notation for specialized classes and redefined associations – we show only the specialization and add the original class and role names from UMM in braces (and bolded) – see Fig. 3 which presents the same specialization as in Fig. 2.

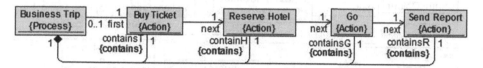

Fig. 3. Business Trip specialization in the alternative notation (Color figure online)

Now let us give a more formal definition of *metamodel specialization*. There may be as many classes specialized from UMM classes as required. But there is a restriction that only a precisely defined set of features enabled by UML *redefines* construct can be applied to inherited from UMM class *properties* – attributes and association ends. The permitted redefinition includes:

- the definition of a default value of an attribute (but not redefining the attribute type and multiplicity) – syntactically the redefinition is ensured by using the same attribute names in the subclass
- for a redefined association end the multiplicity of the association may be redefined (narrowed), explicit redefinition must be used when a different role name is used for a subclass.

No new (non-redefined) attributes or associations can be introduced in the specialization. Default values of attributes are essential here since they determine that a newly created instance of a specialized class will have just these values. A specialization of a UMM class may also be an abstract (fully abstract) one (with the standard UML semantics), if it has a further specialization to non-abstract classes. But concrete classes cannot be specialized further. Specialized classes may have OCL constraints attached.

The goal of these specialization restrictions is to permit only *meaningful* specializations where subclasses are true specializations of the corresponding UMM classes with a similar, but more restricted meaning, thus preserving the intended meaning.

3   Universal Metamodel and Universal Engine

Now when the permitted metamodel specialization has been defined it is time to try to formalize more deeply the intended meaning of a metamodel by adding some precisely defined behaviour to it. We will define this behaviour by means of an executable engine named the Universal Engine (UE) for the given UMM. By definition of this UE we understand a specification how this UE will work on arbitrary specialization of the UMM. In this sense there is only one unique UE for the given UMM.

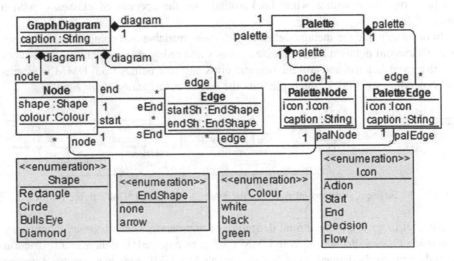

Fig. 4. Universal metamodel for simplified diagram editors (Color figure online)

We will explain the concept of UE on an example – a UMM in Fig. 4 and one of its specializations in Fig. 5. We will explain the functionality of UE on just this specialization, but with the goal to understand how the UE will work on any specialization. Since our main domain in this paper is diagram editor definition, Fig. 4 represents a UMM for a family of very simple editors which are capable of creating just one diagram consisting of nodes and edges, with the structure specified in a specialization.

Nodes can have any shape and fill colour, edges have colour and end shapes. But no text element creation for nodes and edges is present in this simplified version. However one vital element for defining an editor functionality is present here – the Palette together with its elements for creating nodes and edges. Any diagram editor is to be run by a user, but User is not explicitly present in the UMM. Instead, we say that the user can click any palette element – palette node or edge. In response a node or edge instance of the specified kind will be created by UE. Certainly, for a new node the user after the click has to select an empty place in the diagram area, but for a new edge – its start and end nodes. Actually this is all we have to say here on the generic behavior of UE in this simple case, in addition we assume that the editor always starts with an empty diagram with its Palette shown. The details of real behavior of UE for creating a diagram of a specific kind is defined in a specialization of UMM. E.g., only there it is visible what elements will be shown in the palette and what diagram Node specialization will be created by clicking on a PaletteNode specialization. Figure 5 presents one such UMM specialization – a very simple flowchart editor, the custom notation for specialization introduced in Sect. 2 is used there.

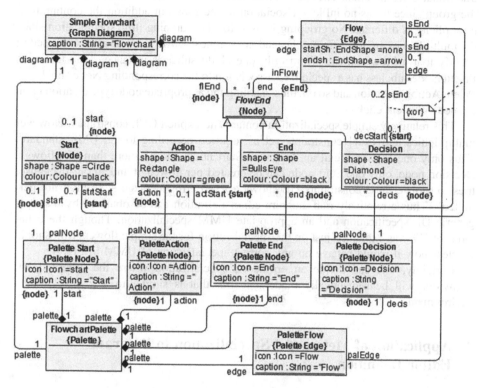

Fig. 5. UMM specialization defining a simple flowchart editor (Color figure online)

The GraphDiagram class is specialized to a concrete diagram kind – SimpleFlowchart. The specialization contains four specialized node kinds – Action, Start, End and Decision as subclasses of the UMM Node class and just one Edge specialization – Flow. Accordingly, the Palette is specialized to FlowchartPalette containing four PaletteNode subclasses (one for each node kind) and one PaletteEdge subclass for creating Flow instances. Default values are assigned to all inherited attributes in subclasses, for node and edge subclasses these are the default style attributes to be used by UE when a node or edge instance is created (note the different shapes for all node kinds). For palette elements different icons and texts are specified accordingly. All UMM associations are redefined as well – using the automatic redefinition where association end names are the same for the superclass and subclass and explicit redefinition otherwise.

To reduce the number of associations in the specialization an abstract subclass is used – the *FlowEnd* class. It groups together the concrete subclasses Action, End and

Decision, which have a common containment association in the SimpleFlowchart diagram and a common association to incoming Flow instances. The fact of specializing the Node class in UMM is also shown in the *FlowEnd* class, the subclasses inherit it. Note that the specialization within the specialized metamodel is shown via the traditional UML notation. The fourth Node subclass – the Start couldn't be included in the group since it has no inFlow association to the Flow, in addition the containment multiplicity is different. No grouping is used for the outgoing flow specification since the multiplicities are too different, instead the simplest UML {xor} constraint is used to specify that a Flow can start from only one Node subclass instance. Note that each PaletteNode subclass has a specialized association to the corresponding Node subclass – PaletteAction to Action and so on, thus enabling the appropriate node type creation upon a palette element click.

This relatively simple specialization contains no explicit OCL constraints. However multiplicities and {xor} constraints act as required according to the UML standard. Thus only one Flow can exit an Action or Start node, and no more than two flows a Decision node, only one Start node can be created per flowchart and so on. If the user tries to violate these constraints UE shows a fixed error message – "Action not permitted". Thus a relatively rich diagram editor definition can be obtained by a simple generic UE specification and an appropriate UMM specialization. Though the behaviour of UE was explained just on the specialization for the simple flowchart, it should be clear how it would behave on any correct specialization of UMM in Fig. 4.

Certainly, one can define an erroneous specialization where the inherited UE behaviour will be semantically inadequate but that is similar to any development environment.

4   Application of Metamodel Specialization to Diagram Editor Definition

The previous section gave a simple introduction into basic concepts of UMM and UE for graphical diagram editor definition. Our goal in this paper is to define a platform for realistic diagram editor definition by means of metamodel specialization. The capabilities of such editors should be similar to our previous editor definition platform TDA [7, 8]. The UMM will provide a general schema for any such editor – be it an editor for flowcharts, for UML class diagrams, for UML activity diagrams etc. The generic behaviour of all such editors will be defined by the *Universal Engine* (UE) operating on UMM. But making the editor behaving just as a Flowchart editor should be made by defining an appropriate specialization of the UMM – then the UE will act as a true Flowchart editor. The UMM will provide a *vision* of such a diagram editor – on what concepts it is operating (see the UMM in Fig. 6). We will consider only diagram editors for pure graphical modelling purposes – without the need to generate some code from the diagram, to run an interpreter on it etc.

The UMM for our diagram editor domain provides a generic data schema on which the behaviour of UE and thus any specialized editor is based. But the behaviour dynamics involves also the *editor user* whose actions actually determine the result. The previous section introduced one element for interaction with the user – the diagram Palette, but there are more.

The possible user actions will not be explicitly captured as UMM classes but they will be tied up to most of UMM classes. The semantics of UE behaviour will be defined just in terms of these actions – what happens if the user clicks a palette element, double-clicks an existing diagram node, enters a compartment value as a text input etc. But there is a strict requirement that all results of a user interaction must be stored as instances of appropriate UMM classes – more precisely, of their specializations.

Now let us explain our vision of such diagram editors and their potential behaviour on the basis of the UMM in Fig. 6. Typically any real diagram editor, including those defined via TDA, contains the concept of Project – a set of related diagrams having a common usage. Therefore we also include Project class in our UMM. The contents of a project has to be somehow visualized – frequently via a tree. However since we want to restrict our visualization facilities, a Project diagram is introduced instead. It contains Diagram seeds – nodes from which the corresponding diagram can be accessed via double-click. Thus a project diagram is a normal graph diagram. The concepts of Graph diagram, Node, Edge and also Palette were introduced already in Sect. 3, only some more attributes are added to these UMM classes. Certainly, we will not present completely all used in practice diagram style attributes, but only the main ones. The main new concept in this UMM is the Compartment – an element in a node or at an edge containing some text. The value attribute of a compartment shows the string really displayed in a diagram. But there are a lot of other attributes specifying the structure of texts, the means for entering them by a user, a way and style of displaying them and so on, in addition these attributes differ for texts in nodes and at lines, therefore we have NodeCompartment and EdgeCompartment classes. A compartment text may have a substructure – e.g., a class attribute text in UML consists of its name, type, default value, modifiers etc., these elements are separated by constant prefixes or suffixes in the common string value, the order of concatenation may be defined by the subCompNo attribute if required. But during the value creation by the user they typically are processed as separate compartments. This structuring in the UMM is supported by the parentCompart – subCompart association, permitting each part to be processed separately as a subcompartment.

Further, the inputContr attribute determines the input control type, which is offered to the user for entering the compartment or subcompartment value. Certainly, the supported types of input controls depend on capabilities of UE, but the minimum list includes simple text input, checkbox for entering Boolean values and listbox or combobox for offering to the user a list of values to select from (in case of combobox a direct value input is also permitted). For both these controls there must be a possibility to define the appropriate value list, therefore the itemList attribute of type Set (String) is added. The default value of this attribute must be set in the specialized compartment class (if listbox or combobox is selected for the compartment input), this value may be a constant set or an OCL expression deriving the set from other diagram elements already created. One more nontrivial input control is multiline input containing rows with the same properties (such as the whole attribute list for a class), there a special CompartmentRow subclass permits to process each line separately and add a new line (a line may be a structured compartment as well).

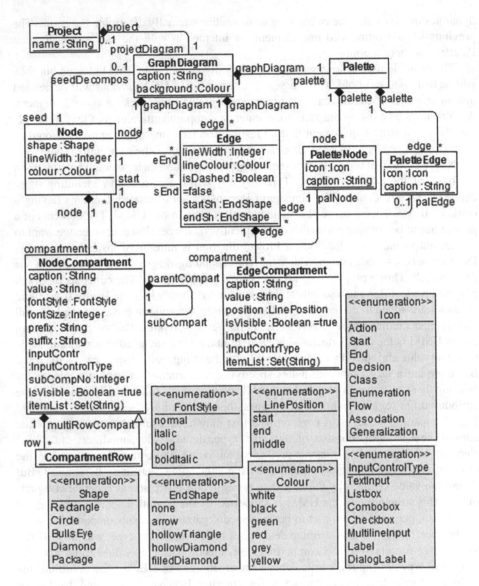

Fig. 6. UMM for real diagram editors (Color figure online).

The whole input process of a compartment is organized by UE in a fixed way, the specialization may configure only a compartment structure, the input control used for a compartment/subcompartment entry and provide the required values. In addition, a standard UML constraint (an OCL expression returning a Boolean value) may be added to the specialized compartment class to check the entered compartment value after the user has completed the input of this compartment. Some examples for these rather complicated specialization features are demonstrated in Sect. 6 on class diagram editor fragments.

We conclude this section by a rather informal description of UE related to this UMM. Upon start the UE permits the user to create a new diagram editor project of the kind defined by the current specialization (a flowchart project, a class diagram project etc.) or open an existing project of this kind. After that the project diagram (either empty or already filled) with its palette is shown. The user can add a new diagram of a supported kind via creating its seed from the palette, or open an existing diagram – by double-clicking on the seed. The diagram is opened together with its palette, then new diagram elements can be added in the manner described in the previous section. But a new possibility is to enter compartment values which are defined in the specialization – the compartment editor (a dialog form) is opened after a new node or edge is created. The compartment editor can be opened also for existing nodes or edges by a double-click. Besides this specialization-related UE behaviour, UE offers some default behaviour to the user – to save a project, to modify the default style of a node or edge, to modify the layout etc.

5   The Real Flowchart Specialization Example

In this section the new possibilities of UMM and UE are demonstrated on a more realistic flowchart editor example – see Fig. 7. The Project class from UMM is specialized to FlowchartProject with just one FlowchartProjectDiagram attached to it. This diagram contains named FlowchSeed nodes from which the corresponding Flowchart diagram instance can be opened, seed nodes can be created by the FlProjectPalette with the sole FlPrPaletteSeed element. In order to have a user-defined name for a seed (and the related flowchart as well), the FlowchNameCompart class (specialized from the NodeCompartment) is associated to FlowchSeed. Only the caption and inputContr attributes with their default values appear in the compartment specialization – other attribute values are not required for this simple case.

The Flowchart node definitions in this specialization are similar to those in Sect. 3, only a name can be created for an Action and a condition text for a Decision – both use simple TextInput fields for value input. The Palette is also similar to that in Sect. 3. But here two edge subclasses are defined – the Flow as in Sect. 3 and the ConditionalFlow which can exit only a Decision node (no more than two instances). A conditional flow has a text attached near to its start – typically Y or N but any other text can be used as well. Therefore in the specialization the class CondValueCompart (a subclass of EdgeCompartment in UMM) is used. A different input control type – Combobox is used there, and for this control also the default value for itemList attribute must be defined – here it is a constant set of strings – Set {"Y", "N"}. The position attribute specifies that the entered text has to be positioned near the start of the edge.

Note also the use of the abstract superclass FlowchEdge in the specialization, with the concrete subclasses Flow and ConditionalFlow. This construct permits to use one common palette edge element for both – it is associated to the superclass and the user has not to think which kind of flow to select – he just clicks the flow in the palette and selects either Action (or Start) or Decision instance as a start node. Now UE is capable

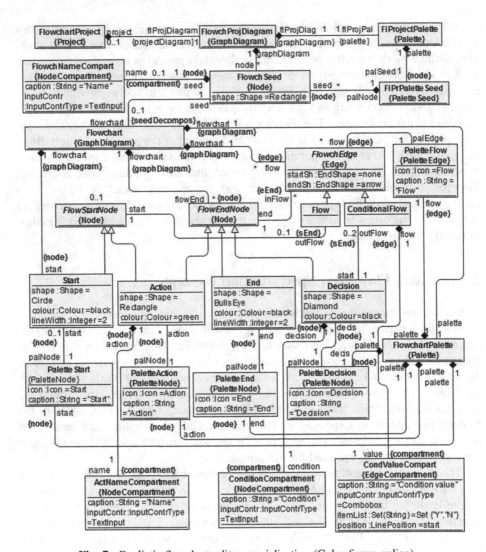

Fig. 7. Realistic flowchart editor specialization (Color figure online)

to infer automatically which subclass of FlowchEdge to create – the specialization explicitly defines which flow subclass can start from the given start node type. Certainly, when using such a construct it must be ensured during the specialization building that only one edge subtype is legal for each start node choice.

Thus this example shows that using only basic UML class diagram constructs such as multiplicity a UMM specialization can define relatively complicated editor behaviour. In order to obtain a more compact specialization, it is assumed that some more compartment style attributes (in addition to those shown in Fig. 6) have their default values set already in UMM, e.g., fontStyle is set to the value normal.

6 Fragments of Class Diagram Example

In this section some basic fragments containing new features for a specialization of the same UMM in Fig. 6 defining a class diagram editor will be given. The functionality of the editor is approximately that used for creating class diagrams (metamodels) used in this paper. The whole specialization can be defined in 3 class diagrams each to be shown in an A4 page in a readable way. The supported node types are Class and Enumeration, but edge types – Association and Generalization. In addition, the UML package mechanism is included, but in a slightly non-standard way – using Package diagrams. The main new elements in this specialization are the use of compartment rows, subcompartments and OCL expressions for default values and constraints. Figure 8 presents a fragment of this specialization showing the Class node and some of its compartments: Class name compartment, IsAbstract compartment and Attribute compartment. The IsAbstractCompartment enables the user to enter the Boolean value specifying whether the given class is abstract or not; this is done via a checkbox control (however the value is stored in the model as a string "true" or "false"). This value is not visible directly in the class node – therefore the attribute isVisible is set to false (so it is specified in the UML standard). Instead, the value is displayed by setting the appropriate style for the class name compartment (*italic* if the class is abstract). To specify this setting, the value of fontStyle attribute of ClassNameCompartment is set by an OCL expression:

```
fontStyle : FontStyle = if self.class.isAbstract
.value.toBoolean() then italic else bold endif
```

(the expression is not shown in Fig. 8 to reduce the box size). The Attribute compartment is to be created by the user via a specific control – MultiLineInput. This control is specially adjusted to creating texts consisting of logically independent lines, such as attributes or operations in a Class node. Therefore the UE provides an independent entry of each line using the CompartmentRow class in UMM which is specialized here to AttributeRow. Further, the line can have a complicated substructure: each attribute has a name, type, default value, multiplicity etc. This is supported in UE by the subcompartment concept (see Sect. 4). Here we have the AttributeName, AttributeType, DefaultValue etc. subcompartments (only the first two are shown in Fig. 8). To specify how the values are to be concatenated to a common string, prefixes or suffixes are used – see the prefix ":" for the type. Each subcompartment can be entered using a specific control – the type is entered using a combobox offering the most typical values (primitive types, all defined enumerations in the model). Therefore the itemList attribute of this compartment is specified by the OCL expression:

```
itemList : String[*] = set{"Integer","Boolean", "String",
"Real"} -> union(Enumeration.allInstances().name.value ->
asSet() )
```

The allInstances OCL construct here iterates over the whole project, thus all Enumeration instances are collected and their name values included in the list.

A similar but more complicated expression can be defined to offer all already existing class names when a class name is to be entered (in order to create occurrences of a class in several diagrams). Similar expressions preparing a typical value list can be defined for other subcompartments as well, e.g., for type-dependent default value prompting.

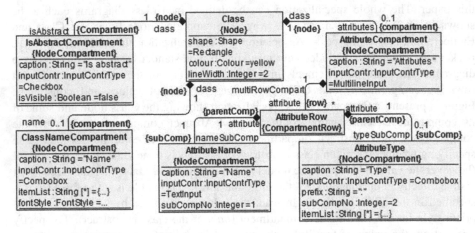

Fig. 8. Fragment of class diagram specialization – class attributes (Color figure online)

Finally, we show an example of OCL class constraint to be used for checking the entered compartment value correctness – the class AttributeName has a constraint specifying that attribute names must be unique for a class:

```
{self.attribute.multiRowCompart.attribute.nameSubComp
.value -> forAll(val | val<>self.value)}
```

Such constraints have to be specified directly for the corresponding compartment class, then UE knows that the constraint has to be evaluated right after the user has exited the corresponding input control.

To conclude, the use of OCL permits to obtain at least the same functionality of a class diagram editor as can be defined using custom transformations at extension points in TDA and close to many commercial UML editors.

7   Implementation Principles

The implementation of metamodel specialization based DSML editor platform really consists of two parts – the Base UE including DSML project management, diagram drawing and layout management, dialog engine with current form building and user input processing and a UMM specialization interpreter. This interpreter has to find the relevant specialization class together with the specialized attributes (default values set), create a correct instance of this class (to be passed to drawing engine for visualization or some other relevant part of Base UE) and find the specialized associations outgoing from this class. These associations determine the other specialized classes to be processed in this step, e.g., the specialized compartments for the given node subclass. In addition, the OCL expressions for default values and constraints must be interpreted. If Eclipse EMF is used as a model repository there is a freely available OCL interpreter. For other repositories the solution used in TDA can be applied – use the Lua/lQuery instead of OCL, this language has sufficient expressive power for defining expressions and is implemented within TDA for several repositories. The Base UE in fact is quite similar to the set of existing component engines in the TDA implementation. Thus the effort for implementing the proposed approach could be lower than that used for TDA, in addition the base components could be reused.

8 Conclusions

The analysed DSML examples show that the proposed approach for DSML editor definition based on metamodel specialization has a number of advantages and is usable in practice. The specialized metamodels which use only basic UML class diagram features and OCL for more complicated situations are natural formalizations of the graphical syntax of the DSML to be defined. Thus such metamodels are sufficiently easy to create and read. To a great degree this fact shows up when compared to the corresponding DSML definitions in the existing TDA platform. For very simple DSMLs such as the simple flowchart where the type instances in TDA can be created using the Configurator without any extension points the efforts are comparable. But for Class diagram editor a significant use of extension points and transformations is required in TDA, with a large effort required to create these transformations due to the complicated runtime metamodel in TDA. Thus the complete DSML definition there is in fact invisible, the type metamodel instances provide only a graphical skeleton of class diagram definition. At the same time the definition using metamodel specialization describes the supported functionality in a very explicit way. Creating of required OCL constraints is also quite straightforward since the specialization directly defines also the runtime metamodel. If more features are to be added, the specialization extension by new specialized classes is also very straightforward. Most probably, if TDA would be created now, the metamodel specialization approach would be used.

Acknowledgements. This work is supported by the Latvian National research program SOPHIS under grant agreement Nr. 10-4/VPP-4/11.

References

1. Eclipse. http://www.eclipse.org
2. Graphical Modeling Framework (GMF, Eclipse Modeling subproject). http://www.eclipse.org/gmf/
3. Obeo Designer: Domain Specific Modeling for Software Architects. http://www.obeodesigner.com/
4. EuGENia Live. http://eugenialive.herokuapp.com/

5. Cook, S., Jones, G., Kent, S., Wills, A.C.: Domain-Specific Development with Visual Studio DSL Tools. Addison-Wesley, Boston (2007)
6. Kelly, S., Tolvanen, J.-P.: Domain-Specific Modeling: Enabling Full Code Generation. Wiley, Hoboken (2008)
7. Barzdins, J., Rencis, E., Kozlovics, S.: The transformation-driven architecture. In: Proceedings of DSM 2008 Workshop of OOPSLA 2008, Nashville, Tennessee, USA, pp. 60–63 (2008)
8. Sprogis, A.: Configuration language for domain specific tools and its implementation. Ph.D. thesis (in Latvian), University of Latvia, Riga (2013)
9. Barzdins, J., et al.: GrTP: transformation based graphical tool building platform. In: Proceedings of MDDAUI 2007 Workshop of MODELS 2007. CEUR Workshop Proceedings, Nashville, Tennessee, USA, vol. 297, 4 p. (2007). http://ceur-ws.org
10. Sprogis, A.: The configurator in DSL tool building. In: Computer Science and Information Technologies, Scientific Papers, vol. 756, pp. 173–192. University of Latvia (2010)
11. Liepins, R.: lQuery: a model query and transformation library. In: Computer Science and Information Technologies, Scientific Papers, vol. 770, pp. 27–46. University of Latvia (2011)
12. Sprogis, A., Barzdins. J.: Specification, configuration and implementation of DSL tool. In: Frontiers of AI and Applications. Databases and Information Systems VII, vol. 249. IOS Press (2013)
13. Rencis, E., Barzdins, J., Kozlovics, S.: Towards open graphical tool-building framework. In: Proceedings of BIR 2011, pp. 80–87. RTU Press, Riga (2011)
14. Pierre, S., Cariou, E., Le Goaer, O., Barbier, F.: A family-based framework for i-DSML adaptation. In: Cabot, J., Rubin, J. (eds.) ECMFA 2014. LNCS, vol. 8569, pp. 164–179. Springer, Heidelberg (2014)

Models of Event Driven Systems

Zane Bicevska[1]([⊠]), Janis Bicevskis[2], and Girts Karnitis[2]

[1] DIVI Grupa Ltd, Riga, Latvia
Zane.Bicevska@di.lv
[2] Faculty of Computing, University of Latvia, Riga, Latvia
{Janis.Bicevskis,Girts.Karnitis}@lu.lv

Abstract. This paper provides the business process modeling approach based on usage of Domain Specific Languages (DSL). The proposed approach allows us to create executable information systems' models and extends the concept of Event Driven Architecture (EDA) with the business process execution description. It lets us apply principles of the Model Driven Development (MDD) in order to create the information system which complies with the model. The proposed approach provides a set of advantages in information systems development, use and maintenance: bridges the gap between business and IT, an exact specification, which is easily to implement into information system, up-to-date documentation etc. The practical experience proves the viability of the proposed approach.

Keywords: Business process modeling · Event driven architecture · Services oriented architecture · Domain specific language

1 Introduction

Model driven development (MDD) is a concept that offers a wide range of advantages for development, maintenance and use of information systems [1, 2]. The main advantages are as follows: (1) MDD provides a higher level of abstraction to describe the business process, and it leads to reduced efforts for error-prone description, meaningful validation and exhaustive testing, (2) MDD bridges the gap between business and IT, (3) MDD captures domain knowledge, (4) MDD results in software being less sensitive to changes in business requirements, (5) MDD provides up-to-date documentation.

To reach the benefits of MDD, there should be used modelling languages and tools allowing accurately describe the system's operation. It means the objects of high abstraction level should be transformable into source code of an information system. The most popular modeling languages like UML [3] and BPMN [4] are able to carry out the task only partially [5]. The languages use high abstraction concepts without linking them to a specific domain objects and views. To achieve a full compliance, the objects of high abstraction level must be linked to the gradual detailing level up to the level of implementation. Many authors propose usage of Domain Specific Languages (DSL) as the solution. For instance, [6] argue that although UML Cis a useful modeling language, it is not an appropriate language for MDD, because UML is designed for documenting and not for programming.

G. Arnicans et al. (Eds.): DB&IS 2016, CCIS 615, pp. 83–98, 2016.
DOI: 10.1007/978-3-319-40180-5_6

As a response to this the MDD community shifted focus to smaller, more specific languages [5, 7]). These DSLs focus on a specific problem area or even a business domain. That ultimately also defines the scope and applicability of these languages. They are by definition not meant to solve a generic issue and hence don't attract a large developer community. The usage of DSLs in the MDD context leads to the requirement the models should be executable. And the challenges are still there - how to define business models and how to make them executable? The essential change is that models can now be directly used to drive software development.

This paper describes a platform for defining of DSLs ensuring executable models. The main focus of the research is devoted to so-called Event Driven Systems (EDS) where the data processing is caused by system's external or internal events. The model of an EDS consists of widely-used modeling artefacts like data object, activity, data object's state, event, time constraints etc. The proposed tools let you:

- to define a new DSL, i.e. it's artefacts and rules for graphical representation of diagrams;
- to create an editor for editing of graphical diagrams in the new DSL;
- to define semantics of the used concepts and the execution of the graphical diagrams.

The EDS represent a class of models where each instance is an executable model in a particular DSL. The models can be transformed into the executable source code either by interpreting the model, or by generating the source code from the model using a code generator.

As examples of EDS serve a wide range of data processing systems used in our daily life – state information systems (population register, vehicle register, business register etc.) as well as commercial solutions (e-shops, CRMs etc.). For example, the population register processes the data of citizens as data objects, and the personal life events (birth, education, occupation, marriage, children etc.) serve as events which change the states of data objects.

The proposed approach differs from the traditional UML and BPMN models as it offers a wider range of artefacts, a precisely defined semantic of model execution and a usage of domain specific concepts in the modelling language.

The present work is a continuation of the previous modeling research [8] with the main focus on the executability of models. The first section describes the purpose of the study – to create a DSL based approach for modelling of executable event–driven information systems. The second to fourth section describe respectively the syntax, the semantics and the pragmatics of the proposed approach. The fifth section describes the related research. The conclusion contains summary about similarities between EDS and MDD approaches.

2   Models of Event-Driven Systems

2.1   Context

In everyday life we use information systems that gather information about certain events and use the information in different ways. For instance, banking systems collect the information about transactions in accounts (income, expenses and fees), hence the account's balance can be calculated from the gathered information at any moment.

The event-driven systems process data objects of a certain type and the processing is response initiated by different events changing the values of data object's attributes. For instance, a records management solution deals with the data object *Document*, and the processing steps of documents can be described by various events - document registration, response preparation, acceptance, sending of response etc. The event-driven system stores the passed events into the data object's *Document* attribute *State* using the values *Registered*, *Prepared*, *Accepted* and *Sent*. This allows you to keep track of each data object and all instances of data objects at the same time by using statuses as results of changed attributes' values of data objects.

The main goal of this research is to design a DSL-based modelling platform for describing of EDS to achieve a better appliance of MDD principles in development of information systems.

2.2 Business Process Modeling Environment DIMOD

Usually business processes of EDS are described using some graphical DSL. As the EDS represents a wide class of information systems, the used DSLs can also be different. Hence it is advisable to use not only one specific graphical editor supporting one concrete DSL but to create your own editor for each used DSL. This can be done if there is a graphical editor building platform that allows not only defining the appropriate DSL editor, but also provides other functions such as checking of model's correctness.

Currently there are several such platforms in use; one of them, called GrTP [9], is produced in Latvia. The business process modeling environment DIMOD, used in this research, was derived from the graphical editor building platform GrTP. Accordingly the DIMOD features are widely affected by those of GrTP.

The business process modeling environment DIMOD is intended:

- to define the DSL using meta-model that is stored into internal tables of the environment's repository. DSL defining parameters may be defined and modified using the separate configuration component Configurator [10]. Once the DSL is defined the corresponding modeling environment (editor) is created for the DSL automatically;
- to create and edit business process diagrams in the DSL. This is usually done by some highly qualified modelling experts in collaboration with domain experts ("clever users");
- to check the business process models' internal consistency. Both IT and domain experts are involved in it;
- to publish the created model/diagrams in web. It allows a wide range of users to use the business model diagrams without installation of the specific modeling environment DIMOD.

DIMOD also includes a feature to call APIs in predefined points for implementation of DSL specific semantics. It provides accessibility of specific functions or external sources as well as read/write access to model's repository.

2.3   Characteristic Aspects of EDS Modeling Languages

Traditionally, there are three views used to describe programming and/or modeling languages:

1. Syntax of the modelling/programming language. The syntax of modelling language is determined by the language meta-model. A program or a model is considered to be syntactically correct if it meets all the conditions included in the meta-model. The syntactical correctness of the model/program does not guarantee that the information according the model is processed exactly in that way which has been desired by the author. Therefore it is not sufficient to have only syntactically correct model to create an information system; it is necessary to define the semantics of the language.
2. Semantics of the modelling/programming language. The semantics of modelling language describes the meaning of concepts used in the language. The semantics can be defined in different ways – by axiomatic mathematical theory or by describing of an abstract machine explaining the language command execution and the meaning of language structures. In this paper the concept of abstract machine is used as it is well-known in the IT industry.
3. Pragmatics of the modelling/programming language. The pragmatics gives recommendations how the language should be applied and describes the methods. The pragmatics is based on practical considerations like language usage costs, development deadlines, availability of support, complexity of the language etc.

The following sections will include analysis of all three aspects of EDS.

2.4   Ontological Model of EDS

A modeling language should be defined in two levels: (1) firstly definitions of the ontological notions should be created, and then (2) the graphical representation is aligned to defined notions. The separation between the used concepts and their graphical depiction is the essential difference between the proposed approach and the traditional way used in many modeling languages where basic language elements are defined by graphical symbols representing basic concepts.

The ontology of the EDS business process model proposed by the authors is shown in the Fig. 1. The meta-model consists of a voluntary number of elements. Every element has at least three predefined attributes – type, name, and identifier – and a voluntary number of additional attributes.

Besides the five predefined types of elements the model may contain a voluntary number of additional types of elements. The predefined types are as follows:

- object represents external donors/receivers of the information as well as sources/targets of data; some examples of such elements are messages, documents, records in data bases, legislative acts, etc.

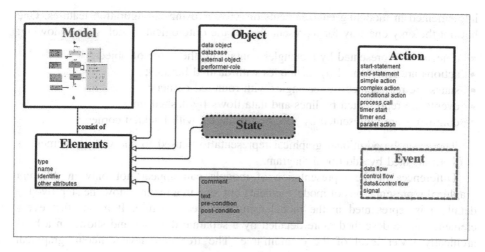

Fig. 1. Ontological meta-model of EDS

- action represents activities with data objects such as start/end of processes, execution of process's steps, conditional execution, parallel activities, control of deadlines, etc.
- state describes corteges of attributes' values available for the data object to be processed. The corteges of values are equivalence classes with objects that are to be processed similarly during the specific process. For instance if the document processing should be modeled it is important to identify all documents being in states Received, Registered, Processed etc. It can be done using the attribute Document's state.
- event represents data flows that are received/sent by the process form/to external information sources/targets and control flows representing the sequence of process steps, and other types of flows.
- comment contains informal description about activities processing, states and flows.

The set of notions described above is sufficient to apply business process models for many uses.

3 Syntax of EDS Models

3.1 Representation of EDS

Depending on the demands of DSL users and features of the DSL supporting environments, the above described business processes can be represented by diagrams collocated in a tree-like structure. Each diagram may contain a voluntary number of elements. Actions and states will be specified in a separate diagram, and it will be represented in a lower level of the diagram tree under the respective element.

The elements described above may be represented using different graphical elements, like boxes, lines, pictures etc. The representation of elements will be implemented in modeling environments directly or using configuration features. One but not the only one way for representation of the ontological model is the following:

- objects are represented by rectangles containing the name of object;
- actions are represented by rectangles with dashed borders;
- states are represented by rectangles with rounded corners;
- events are represented by lines and data flows by dashed lines;
- comments are represented by specific rectangles with buckled corners.

There can be additional graphical representations used to show if an element is in-depth detailed by additional diagram.

Differences in the representation of models can appear not only in different graphical representations of model elements but also in a manner how the step-by-step detailing is represented in the model using the tree-structure. It means that every element can be described more detailed by a separate diagram and shown in a hier-archically lower level of the diagram-tree. The tree can have different graphical representations.

3.2 DSL Example

One example of EDS is the working time tracking (WTT) system, and it was used to show the approach of DSL definition with consecutive creating of the modelling environment in DIMOD.

The syntax of the DSL used for modelling the WTT business processes is described by the meta-model in the Fig. 2.

There are two main categories of elements (objects) in the DSL: (1) source elements (*From-Element* in the Fig. 2) are those where the *dataFlow* and *controlFlow* events are started from, (2) target elements (*To-Elements* in the Fig. 2) are those where the events

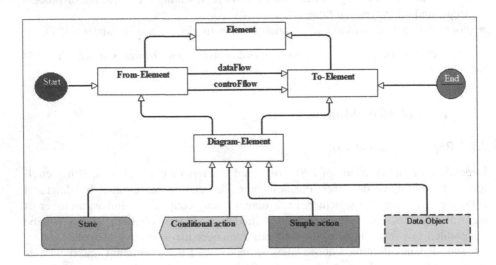

Fig. 2. DSL meta-model for modelling of WTT business processes [11]

are finished. Elements of graphical diagramms are described as *Diagram-Element*, and they in turn can be either *State, Conditional action, Simple action,* or *Data Object* (Document or Screen form).

An example of the diagram created by DIMOD generated tool describing the WTT business process for time reporting and accepting is given in the Fig. 3.

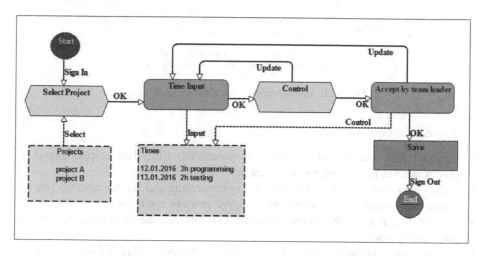

Fig. 3. An example of WTT business process diagram

The process starts with the Sign In event. Then the conditional action Select Project is available where users can specify the instance of the data object Project for which the spent working time will be reported. The project's selection is finished with an event OK that leads the data object into the state Time Input. Users can enter working time data (data object's parameters are changed) and finish the session by initiating the event OK. The event transfers the entered data to the action Control. If any syntactical errors are discovered, the event Update leads the data object back to the state *Time Input*. If the entered data is syntacticaly correct, the event *OK* transfers data objects to team leader for accepting the entered working time information on the substance. The action *Save* stores the entered data and the process is finished.

4 Semantics of EDS

4.1 Memory of Process Execution

Let us define the semantics of process execution by using a "hypothetical" machine (see Fig. 4). It consists of (1) a memory that stores instances of data objects and instances processes, and (2) an engine that is able to execute particular process steps/statements consecutively. Data object instances are stored in the data objects part and process instances – in the processes part of the memory.

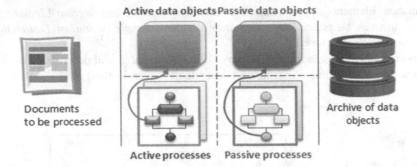

Fig. 4. Business process execution memory [12]

The data objects part contains:

- prepared data objects (Documents to be processed) - the data objects should be processed, but processing has not been started yet
- Active data objects - the data objects are in processing and are not finished yet; the active data objects have corresponding active process instances
- Passive data objects - the processing is finished, but the data objects can be used by other processes
- archived data objects (Archive of data objects) - the processing is finished, and the data objects can't be used by other processes anymore

The processes part of the memory stores instances of processes that were created by deriving the instances from the process diagrams (process descriptions) and by linking them to the data objects to be processed. There are two kinds of process instances:

- Active processes– the process instances are in processing and are not finished yet; the active processes have corresponding active data objects instances;
- Passive processes - the processing is finished; nevertheless it is advisable to save the passive process instances for repeated processing if such a need would arise.

4.2   Process Initiation and Execution

The process execution starts with user login into the information system. Let us assume the user identification is implemented in a traditional way, i.e., by entering login name and password. It is important only that the information system, in result of electronical identification, has granted to the particular user the rights to execute certain business processes and access certain data objects.

The first step of the business process execution is to let the user choose the business process that should be executed and link the process to the desired data object. There are two different cases possible: (1) an initiating of a new process, and (2) a continuation to execute an already initiated process.

If a new process is initiated, a new process instance is created by copying the process description. During the execution the recently created process instance will be linked to the data objects that should be processed. Technically the execution of the process instance is activated, for example, by user clicks on menu control.

To continue the execution of a process that has already been initiated before, the engine finds the process instance in the active process part of the memory, and the process execution is started/continued from the point in the process model where it was interrupted in the previous execution session. In addition, the user can see the executed steps in the process diagram. Technically it may be implemented as an individual user's activity that acts according to the to-do-list principle – the user can choose between all active (unfinished) process instances that are attributable to him/her.

After the process execution is initiated, the user can manage the execution by selecting one or several steps of the process. The process execution can be stopped (postponed) by the user, and the active process then will be stored into the memory.

Processes are executed step by step. The steps can be processed manually, semi-automatic, or automatic. Manual steps are initiated, for example, by the user clicking on the control-window buttons. Semi-automatic steps are also initiated by the user but, in contradiction to manual steps, the execution is completed by certain user's activities. For example, within the step a screen form is opened with editable fields, the user fills in the necessary data and presses the button OK. Automatic execution is performed without privity of user. For example, data object saving in the database is an automatic operation.

4.3 Visualization of Data Objects and Process Execution

Data objects are represented in screen forms, and one screen form describes one or several data objects. A screen form can contain many fields to be added/edited/deleted. It can also serve for selecting of data objects.

There are some improvements practiced for easier identifying and specifying of data objects: (1) show attribute values in screen forms to let users recognize the necessary objects, (2) allow filter data objects in screen forms to narrow the set of object instances, (3) colour fields in screen forms according to the data objects instance status (for instance, the corrigible fields could be shown in a different colour). Furthermore, some data columns or fields could be disabled (not displayed at all) if the user does not have rights to access the data objects. Development of screen forms is relatively simple, and it can be done using traditional methods.

Business process diagrams are suitable for visualization of process execution. The process execution can be represented by colouring the already executed activities and control flows in a different colour. In the general case, only those elements of business process diagrams will be coloured which have been involved in the processing of the concrete data object. The shaded track allows identify the executed steps and continue process execution from the point in the process model where the execution was interrupted in the previous session.

5 Pragmatics of EDS Models

5.1 Implementation of Direct Business Process Execution

The development of information systems with built-in business process interpreter (execution engine) differs significantly from the development of the ordinary information systems. Instead of developing the windows with popup menus, control buttons and other interface elements for process managing, graphical diagrams with process descriptions/definitions should be available for the users to ensure the possibility to manage processes directly, without usage of menus. The process execution must be visualized in the process diagram allowing users to monitor the performed actions. The traditional application development has been replaced by creating the diagrams describing business processes.

Usually the business process modeling is carried out by a small number of users, so-called "modelers". It means the modeling environments and tools must not compulsory be web applications. For instance, the modeling environment DIMOD is currently implemented as a desktop-application that is available in modelers' workplaces.

Ordinary users do not create models they use them in a read-only mode. These users need an access to business process descriptions via web because they will use the models for the direct business process execution. The DIMOD solves it by exporting business process diagrams to the html/SVG format (SVG - Scalable Vector Graphic). As a result, a logical copy of model's repository is obtained that is accessible from an internet browser, and all navigation functions in diagrams and their elements are still there.

The appearance of process diagram's elements can be easily changed in html/SVG. For instance, the colour of an element can be switched by editing element's parameters in the html/SVG statements. It can be done by a special routine that finds an element by its SVG-identifier and updates the corresponding html/SVG statement. After refreshing of the diagram the element's appearance will be changed.

There is also an external identifier necessary to maintain the connection between events of graphical element processing and the corresponding software routines that should be developed in some programming language, for instance, Java. The external identifier ensures the activation of separate steps of business processes, for example, by a mouse double-click.

5.2 BiLingva and DIMOD

Authors of the paper have developed and used in practice a domain-specific language BiLingva [11] which is suitable for modeling of EDS. The used DSL BiLingva offers a full range of EDS elements - object states, process steps (events), execution sequences (flows) as well as decision points (conditions), process branching, time intervals, parallel processing.

Each use of DSL consists of several steps. In the first step the DSL for process modeling is defined in cooperation with organization's business specialists. Definition of DSL is based on some pre-defined DSL complemented with additional attributes and industry specific input/output documents. In practice definition of DSL simultaneously implies the definition of DSL support environment, for example DIMOD.

In the next step, the organizational business model is built by using DSL in the modeling environment.

In the final step the models created using DSL, for instance BiLingva, can be transferred into EDS data objects (elements) and stored in the EDS database automatically. If the EDS has a built-in interpreter or uses some external interpreting engine, the business processes can be executed according to their descriptions (models) in DSL.

The EDS approach differs from the MDA approach crucially. The MDA demands an information system should be generated from the model. The EDS approach proposes to develop information systems in the traditional way (including the required functionality as well as non-functional features - usability, security, performance, etc.) with the difference that the event-driven (workflow) execution is implemented as a separate and individual component which uses business process descriptions in a specific way.

The modeling-based information systems' development techniques are not widely known among business professionals. Business analysts with the basic modeling knowledge can create only informal models (informal description models for better understanding of business processes). Just after the level of DSL knowledge is increased they are able to create formal models (formal models defining the business processes). In practice, it may take up to 2 to 4 years for different organizations to learn modeling techniques.

Viability of the proposed EDS approach using BiLingva has become clear in practice, as this approach is being used in a number of information systems in Latvia for more than 10 years [8]. Particularly striking was the users' positive attitude towards graphical process diagrams. In a short period the diagrams became the user guides and substituted the text documents used before, as the diagrams are more accurate and explain system's operation to the users more comprehensively.

6 Related Work

Let's look at four research directions that are the most related to the EDS research – Event Driven Architecture (EDA), Service Oriented Architecture (SOA) 2.0, workflow management systems and executable UML.

6.1 EDA

EDA is a distributed information systems architectural style [13, 14] characterized by the following:

- software processes are completely decoupled from each other,
- processes only depend upon events,
- events are published in a standard, consistent manner,
- software processes can react to different types of events or different contents of events autonomously.

EDA approach spends the main attention to recording of events in external or internal business process environment, therefore making them accessible for processing routines. Descriptions of the event initiated business processes are not discussed closer. Events are characterised by attributes (name, type, type-specific attributes).

The EDS approach deals with both events and the processes that handle the events. Using the formalisation described in the previous sections, events can be defined by two kinds of flows: (1) control flow, and (2) data flow. The control flow describes the sequence of events which initiate the execution of processes or execute process' steps. The completing the execution of one process's step which is followed by the execution of the consecutive step is also considered to be the part of the control flow. The data flow stands for events which describe actions with data objects (read, write, send, etc.).

An EDS model is some kind of EDA generalisation as it makes possible to describe not only the event control and data flows, but also the business processes and the processed data objects.

6.2   SOA

SOA [15] is defined as a distributed software architecture where self-contained applications expose themselves as services, which other applications can connect to and use. SOA applications claim to be self-describing, discoverable, as well as platform- and language-independent.

SOA 2.0, also called advanced SOA or event-driven SOA, is the next generation of SOA that focuses on events, inspired by EDA. SOA 2.0 enables service choreography, where each service reacts to published events on its own, rather than being requested to do so by a central orchestrator.

Both approaches (SOA and EDA) are considered to be complementary [16] and often one architecture is considered to be a subset of the other, depending on the viewpoint of the architect. They both are evolved from different cultures, but have come to embrace similar principles. EDA evolved from the MessageOrientedMiddleware (MOM) with implementations such as TIBCO/Rendezvous [17] or MQSeries [18], whereas SOA evolved from the DistributedObjects with implementations such as CORBA or COM [19]. They both seem to learn a lot from each other.

The SOA [20] and EDS interrelation is similar to those of EDS and EDA. The EDS focuses on the execution of initiated processes whereas the SOA describes initiation and processing of independent services without deeper analysis of internal structures and operations of the services.

6.3   Workflow Management Systems

According to the Workflow Management Coalition [21], a workflow represents "the automation of a business process during which documents or tasks are passed from one participant to another for action, according to a set of procedural rules". Already now there are information systems running in the interpretative mode according to the business process model. The most popular of them are workflow management systems.

Workflow management systems are usually characterized by concepts such as tasks (process's steps) which are accomplished by different users processing data objects. Ten features are characteristic for workflow systems, the most important are: graphical workflow representation, data object representation in screen forms, generated reports, role-based accessibility. But the main focus in the commercial workflow management systems is on transferring of tasks among different users. It means the solutions ensures recording of internal and external events and delivering of data objects to the executing processes for completion of defined tasks.

In the comparison of workflow management systems the JIRA [22] is recognized as the most popular [23]: "JIRA is the workflow management tool for teams planning and building great products." As the second most popular workflow management system is mentioned KiSSFLOW [24].

The EDS approach not only includes describing of workflow management systems but also provides a number of supplementary features. It confirms that workflow systems are a subset of the EDS approach.

6.4   Executable UML

Numerous studies have claimed [1] the models should be dynamically executable. According to the researches, the descriptions of business processes should be detailed so clearly and consistently that the processes can be executed automatically without human participation. Even if the processes would be executed by people, the business process descriptions should be unequivocal and definite. The goals set by authors of the executable UML are the same as those of the EDS approach.

A comprehensive overview of the executable UML research roadmap is given at [25]. Two research fields of executable UML deserve more attention: (1) Foundational UML (fUML), and (2) Executable and Translatable UML (XTUML).

Foundational UML (fUML) is an executable subset of standard UML that can be used to define, in an operational style, the structural and behavioural semantics of systems. Implementations of fUML as Open Source are given at [26].

fUML is a subset of the standard UML for which there are standard, precise execution semantics. This subset includes the typical structural modeling constructs of UML, such as classes, associations, data types and enumerations. But it also includes the ability to model behaviour using UML activities, which are composed from a rich set of primitive actions. A model constructed in fUML is therefore executable in exactly the same sense as a program in a traditional programming language, but it is written with the level of abstraction and richness of expression of a modeling language.

A system is composed of multiple subject matters, known as domains in fUML terms. They are used to model a domain at the level of abstraction of its subject matter independent of implementation concerns. The resulting domain model is represented by the following elements:

- The domain chart provides a view of the domain being modeled, and the dependencies it has on other domains.

- The class diagram defines the classes and class associations for the domain.
- The state chart diagram defines the states, events, and state transitions for a class or class instance.
- The action language defines the actions or operations that perform processing on model elements.

Executable and Translatable UML (XTUML) [27] is a subset of the UML endowed with rules for execution. With an executable model, you can formally test the model before making decisions about implementation technologies, and with a translatable model, you can retarget your model to new implementation technologies. The UML specification defines the "abstract syntax" of UML diagrams but provides few rules on how the various elements interact dynamically.

XTUML incorporates well-defined execution semantics. Objects execute concurrently, and every object is in exactly one state at a time. An object synchronizes its behaviour with another object by sending a signal interpreted by the receiver's state machine as an event. Each procedure consists of a set of actions, such as a functional computation, a signal send, or a data access. The application model therefore contains the decisions necessary to support execution, verification, and validation, independent of design and implementation. At system construction time, the conceptual objects are mapped to threads and processors. The translator's job is to maintain the desired sequencing (cause-and-effect) specified in the application models, but it may choose to distribute objects, sequentialize them, duplicate them, or even split them apart, as long as application behaviour is preserved.

XTUML has been used on over 1,400 real-time and technical projects, including life-critical implanted medical devices [27]. Recent studies Alf [28] and PSCS [29] are devoted to further research on extensions of executable UML.

The EDS approach offers all fUML and XTUML features, and in addition:

(1) EDS is based on the DSL capabilities and offers a wider range of domain-oriented artefacts than UML, for example pre-defined process control deadline artefacts, internal and external objects used in Data Flow Diagramms.
(2) EDS allows using features of Finite State Machine (FSM) and UML activity diagrams in one common diagram. It lets to use different modeling approaches even on the same graphical diagram – by choose the data object processing can be either described as a state transition (FSM approach) or by data object activities (UML approach). Because the both styles are equivalent [11] the choose of the modeling approach can be left to the modeller; this cannot be assured using BPMN or executable UML approaches.

As a problem for the EDS approach is recognized the necessity of high qualification staff being able to define a new DSL. If the same problem would arise in a project where executable UML is used, there would be possible to switch to the standard UML.

7   Conclusions

This research was devoted to analysis of the EDS modelling/development approach that bases on the usage of DSL. The developed business process modeling environment DIMOD provides all the benefits of MDD for development, usage and maintenance of information systems. It provides:

- defining of new DSLs offering a wide range of artefacts and domain knowledge integration into DSL;
- creating of graphical editors for defining/editing of business process models in the new DSLs;
- usage of DSL resulting in less error-prone description, domain based meaningful validation and exhaustive testing of system;
- usage of DSL meta-model to define the DSL syntax being less sensitive to changes in business requirements;
- detailing of the business process model to the level that the model becomes either executable by interpreter or an executable source code can be generated from the model;
- up-to-date documentation of business processes as a DSL model in a graphical form.

Experience drawn from practical use of DSL in deployment and exploitation of many information systems has contributed to the validity of the proposed approach, thus stimulating new research activities on use of DSL in technologies, in particular, integration of business model and information system, integration of business model into heterogeneous systems, technologies for business models transfer to information systems, new approaches to business model use, correctness of business models and other research directions.

Acknowledgements. This work was supported by the Latvian National research program SOPHIS under grant agreement Nr.10-4/VPP-4/11.

References

1. Haan, J.D.: 15 reasons why you should start using Model Driven Development (2009). http://www.theenterprisearchitect.eu/blog/2009/11/25/15-reasons-why-you-should-start-using-model-driven-development/2009
2. Haan, J.D.: Opening up the Mendix model specification & tools ecosystem (2015). http://www.theenterprisearchitect.eu/blog/2015/10/30/open-mendix-model-specification-and-tools-ecosystem/2015
3. Unified Modeling Language. http://www.uml.org
4. Business Process Model and Notation. http://www.bpmn.org
5. Haan, J.D.: 8 Reasons Why Model-Driven Approaches (will) Fail (2008). http://www.infoq.com/articles/8-reasons-why-MDE-fails
6. Greenfield, J., Short, K., Cook, S., Kent, S.: Software Factories - Assembling Application with Patterns, Models, Frameworks and Tools. Wiley Publishing, Hoboken (2004)

7. Haan, J.D.: Opening up the Mendix model specification & tools ecosystem (2015). http://www.theenterprisearchitect.eu/blog/2015/10/30/open-mendix-model-specification-and-tools-ecosystem/
8. Karnitis, G., Bicevska, Z., Cerina-Berzina, J., Bicevskis, J.: Practitioners approach to business processes modelling. In: Frontiers in Artificial Intelligence and Applications Databases and Information Systems VIII -Selected Papers from the Eleventh International Baltic Conference, DB&IS 2014, pp. 343–356 (2014)
9. Barzdins, J., Zarins, A., Cerans, K., Kalnins, A., Rencis, E., Lace, L., Liepins, R., Sprogis, A.: GrTP: Transformation Based Graphical Tool Building Platform (2014). http://sunsite.informatik.rwth-aachen.de/Publications/CEUR-WS/Vol-297/
10. Sprogis, A.: Configuration language for domain specific modeling tools and its implementation. Baltic J. Modern Comput. **2**(2), 56–74 (2014)
11. Ceriņa-Bērziņa, J., Bičevskis, J., Karnītis, Ģ.: Information systems development based on visual domain specific language BiLingva. In: Szmuc, T., Szpyrka, M., Zendulka, J. (eds.) CEE-SET 2009. LNCS, vol. 7054, pp. 124–135. Springer, Heidelberg (2012)
12. Bičevskis, J., Bicevska, Z.: Business process models and information system usability. Procedia Comput. Sci. **77**, 72–79 (2015). ScienceDirect
13. Michelson, M.B.: Event-Driven Architecture Overview. http://www.omg.org/soa/Uploaded%20Docs/EDA/bda2-2-06cc.pdf
14. Using Events in Highly Distributed Architectures. https://msdn.microsoft.com/en-us/library/dd129913.aspx
15. Theorin, A., et al.: An Event-Driven Manufacturing Information System Architecture. https://mediatum.ub.tum.de/doc/1253955/1253955.pdf
16. EDSOA: An Event-Driven Service-Oriented Architecture Model For Enterprise Applications (2010). http://www.cluteinstitute.com/ojs/index.php/IJMIS/article/view/839/823
17. TIBCO Rendezvous Concepts. https://docs.tibco.com/pub/rendezvous/8.3.1_january_2011/pdf/tib_rv_concepts.pdf
18. MQSeries An Introduction to Messaging and Queuing. ftp://software.ibm.com/software/mqseries/pdf/horaa101.pdf
19. CORBA BASICS. http://www.omg.org/gettingstarted/corbafaq.htm
20. Welke, R., Hirschheim, R., Schwarz, A.: Service-oriented architecture maturity. Computer **44**(2), 61–67 (2011). doi:10.1109/MC.2011.56
21. Workflow Management Coalition Homepage. http://www.wfmc.org
22. Top Workflow Management Software Products. http://www.capterra.com/workflow-management-software/
23. JIRA Homepage. https://www.atlassian.com/software/jira
24. KiSSFLOW. https://kissflow.com/process_playbook/workflow-management-system-10-must-have-features/
25. Seidewitz, E.: Executable UML Roadmap (2014). http://www.slideshare.net/seidewitz/xuml-presentation-140917-executable-uml-roadmap
26. Lockheed Martin/Model Driven Solutions. http://modeldriven.github.io/fUML-Reference-Implementation/
27. Mellor, J.S.: Executable and Translatable UML (2010). https://web.archive.org/web/20100209114705/, http://embedded.com/story/OEG20030115S0043
28. The New Executable UML Standards: fUML and Alf By Jordi Cabot (2011). http://modeling-languages.com/new-executable-uml-standards-fuml-and-alf/
29. Precise Semantics Of UML Composite Structures™. http://www.omg.org/spec/PSCS/

DSML Tool Building Platform in WEB

Arturs Sprogis$^{(\boxtimes)}$

Institute of Mathematics and Computer Science, Riga, Latvia
arturs.sprogis@lumii.lv

Abstract. The paper discusses how to build DSML tool building platform in WEB. Previously this was not possible due to the limitations of the browsers to render graphical diagrams but the technologies have evolved and currently the limitations are eliminated. Basically, the platform consists of three components – Presentation, engine, Interpreter and the Configurator. The paper gives an explanation what are the tasks for each of the component and how they interact with each other. To demonstrate a tool building process, a building of a simple flowchart editor is presented.

Keywords: DSML tools · Web development · Models

1 Introduction

In the paper a web-based domain specific modeling language (DSML) tool building platform is proposed. Typically, DSML tools are more suitable for modeling tasks than universal language tools because DSML tools have a tool specific functionality for each application area as well as domain specific concepts are used in the tool's language [1]. But the main problem in the field of DSML tools is that it takes a lot of effort to program a single DSML tool from scratch [2].

DSML tool building problem is not a new topic, and there already are a number of DSML tool building platforms, for instance, MetaEdit [3], Eclipse GMF [4], Obeo-Designer [5], Graphiti [6]. However, these platforms are desktop applications but currently the majority of the applications are built as web applications since cloud based systems have a number of advantages. Firstly, the cloud services allow accessing data from anywhere in the world and no desktop installations on each user's computer is needed. Secondly, the collaboration is made easier and it can be performed in real-time. In case of DSML tools, it means that a user may edit a diagram and the rest of the collaborators would receive the changes immediately, thus eliminating users messing with complex multi-user mechanisms. Or we can go further, and change the tool "on the fly" that would make the shipping of the latest tool version immediate and would avoid users from performing manual updates. Thirdly, the system can be used on any device including mobile devices and tablets. Fourthly, there is no installation needed, users just have to signup. And finally, the application can embed other websites or use external web services.

Previously DSML tool building in the web was poor due to the technological limitation, for instance, browsers did not support HTML canvas element or poorly supported SVG that is the core to display graphical diagrams. In addition, there are

© Springer International Publishing Switzerland 2016
G. Arnicans et al. (Eds.): DB&IS 2016, CCIS 615, pp. 99–109, 2016.
DOI: 10.1007/978-3-319-40180-5_7

other issues, namely, the desktop and web application architecture is greatly different. In case of web applications, the data are stored on the server but the end-user works with the application through browser on the client. This causes a number of problems for application developers and the most common problems are - data transfer management, respectively, when and what data has to be sent to the client, data synchronization in real-time, data transfer security, data access rights, live html rendering and user management. To implement all of this properly, it takes a lot of work. Fortunately, during the past years the technologies have significantly developed and there are a number of frameworks and libraries like Meteor [7], Derby [8], React [9], Angular [10], Ember [11], etc. that reduce the complexity of web application development helping developers building fancy and complex applications, including DSML tools.

The paper has the following structure. Section 1 explains the architecture and the basic concepts of the platform. Section 2 explains Presentation engine and Presentation model in more details. Section 3 explains Interpreter and Type model. Section 4 explains Configurator and its graphical language. Section 5 explains some implementation details about the platform. Section 6 gives an example of how a simple flowchart tool can be built in the platform. Section 7 gives a related work, and finally the Sect. 8 concludes the paper.

2 Architecture and Basic Concepts

The overall architecture of the platform is presented in Fig. 1. The Presentation engine renders the graphical diagram and displays it on the screen for the end-user. Naturally, the interaction between a user and the tool happens via the screen as well as mouse clicks and keyboard. The Presentation engine receives end-user's actions such as create new element, resizing, or deleting. Then the control is passed to the so-called Interpreter – a component that performs the actual processing of users actions. The logic on how to process the user's actions is stored in the Type model, and the Interpreter exploits this information to process the user's actions. However, the state of the diagram is stored in the Presentation model that is used by the Presentation engine to render the diagrams. Thus, if the Interpreter has to change the diagram, the Presentation model is updated and the Presentation engine re-renders the diagram on the screen for the end-user.

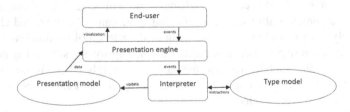

Fig. 1. Overall architecture

To be more specific on event handling process, we will examine a small example by demonstrating how to create a new box in the diagram. To create a new box, the user clicks into the graphical editor's palette and draws a rectangle in the diagram to indicate size and position of the box. Then, the Presentation engine classifies the event type and then passes the control to the Interpreter. The Interpreter "decides" what to do, and, in case of new box, it does the following:

- Performs checking if the given element can be created (count restrictions, etc.)
- If it is allowed, selects the default attribute values (if there are some)
- Creates appropriate instances of the new box in the Presentation model
- Passes the control back to the Presentation engine

Basically, the Interpreter is the "brains" of the platform by managing a procession of end-users actions and the state of the Presentation model accordingly.

3   Presentation Engine

Presentation engine has two tasks. The first task is to render diagrams according to the Diagram model data, and the second is to handle user events.

Figure 2 represents the Presentation model. The overall idea is that tools may contain several projects (class Project), and each project may contain several diagrams (class Diagram) and each diagram consists of number of graphical elements (class Element) – lines (class Line) and boxes (class Box). Each element may have a number of labels (class Compartment) to store element's properties.

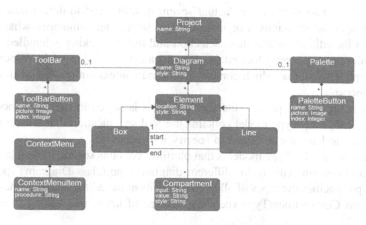

Fig. 2. Presentation model

In more detail, class Project has a property *name* that stores project name. Class Diagram contains two properties – *name* and *style*. The property *name* stores the diagram name; *style* describes how to render the diagram. Class Element has two properties – *style* and *location*. The property *style* describes how to render the element in the diagram; *location* contains the element coordinates. However, class Line has its own properties – *start* and *end*, to indicate that lines always has start and end element.

Class Compartment has three properties – *style*, *input*, and *value*. The property *style* describes how to render the attribute in the diagram; *input* stores the value entered by the end-user and *value* stores the label's representation value. The purpose for *input* is to contain the entered value whereas for *value* to contain the representational value. For instance, if we want to add a prefix "(" and suffix ")" to the attribute value "x", *input* will contain "x" and *value* will contain "(x)".

To allow end-user interact with the editor, we have classes for controls. The class Palette corresponds to the palette control in the diagram, but the class PaletteButton corresponds to the palette control button allowing to create new elements. Class PaletteButton has three properties – *name*, *procedure* and *index*. The property *name* stores the item name; *procedure* stores the name of the procedure that will be executed if the end-user selects the item; the *index* stores the sequential number of the button in the palette. Similarly, the class ToolBar corresponds to the toolbar, but the class ToolBarButton corresponds to the toolbar item.

The class ContextMenu corresponds to the context menu, but the class ContextMenuItem corresponds to the context menu item. The class ContextMenuItem has two properties – *name* and *procedure*. The property *name* stores the item name; *procedure* stores the name of the procedure that will be executed if the end-user selects the item.

4   Interpreter

The Interpreter's task is to do the "actual" event handling and to define how the tool behaves in specific situations. For instance, the Interpreter influences what will be displayed in the editor's palette, how selection and mouse clicking is handled, if some special rendering is necessary for certain elements and so on. It is also the responsibility of the Interpreter to define which context menu and context buttons are available in a certain scenario.

The logic on how to process the user's actions is stored in the Type model (see Fig. 3). Basically, the Type model contains the tool specification that is later used by the Interpreter to handle events and to specify tool behavior.

The idea behind the Type model is that platform contains several tools (class Tool), and each tool contains editors for different diagram type (class DiagramType). Each diagram type specifies the types of allowed elements in the editor (class ElementType). And the class CompartmentType specifies the type of labels that belong to each element type.

The property *name* of classes Tool and DiagramType store the language name. The class ElementType has two properties – *name* and *isAbstract*. The attribute *name* stores the name of the element; *isAbstract* stores if the element type is abstract, respectively, if the element type is abstract it means it does not specify any element in the editor but is used to build element type hierarchy (similar to abstract class in UML class diagrams). The model allows specifying languages containing two types of elements – boxes and lines. In addition to specify a line type, we have to specify which type of element has to

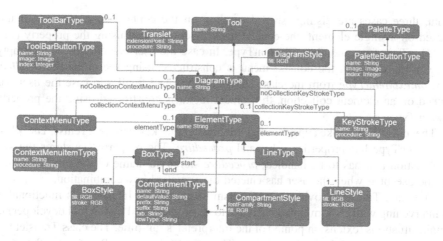

Fig. 3. Type model

be the start element of the line and which element has to be the end element, and in the model this information is described by the class LineType and its properties *start* and *end*.

The class CompartmentType has six properties – *name, defaultValue, prefix, suffix, rowType* and *tab*. The property *name* stores the element's attribute name; the *defaultValue* stores the default value of the attribute, and *prefix* and *suffix* corresponds to prefix and suffix of element's attribute value. For example, if the *prefix* is "≪" and the *suffix* is "≫", then the attribute value built by the Interpreter will be in the form - "≪ this is an attribute value ≫". The property *rowType* stores the widget name to enter attribute values, for instance, the property values can be input, textarea, selection, etc.; *tab* stores the tab name on which the widget is placed.

To specify the style for the editor, model contains style classes. Class *DiagramStyle* and its properties store the initial diagram style that is set by the Interpreter when a new diagram is created. Similarly, classes BoxStyle and LineStyle store the initial style for the boxes and lines, but the class CompartmentStyle stores the initial style for the attributes.

Graphical editors have interface controls as well, and the Type model contains the corresponding classes. The classes PaletteType and PaletteButtonType is used to generate palette and its buttons in the editor by the Interpreter. The class PaletteButtonType has properties – *name, image* and *index*. The property *name* stores the palette button name; *image* stores the palette button image and the *index* stores the sequential number of the button in the palette.

Similarly, the classes ToolBarType and ToolBarButtonType is used to generate the toolbar buttons. The ToolBarButtonType has properties – *name, image* and *index* having the same meaning as in case of PaletteButtonType.

Classes ContextMenuType and ContextMenuItemType are used to generate context menus. The class ContextMenuItemType has properties – *name* and *procedure*. The property *name* stores the name of the menu item; *procedure* stores the name of the procedure that will be executed when the user selects the item. To display the context menu, three cases are distinguished depending on the context. In case the user has clicked on a single element, the context menu is constructed using the property *contextMenuType* from the class ElementType. In case the user has clicked on an empty place in the diagram, the context menu is constructed using the property *noCollectionContextMenuType* from the class DiagramType. And finally, in case the user has clicked on an element collection, the context menu is constructed using the property *elementType*.

The class KeyStrokeType is used to handle the key stroke events. The class KeyStrokeType has properties – *key* and *procedure*. The property *key* stores the key combination that has to be handled; *procedure* stores the name of the procedure that will be executed when the user has entered the specified key combination.

The class Translet provides a mechanism to extend the Interpreter's functionality by intervening with custom functions (have to be programmed by the tool developer) in certain situations (extension points) of the Interpreter's run-time. The class Translet has properties *extensionPoint* and *procedure*. The property *extensionPoint* stores the name of the extension point; *procedure* stores the procedure name that will be executed by the Interpreter if the extension point is specified.

5   Configurator

As mentioned in the previous section, the Type model contains the tool specification and basically the tool is ready if we have the specification. Thus, the problem is to create the instance of the Type model. A straightforward approach would be to manually enter the tool specification using some textual format, for instance, JSON or XML. However, this approach would be error-prone and slow because the tool developer must enter rather big amount of data and there is no validation if the entered data is correct. To solve this problem, we have created a special tool called Configurator.

The task of the Configurator is to provide an interface for the tool developer to enter the tool specification. The Configurator's interface is partially graphical, that is, it allows creating box and line specifications as graphical elements in the diagram and representing their default style in the target editor by their appearance. But element property values are entered through dialog forms. The advantage of this approach is that when creating a new object the default values are automatically created and interface avoids entering data in the unsupported format.

Figure 4 represents the graphical representation of the specification of the simplified flowchart editor (gives an overview, not the full specification of the editor). Altogether there are six boxes and seven lines in the diagram, however boxes "Out" and "In" are abstract types, thus they do not specify any editor's element but are used to make element hierarchy. And only one line of seven connecting boxes "Out" and "In" specifies an element in the editor because the rest of the lines specify that elements "Out" and "In" are super types, thus making its subtypes to inherit the allowed outgoing and incoming lines.

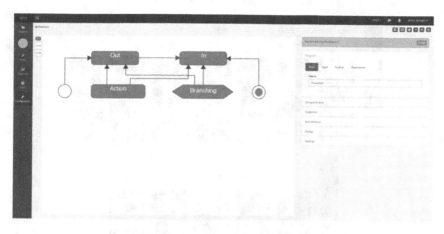

Fig. 4. Configurator's screenshot

6 Implementation Overview

To implement the platform, we have to build a system that unifies all the presented components, and, since the platform is web based, we had to implement additional things like user account management, data transfer between server and client, and user access rights to the data as well.

To implement our tool building platform we have chosen Meteor. The choice on behalf of Meteor is for a number of reasons. It is a pure JavaScript framework; it is full stack framework that natively integrates with widely used database MongoDB [12] it automatically propagates data changes to clients without requiring the developer to write any synchronization code allowing real-time collaborations (reactivity); it supports mobiles; it comes with thousands of plug-in libraries including many of the popular, and finally there is growing community posting questions, answering questions, bloging, etc.

Basically, Meteor removes the need to program the web and the communication layers that is very helpful to build the environment for the platform, however, we still have to implement all the components by ourselves. The Presentation engine is implemented from the scratch using KonvaJS (HTML5 canvas library) [13]. The Interpreter is implemented as a universal event processing layer on top of the Type model. The Configurator is implemented by applying the proposed methodology for DSML tool building. Namely, it has been specified by the Type model instance, the events are processed by the Interpreter and the graphical representation of the model is render by the Presentation engine, thus the Configurator is a DSML tool that allows specifying other DSML tools.

7 Example

To demonstrate the tool specification in more details, we will look at a simple flowchart editor. The proposed flowchart language contains four box type elements – "Start", "Action", "Branching" and "End", and one line type element – "Flow". According to

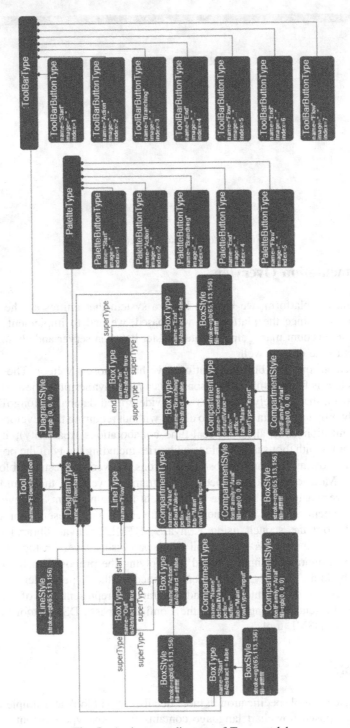

Fig. 5. An instance diagram of Type model

the proposed methodology, to build a flowchart editor we have to enter a specification in the Configurator, and the graphical model representing the flowchart editor's configuration is presented in Fig. 4.

However, what actually happened "behind the scene" is presented in Fig. 5, respectively, the Configurator created the instance of Type model saving us from low-level instance manipulations as well as avoiding entering the incompatible instance of the Type model.

When we have specified the tool, we can create a flowchart diagrams. Figure 6 presents a flowchart diagram.

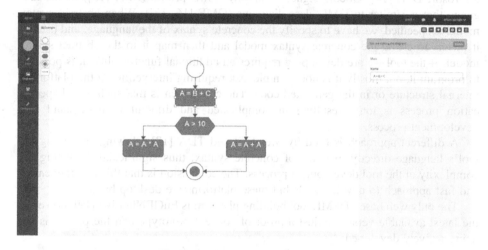

Fig. 6. A flowchart editor diagram

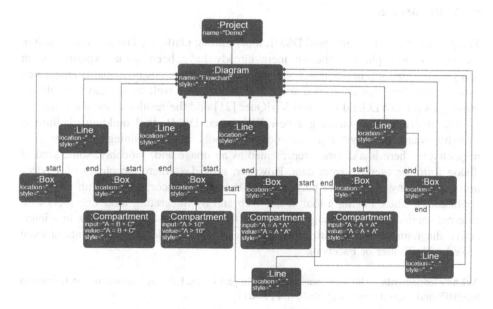

Fig. 7. An instance diagram of Presentation model

Figure 7 presents flowchart editor diagram represented as an instance of Presentation model.

8   Related Work

There is a number of existing DSML tool building platforms and typically they use one of two approaches. The most commonly used approach requires, at first, to specify an abstract syntax model of the language using UML class diagrams (Eclipse GMF, Microsoft DSL [14], ObeoDesigner, Graphiti), UML profile (RSA [15]) or the language that is similar to UML class diagrams (GME [16, 17]). When abstract syntax model is specified, we have to specify the concrete syntax of the language, and usually it means to create the concrete syntax model and then map it to the abstract syntax model. If the tool we are developing requires an additional functionality, it is possible to program it, but mostly it is not a simple task requiring intervening in the platform's internal structure or in the generated code. The conclusion is that such a tool specification process is long, resulting in complicated and difficult to understand tool development process.

A different approach is used by MetaEdit and TDA [18] allowing specifying the tool's language directly in terms of concrete syntax, thus significantly reducing the complexity of the tool development process. The conclusion is that this is a convenient and fast approach to develop tools but these platforms are desktop-based.

The only web based DSML tool building platform is EuGENia Live [19], however, the latest available version is just a proof of concept prototype and the project is not being actively developed.

9   Conclusions

The paper presented a web-based DSML tool building platform. The platform is still in the development phase, although there already have been some experiments on applying the platform in practice. For example, all the diagrams in the paper are created using a tool that was developed using the platform as well as we have developed prototypes for OWLGrEd [20] and ViziQuer [21] and the results seem promising.

The motivation on building a new Web-based DSML tool building platform is straightforward. Mostly graphical modeling tools allow building static models, respectively, there is a diagram represented as an image and, probably, some kind of dialog form for users to enter data. However, nowadays there are plenty of services online and many of them are purposely made open for incorporation with other web applications. Thus, having a tool in the Web and by integrating modeling tools with external services we can change these tools from being just static images into interactive diagrams extended with videos, images, links to relevant documents or even linked with Twitter or Facebook.

Acknowledgements. This work was supported in part by the Latvian National research program SOPHIS under grant agreement Nr.10-4/VPP-4/11.

References

1. Kelly, S., Tolvanen, J.P.: Domain-Specific Modeling: Enabling Full Code Generation, p. 448. Wiley, Hoboken (2008)
2. Cook, S., Jones, G., Kent, S., Wills, A.C.: Domain-Specific Development with Visual Studio DSL Tools. Addison-Wesley, Boston (2007)
3. MetaCase - MetaEdit+ Modeler DSM Tool. http://www.metacase.com/mep/
4. Eclipse Modeling. http://www.eclipse.org/modeling/gmf/
5. Obeo Designer: Domain Specific Modeling for Software Architects. http://www.obeodesigner.com/
6. Graphiti Home. http://www.eclipse.org/graphiti/
7. Meteor. https://www.meteor.com/
8. Derby. http://derbyjs.com/
9. React - A Javasctipt library for building user interfaces. https://facebook.github.io/react/
10. Angular - Superheroic JavaScript MVW framework. https://angularjs.org/
11. Ember - A framework for creating ambitious web applications. http://emberjs.com/
12. MongoDB for GIANT Ideas. https://www.mongodb.org/
13. Konva – JavaScript 2d canvas framework. http://konvajs.github.io/
14. Visualization and Modeling SDK (DSL Tools). http://code.msdn.microsoft.com/windowsdesktop/Visualization-and-Modeling-313535db
15. RSA. http://www.ibm.com/developerworks/rational/products/rsa/
16. Davis, J.: GME: the generic modeling environment. In: OOPSLA 2003: Companion of the 18th Annual ACM SIGPLAN Conference on Object-Oriented Programming, Systems, Languages, and Applications (2003)
17. GME: Generic Modeling Environment | Institute for Software Integrated Systems. http://www.isis.vanderbilt.edu/Projects/gme
18. TDA. http://tda.lumii.lv/welcome.shtml
19. EuGENia Live. http://eugenialive.herokuapp.com/
20. OWLGrEd – Editor for Compact UML-style OWL Graphic Notation. owlgred.lumii.lv
21. ViziQuer – Structured Semantic Data Search Tool. viziquer.lumii.lv

Decision Support Systems and Data Mining

Algorithms for Extracting Mental Activity Phases from Heart Beat Rate Streams

Alina Dubatovka[✉], Elena Mikhailova, Mikhail Zotov, and Boris Novikov

Saint Petersburg State University, 7-9, Universitetskaya nab.,
St. Petersburg 199034, Russia
alina.dubatovka@gmail.com, borisnov@acm.org,
{e.mikhaylova,m.zotov}@spbu.ru

Abstract. The paper presents algorithms for automatic detection of non-stationary periods of cardiac rhythm during professional activity. While working and subsequent rest operator passes through the phases of mobilization, stabilization, work, recovery and the rest. The amplitude and frequency of non-stationary periods of cardiac rhythm indicates the human resistance to stressful conditions. We introduce and analyze a number of algorithms for non-stationary phase extraction: the different approaches to phase preliminary detection, thresholds extraction and final phases extraction are studied experimentally. These algorithms are based on local extremum computation and analysis of linear regression coefficient histograms. The algorithms do not need any labeled datasets for training and could be applied to any person individually. The suggested algorithms were experimentally compared and evaluated by human experts.

Keywords: Pattern recognition · Signal processing · Mental activity phases · Data stream · Linear regression · Phase extraction

1 Introduction

There is a large amount of works describing the usage of heart rate variability (HRV) measures for monitoring of physiological arousal, attention, stress and general cognitive workload of operators in the process of real or simulated professional work [13,18].

Still, the majority of time- and frequency domain measures of HRV that have been used are coarse-grained and do not enable identification of moment-to-moment changes of operator mental state in the course of activity. However in many cases it is important not so much to measure the generalized level of physiological arousal of an operator during the performance of professional tasks, as to identify the moments in time when arousal sharply increases, which can be caused both by objective factors, such as increasing task demands, and by

A. Dubatovka—The work of the first author is partially supported by the Google Anita Borg Memorial Scholarship 2015.

G. Arnicans et al. (Eds.): DB&IS 2016, CCIS 615, pp. 113–125, 2016.
DOI: 10.1007/978-3-319-40180-5_8

psychological factors, such as decision-making difficulties [5,14]. It is known, that people respond to the increase of cognitive workload with increasing of heart rate (HR) and reduction of heart rate variability (HRV) [13,18].

These periods of sudden changes of HR and HRV parameters are usually called the non-stationary (transitive, unsteady) phases (NSPh), in contrast to stationary periods, that are characterized by stability of HR and HRV parameters over time [16]. In a number of studies it has been shown, that time and peak characteristics of non-stationary phases, registered under physical or mental stress, can be considered as informative indicators of stress resilience [14,16]. Meanwhile, absence of reliable algorithms for detection and analysis of non-stationary periods of a heart rate essentially complicates the progress of studies in this area.

Technically, the problem may be characterized as "small data", rather than BIG DATA. Specifically, this means that the amount of available data is relatively small; hence, it is hard to expect high performance on any machine learning algorithm. Further, the differences between phases to be detected may be lower than differences between individuals. Thus, we have to estimate parameters of algorithms (such as thresholds) on the fly, based on incoming data stream. Several attempts to use known techniques (such as change point detection etc.) were tried, and everything failed.

The approach of this work is to identify high-level properties of data during a state of interest. These properties are somewhat fuzzy. To improve performance, a cascade of algorithms has been used: after preliminary identification of phase intervals, a set of thresholds is calculated, and then the intervals are finalized based on these thresholds.

The contributions of this paper include:

- A number of algorithms for extraction of high-level mental activity phases from a stream of low-level sensor readings (such as heart beat rate)
- A comparative analysis of performance of the proposed algorithms.

The rest of the paper is organized as follows. Section 2 contains a description of the problem and outlines the approach. Section 3 outlines the algorithms, Sects. 4 and 5 describe the experimental environment and results, respectively. Finally, Sect. 6 contains a brief overview of related work. Section 7 summarizes the results of the research.

2   The Problem and Approach

The data used for our computational experiments contain recorded streams of heart beat rate readings in a training center. Each file contains a recording for one individual during approximately half-hour training session. Each session consists of several periods when trainees were performing certain tasks (sometimes called work intervals below) interleaved with relaxation. The heart beat rate is recorded 4 times per second. Under the terms of the study, the moments of the beginning and end of work are known.

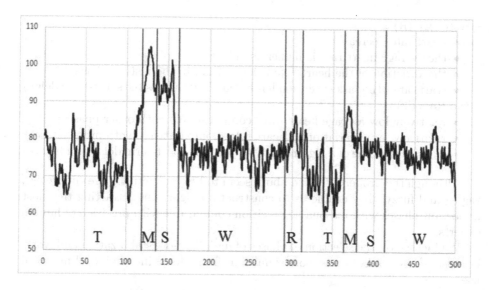

Fig. 1. Example graph the subject heart rate with manually selected phases

Psychologists found that during the execution of work and rest, the subject passes through phases of Mobilization, Stabilization, Work, Recovery (or rehabilitation) and the Rest. It is important to note that the phases of mobilization, stabilization and work are the steps of task execution, while the recovery phase and rest belong to repose stage. Figure 1 shows the heart rate of one subject with manually marked time intervals corresponding to the above phases.

The following high-level properties of heart beat rate stream were identified during preliminary analysis performed together with a team of psychologists [19, 20]:

- Mobilization (M)
  - heart rate stable increases (noticeable in comparison with normal fluctuations)
  - starts at task receiving (with possible short-term delay or advance)
  - the average value of the heart rate is higher than in the resting phase
  - the variation of the heart rate is lower than in the resting phase
- Stabilization (S)
  - the heart rate decreases
  - the average value of the heart rate is higher than in the resting phase
  - the Variation of the heart rate is lower than in the resting phase
- Work (W)
  - the average value of the heart rate is higher than in the resting phase
  - the variation of the heart rate is lower than in the resting phase
  - Ends at the time of the termination of task execution (with possible short-term delay)

– Recovery (R)
  • heart rate decreases
  • the average heart rate is higher than in resting phase
  • the variation of the heart rate is lower than in the resting phase
  • starts after the task execution has finished (with possible short-term delay)
– the Rest (T)
  • relatively low average heart rate (comparatively with other phases)
  • relatively large variation in heart rate (comparatively with other phases)
  • Ends at the time of assignment (with possible short-term delay or advance)

Although the properties listed above seem to be (and actually are) imprecise, vague, and fuzzy, they enable us to construct an algorithm extracting the most important phases with reasonable precision comparable with precision of human experts.

We focused on the selection of phases of mobilization, stabilization and work, because of task execution is more significant from the point of view of practical use.

3   Methodology

Our techniques for phase boundary consist of three main stages: primary allocation interval, the automatic determination of thresholds and refinement of intervals. A notable feature of our approach is its ability to automatically adapt to every particular person without any prior learning on other people; to take into consideration physiological characteristics of different people. Figure 2 shows schematically the main stages of the algorithm and alternatives for their implementation used in the experimental evaluation. Thus, we had 3 alternatives for each of the first two stages and 2 — for the last stage of the method.

3.1   Preliminary Interval Extraction

From all source points algorithm selects the most suitable for the phase boundary role, that is, the entire original time period is separated into primary segments for further processing.

No Straight. The initial allocation interval is trivial, every point equally applies to the role of the phase boundary and is considered in the final stage.

Uniform partition. The initial line segment is divided into a sufficiently large number of intervals of equal length — the primary intervals.

Local extremes. Algorithm finds extremum inside the symmetric "window" of some length as local extremum. In the described experiments, the "window" length equals 21 (10 samples left and 10-right). Obtained local extremums were considered subsequently as the initial ends of the intervals.

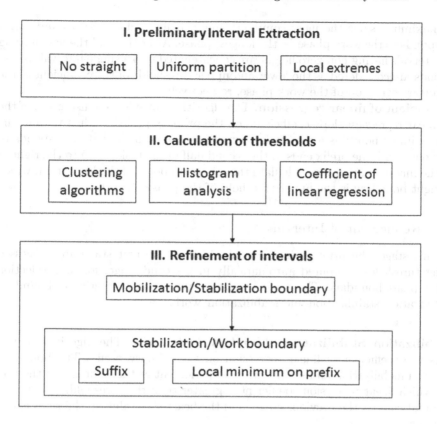

Fig. 2. Scheme of a three-stage algorithm

3.2   Calculation of Thresholds

The goal of the second stage is to calculate thresholds to be used in the last stage as discriminators. Than is, the thresholds refine the imprecise properties of mental phases based on data specific for each trainee.

For example, it is necessary to find out which values of the slope of the linear regression should be considered as a significant increase (characterizing mobilization phase), and which values should be considered as slow descent, i.e., stabilization. To do this, the algorithm determines the lower and upper (sometimes — only the upper) thresholds of the coefficient of the linear regression of the work phase as it is characterized by a near zero coefficient of linear regression.

Clustering algorithms. The algorithm calculates the lower and upper thresholds of the linear regression coefficient with clustering [7] values for all primary prefixes (the prefix with ends in a selected at the first stage points) clusters. The boundaries between the obtained clusters are upper and lower thresholds.

Histogram analysis. The algorithm starts with building a histogram on values of the coefficient of the linear regression on all primary prefixes. Then we find maximum, since the most common value usually is close to zero and corresponds to the work phase — the longest phase. After that, all the neighboring intervals having non-zero height are concatenated into a single interval, whose ends are considered as the lower and upper thresholds of the coefficient of the linear regression of the work phase, respectively.

Coefficient of linear regression. The algorithm constructs a histogram of the linear regression slope coefficients on the primary prefixes. Then we find the maximum point; as a rule, it is a small positive number that corresponds to "long" with the prefix ends in the second half of the task. Then to the right of the maximum point we calculate the minimum point of the histogram, which right boundary is the upper threshold of the phase of work.

3.3 Refinement of Intervals

At this stage, the primary intervals obtained at the first stage are processed using thresholds computed automatically in a second stage, for final selection of the phase boundary. This stage consists of two steps: extraction of boundary mobilization–stabilization and stabilization–work.

Mobilization–Stabilization Boundary Detection. The algorithm calculates the coefficient of linear regression on the primary prefix. The boundary between mobilization and stabilization is carried out at the right end of the prefix, which linear regression coefficient is greater than the threshold, determined automatically at the previous stage, and the heart rate value on the right end is maximal.

Let PI be a set of preliminary extracted points, $\beta_{\text{prefix}}(x)$ — linear regression coefficient for the first x points in seria and $HR[x]$ — the heart rate value in the specified point. Denoting the threshold determined in the second stage as θ, we can determine the boundary between mobilization and stabilization as

$$\underset{x \in \text{PI} \;\&\; \beta_{\text{prefix}}(x) \geq \theta}{\operatorname{argmax}} \quad HR[x].$$

Stabilization–Work Boundary Detection

Suffix. The algorithm calculates the coefficient of the linear regression on the suffixes with the left ends at the points obtained at primary selection of the intervals. The intended separation between stabilization and work is done on the left border of the suffix of the coefficient of linear regression which has the minimum modulo.

If PI is a set of preliminary extracted points and $\beta_{\text{suffix}}(x)$ is linear regression slope coefficient for the last x points in seria, then stabilization–work boundary is

$$\underset{x \in \text{PI}}{\operatorname{argmin}} |\beta_{\text{suffix}}(x)|,$$

Local minimum on prefix. The algorithm calculates the coefficient of linear regression on primary prefix. We select the boundary of stabilization phase in the right end of the prefix, if the value of the coefficient is a local minimum among the coefficients of linear regression on the received prefixes. If we have several local minima, we select the leftmost inside the interval of the task executing (since minimum before working probably predates the growth of the HR during mobilization phase and refers to the resting phase).

Denote a set of preliminary extracted points as PI and the linear regression coefficient for the first x points in seria as $\beta_{\text{prefix}}(x)$. In this case we can compute the boundary using the following formula:

$$\operatorname*{argmin}_{x \in \text{PI}}{}_{\text{loc}} \beta_{\text{prefix}}(x).$$

The techniques outlined above can be used on different modes depending on the amount of data used for calculation of thresholds. If the data stream is recorded, the entire file may be used in off-line mode. To process actual stream, a window must be specified for on-line mode. Note, however, that the window size must be at least greater than the duration of a task.

The off-line mode is expected to produce better results and is applicable in practical cases such as evaluation of trainees after a training session. Of course, the algorithms cannot be used for automatic evaluation of trainees, they only can find out who of trainees requires more attention of trainer or evaluator.

4   Experimental Evaluation

The purpose of the experimental study of these algorithms is to obtain information about the results applicability based on expert estimates. As a numerical characteristic of the algorithm quality, we used precision, defined as the proportion of tasks with properly selected phases.

All experiments were performed in off-line mode.

We evaluated algorithms with data containing the heart rate readings of 63 operators, each operator completed 5 tasks of MATB [4] and FTP [15] computerized aviation simulation tests. Heart rate was recorded 4 times per second, and the entire test lasted 1375 seconds, that means we had 5500 points for each person.

Data were obtained on three groups of operators (trainees):

- E0 — 15 specially trained "experts" having extensive experience in performing tasks;
- U0 — 25 "newcomers" who carried out tasks for the first time;
- U1 — 23 "newbie after training", a little trained but not specially prepared.

Table 2 contains mean, variance and range of data across these groups.

4.1 Preliminary Experiment

Because of variability at all stages of the proposed approach the number of combinations of algorithms is quite large, we must do initial testing on a smaller amount of data to reduce many of the methods discussed in further experiments. During these experiments, we assessed all possible combinations of realizations of the stages of the above algorithm applied to data of the first experiment all subjects from the group of experts to identify the most successful approaches for further work.

At the preliminary stage the data of experts were used, since this group has most clearly expressed phase activity due to the fact that they are the most trained of all operators and less prone to stress. This makes it easy to evaluate the results of the algorithms — the initial phase is usually visible to the layperson. So it's enough quite a visual analysis of graphics with a dedicated algorithm phases, to evaluate results without any additional expert assessments.

4.2 The Main Experiment

Based on preliminary studies, the following combinations of the algorithms showed best results:

1. No straight + local minimum on prefix:
   primary selection of the intervals is not performed, the boundary mobilization–stabilization is determined with the threshold value of the coefficient of the linear regression is automatically determined by the method of coefficient of linear regression, and the boundary stabilization–the work is by finding the local minimum of the regression coefficient on the prefixes;
2. Local extrema + suffixes:
   the primary ends are at the points of local extrema, boundary mobilization–stabilization is determined with the threshold value of the coefficient of the linear regression is automatically determined by the method of coefficient of linear regression, and the boundary stabilization–work — by means of suffixes;
3. Local extrema + local minimum on prefix:
   The primary ends are at the points of local extrema, boundary mobilization–stabilization is determined with the threshold value of the coefficient of the linear regression is automatically determined by the method of coefficient of linear regression, and the boundary stabilization–the work is by finding the local minimum of the regression coefficient on the prefixes.

In this experiment, the algorithms were applied to all available data: five assignments of examinees for each of the three groups — using expert assessments for more detailed analysis of methods.

5   Results of Experiments

Figure 3 show a graphical examples of the algorithms "No straight + local minimum of the prefix", "Local extrema + suffixes" and "Local extrema + local minimum on prefix" on three seria of data from different people. It is rather easy

No straight + Local minimum on prefix

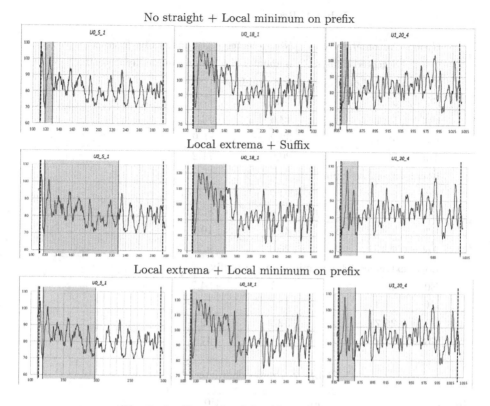

Local extrema + Suffix

Local extrema + Local minimum on prefix

Fig. 3. An Example of the Algorithm Output

to see, that the algorithm "No straight + local minimum of the prefix" produced the best results in our experiments.

Table 1 shows the proportion of selected phase among all experiments, the proportion of phases, where manual selection failed, and the share allocated among the marked phases of the experiments.

It should be noted that for some subjects, the experts were not able to identify phases in some tasks, so the table shows the estimation accuracy with and without these tasks. Table 2 contains mean values of length and linear regression coefficient for mobilization (M), stabilization (S) and work (W) phases that were identified by experts.

6 Related Work

The problem of separation of the phases of mental activity belongs to the class of tasks in the study of brain activity, most of which are investigated on the basis of data analysis ECG, EEG, heart rate [17]. There are several different approaches to processing the above data and other streaming data: a wavelet transformation, machine learning techniques, finding points of change. The choice of approach

Table 1. Results of the algorithms

Group	Automatically detected phases %	Not detected manually %	Automatically detected among manually detected%
No straight + Local minimum on prefix			
E0	57.3	14.7	67.2
U0	54.4	13.6	63
U1	58.3	32.2	85.9
Total	56.5	20.6	71.2
Local extremes + Suffix			
E0	8	14.7	9.4
U0	7.2	13.6	8.3
U1	7	32.2	10.3
Total	7.3	20.6	9.2
Local extremes + Local minimum on prefix			
E0	46.7	14.7	54.7
U0	34.4	13.6	39.8
U1	35.7	32.2	52.6
Total	37.8	20.6	47.6

Table 2. Phases characteristics

Group	E0	U0	U1
Length of mobilization phase	13.29 ± 5.40	14.04 ± 7.84	12.46 ± 6.48
Length of stabilization phase	13.87 ± 6.53	15.96 ± 9.24	11.70 ± 5.32
Length of work phase	159.99 ± 9.46	156.83 ± 14.06	161.98 ± 9.71
Linear regression coefficient on mobilization phase	1.24 ± 1.10	1.48 ± 1.48	2.01 ± 1.65
Linear regression coefficient on stabilization phase	-1.28 ± 0.79	-1.33 ± 0.99	-1.85 ± 1.21
Linear regression coefficient on work phase	-0.0030 ± 0.0246	-0.0027 ± 0.0269	-0.0036 ± 0.0206

is determined by the nature of processed data and the features of the task. In our problem the techniques listed above are not applicable due to high amount of noise, relatively small amount of data, and too high variability of data.

Specifically, for example, data processing of ECG and EEG is typically associated with the wavelet transform [6,11,12]. But such a transformation cannot be applied to heart rate due to the rapidly oscillating data, and also because of errors and failures in data that is highly sensitive Fourier transform.

Machine learning algorithms, most common approaches to data analysis, can be applied to the study of mental processes, for example, to classify brain activity [3,6]. However, all used methods are examples of learning and, therefore, require a large amount of data marked in advance, which makes them not applicable in this problem because the nature of the behavior of heart rate varies between individuals and depending on the type of the job.

At the first glance, the algorithms for points of change (change point detection) [1,8,9] could be helpful in the task of mental activity phase detection (since the behavior of the data at the boundary of two phases is changed). However, they are designed for more or less smooth behavior of the trend that is incorrect for this data. Attempt to use these algorithms in our problem did not give meaningful results.

7   Conclusion

In this work, we proposed various algorithms for processing streaming data containing measurements of heart rate (HR) in performing the work, to highlight the phases of mental activity taking into account individual psychophysiological characteristics of each subject. We proposed a three-stage algorithm to solve this problem without pre-labeled data with only General knowledge about mental phases. The algorithm allows to distinguish the phase of mental activity considering the physiological and psychological characteristics of each person.

A comparison of several variants of this scheme was done by running algorithms on real data with subsequent assessment of their quality by specialists in the subject area. Best results were achieved by a method based on the analysis of the behavior of the coefficient of the linear regression, approximating the heart rate, on the prefix.

The proposed allocation algorithms phase metal activity are useful for many important practical applications, including:

- systems of automatic monitoring of mental status of different professional groups (air traffic controllers, pilots, etc.) when performing critical operations;
- adaptive human-machine interfaces and cognitive support systems of human activity;
- methods of assessing the level of fitness of personnel and their resistance to influence of factors of physical and mental stress;
- systems of diagnostics of disorders of mental activity in various neuropsychiatric diseases;
- online medical information assistance services [10];
- healthcare systems [2].

References

1. Adams, R.P., MacKay, D.J.: Bayesian online changepoint detection. Cambridge, UK (2007)
2. Balandina, E., Balandin, S., Koucheryavy, Y., Mouromtsev, D.: Iot use cases in healthcare and tourism. In: 2015 IEEE 17th Conference on Business Informatics (CBI), vol. 2, pp. 37–44, July 2015
3. Cinaz, B., Arnrich, B., Marca, R.L., Troster, G.: Monitoring of mental workload levels during an everyday life office-work scenario. Pers. Ubiquit. Comput. $17(2)$, 229–239 (2013). http://dblp.uni-trier.de/db/journals/puc/puc17.html
4. Comstock, J.: The Multi-attribute Task Battery for Human Operator Workload and Strategic Behavior Research. NASA Langley Research Center, Hampton (1992). https://books.google.ru/books?id=JlY3AQAAMAAJ
5. Driskell, J., Salas, E., Johnston, J.: Making and performance under stress. In: Military Life: The Psychology of Serving in Peace and Comba, vol. 1, pp. 128–154 (2006). Military Performance
6. Gupta, A., Agrawal, R.K., Kaur, B.: A three phase approach for mental task classification using EEG. In: Gopalan, K., Thampi, S.M. (eds.) ICACCI, pp. 898–904. ACM (2012). http://dblp.uni-trier.de/db/conf/icacci/icacci2012.html
7. Hartigan, J.A., Wong, M.A.: A k-means clustering algorithm. JSTOR Appl. Stat. $28(1)$, 100–108 (1979)
8. Inclan, C., Tiao, G.C.: Use of cumulative sums of squares for retrospective detection of changes of variance. J. Am. Stat. Assoc. $89(427)$, 913–923 (1994). http://www.jstor.org/stable/2290916
9. Killick, R., Eckley, I.A.: Changepoint: an R package for changepoint analysis. J. Stat. Softw. $58(3)$, 1–19 (2014). http://www.jstatsoft.org/v58/i03/
10. Korzun, D.G., Borodin, A.V., Timofeev, I.A., Paramonov, I.V., Balandin, S.I.: Digital assistance services for emergency situations in personalized mobile healthcare: smart space based approach. In: 2015 International Conference on Biomedical Engineering and Computational Technologies (SIBIRCON), pp. 62–67, October 2015
11. Malhotra, V., Patil, M.K.: Mental stress assessment of ECG signal using statistical analysis of bio-orthogonal wavelet coefficients: part-2. Int. J. Sci. Res. (IJSR) $2(12)$ (2013). http://www.ijsr.net/archive/v2i12/MjYxMjEzMDE=.pdf
12. Malhotra, V., Patil, M.K.: Mental stress assessment of ECG signal using statistical analysis of bio-orthogonal wavelet coefficients: part-2. Int. J. Sci. Res. (IJSR) $3(2)$ (2014). http://www.ijsr.net/archive/v3i2/MDUwMjE0MDE=.pdf
13. Mulder, L., de Waard, D., Brookhuis, K.: Estimating mental effort using heart rate and heart rate variability. In: Stanton, N., Hedge, A., Brookhuis, K., Salas, E., Hendrick, H. (eds.) Handbook of Human Factors and Ergonomics Methods. CRC Press, Boca Raton (2004)
14. Novikov, V.S., Stupakov, G.P., Lustin, S.I., et al.: In: Novikov, V.S. (ed.) Physiology of Flight Work. Nauka, St. Petersburg (1997)
15. Petrukovich, V.: Technology for assessing the flight navigator's capacity to operate with numerical information in the spatial pattern structure. Vestnik Baltiyskoi pedagogicheskoi akademii (Bull. Baltic Pedagogical Acad.) 34, 83–90 (2000)
16. Sapova, N.: Complex evaluation of heart rhythm regulation during measured functional loads. Fiziologicheskii zhurnal SSSR imeni I. M. Sechenova (Sechenov Physiol. J. USSR) $68(8)$, 1159–1164 (1982)

17. Task Force of the European Society of Cardiology the North American Society of Pacing Electrophysiology: Heart rate variability standards of measurement, physiological interpretation, and clinical use. Circulation **93**(5), 1043–1065 (1996). http://circ.ahajournals.org/content/93/5/1043.full. hRV autonomic risk factors
18. Veltman, J., Gaillard, A.: Physiological workload reactions to increasing levels of task difficulty. Ergonomics **41**(5), 656–669 (1998)
19. Zotov, M., Forsythe, J., Petrukovich, V., Akhmedova, I.: Physiological-based assessment of the resilience of training to stressful conditions. In: Schmorrow, D.D., Estabrooke, I.V., Grootjen, M. (eds.) FAC 2009. LNCS, vol. 5638, pp. 563–571. Springer, Heidelberg (2009). http://dblp.uni-trier.de/db/conf/hci/hci2009-16.html
20. Zotov, M.V., Petrukovich, V.M., Akhmedova, I.S., Palamarchuk, N: Optimization factors of regulation of cognitive activity during training. Vestnik Sankt-Peterbugskogo universiteta (Saint Petersburg State Univ. Bull.) **12**(2), 17–31 (2011)

Scheduling Approach for Enhancing Quality of Service in Real-Time DBMS

Fehima Achour[1](✉), Emna Bouazizi[1], and Wassim Jaziri[1,2]

[1] MIRACL Laboratory, Higher Institute of Computer Science and Multimedia,
Sfax University, Sfax, Tunisia
fehima.achour@gmail.com, emna.bouazizi@gmail.com, jaziri.wassim@gmail.com
[2] College of Computer Science and Engineering, Taibah University,
Almadimah, Kingdom of Saudi Arabia

Abstract. Applications are increasingly characterized by manipulating large amounts of data and by time constraints to which are submitted data and treatments. RTDBMSs (Real-Time DataBase Management Systems) are an appropriate formalism to handle such applications. However, a RTDBMS often goes through overload periods following the unexpected arrival of user transactions. During such periods, transactions are more likely to miss their deadlines and that directly affects the QoS (Quality of Service) provided to users. Thus, our work is to propose a new scheduling protocol to optimize the execution of transactions without exceeding their deadlines. It consists on assigning priorities to transactions based both on their deadlines, their arrival dates and their priority levels defined by users. Also, we show that our approach can maximize the number of successful transactions, in particular those classified as critical for users. The obtained results are compared with conventional scheduling approaches.

Keywords: Real-Time Database Management Systems · Transaction · Scheduling · Feedback control · Quality of Service

1 Introduction

In current real-time applications, the main goal is to efficiently handle large amounts of data while meeting the time constraints imposed on them. Indeed, a conventional DataBase Management System (DBMS) is suitable for an efficient management of large amounts of data. However, such a system does not consider time constraints. On the other hand, real-time systems have appropriate mechanisms for the management of these time constraints but not for large volumes of data. Thus, Real-time DBMSs (RTDBMSs) have emerged.

We can define a RTDBMS as a combination of a conventional DBMS and a real-time system. Its purpose is both to handle large amounts of data while maintaining their logical consistency and to ensure the compliance with time constraints [1]. When the transactions submitted by users arrive at varying frequencies, such a system goes through instability periods in which the system

© Springer International Publishing Switzerland 2016
G. Arnicans et al. (Eds.): DB&IS 2016, CCIS 615, pp. 126–135, 2016.
DOI: 10.1007/978-3-319-40180-5_9

becomes overloaded. During overloading periods, there is a lack of RTDBMS resources, so that transactions greatly miss their deadlines. Then, it is necessary to better manage the available resources in order to ensure a good Quality of Service (QoS) by satisfying the most important transactions face the required constraints.

As it is essential to any execution process of transactions, the scheduling is addressed in several research projects. It is useful for determining the time in which transactions must be running in the system. When receiving a transaction, the scheduler decides to run it immediately or to delay it or to completely reject the transaction. In a real-time context, such a mechanism would greatly help the transactions to meet their time constraints and then improving the QoS provided for users. In RTDBMSs, the QoS management has been widely discussed. The works included in this guidance are mostly based on feedback control scheduling techniques [2,3]. Among these works, we mention the work of Katet et al. [4] where a feedback control architecture has been used to take into account the real time derived data. This same architecture was adapted by Zeddini et al. [5] for managing QoS in multimedia systems. It is also used in [6] for QoS management in real-time spatial big data.

In this paper, we aim to enhance the QoS in RTDBMSs by satisfying the maximum of important user queries. We are interested in transactions scheduling by proposing a new protocol combining three criteria for assigning priorities to the transactions. This allows in particular to take into account the criticality levels of transactions defined by the concerned users. Thus, we ensure that the maximum of transactions qualified as critical for users to be executed without missing their deadlines. Thereafter, we apply the proposed protocol in a RTDBMS architecture based on feedback control scheduling which is suitable to maintain a robust system behaviour facing instability periods.

The remaining of the paper is organized as follows. In Sect. 2, we present the RTDBMS model that we use in our approach. Thereafter, an overview of scheduling protocols on which we base our work is given in Sect. 3. Section 4 describes the new scheduling protocol we propose and whose performance is evaluated at Sect. 5. We conclude this work in Sect. 6 by briefly discussing our approach and presenting our future work.

2   A RTDBMS Model Based on Feedback Control

To fight against instability phases in RTDBMSs due to unpredictable arrival of user transactions, a feedback-based model was proposed, and which is considered as the basic model for QoS managing in RTDBMSs.

2.1   Data Model

The database that we consider in our model contains real-time and non real-time data. The real-time data are sensor data that describe real world entities. They must be regularly updated to reflect more accurately the real world state.

Each real-time data object has a validity interval beyond which it becomes outdated and then unusable. It also has a timestamp which is the date of its last update. Real-time data can have a deviation from its value in the real world called the *Data Error* (*DE*). The *DE* should not exceed a threshold that we denote *MDE* (*Maximum Data Error*). The non real-time data are those found in conventional databases and which do not dynamically change with the time.

2.2   Transactions Model

As for data, in a RTDBMS, real-time transactions have to meet the time constraints to which are associated. In our model, we consider firm deadline transactions [7] where if a transaction misses its deadline, it will be aborted and becomes useless for the system. We distinguish two real-time transaction types according to the accessed data object: update transactions and user transactions. Update ones are those performed periodically to refresh the real-time data. However, the user transactions, which are aperiodic, consist of a set of read operations on both real-time and non real-time data, and of write operations on only non real-time data.

2.3   Performance Metric

The basic performance criterion considered in our model is the satisfaction rate of the system users. It represents the *Success Ratio* (*SR*) of transactions that are qualified as critical for users. It is defined as a QoS parameter measuring the percentage of transactions that meet their deadlines and having the following formula:

$$SR = 100 \times \frac{\#Timely}{\#Late + \#Timely}(\%) \tag{1}$$

Where #Late and #Timely respectively denote the number of critical transactions that have missed their deadlines and the number of successful critical transactions.

2.4   QoS Management in RTDBMSs

The QoS is fundamental to assess the RTDBMS performance. In [8], the authors proposed an architecture using a feedback-based control scheduling for QoS managing in a RTDBMS. It is called *FCSA* (*Feedback Control Scheduling Architecture*) and it allows to control the RTDBMS behaviour, making it more robust during instability periods. Figure 1 shows the *FCSA* we briefly describe in the following.

The admission controller is responsible for filtering user transactions. Its functioning is controlled by the feedback loop in order to manage the system instability periods. The number of accepted transactions depends on the system state informed by the monitor and on QoS parameters specified by the

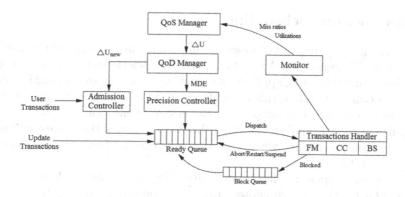

Fig. 1. RTDBMS architecture based on feedback control [8].

DBA^1. The accepted transactions are sent to the ready queue to be executed by the transactions handler. The transactions handler consists of a freshness manager (FM) checking the freshness of real-time data before being accessed by transactions, a concurrency controller (CC) resolving accessing data conflicts appearing between transactions, and a base scheduler (BS) determining the execution time of each transaction. The monitor is responsible for measuring system performance based on the execution results provided by the transactions handler. These measurements belong to the feedback loop and are sent to the QoS manager to inform it about the system state. Using these values and the system reference parameters set by the DBA, the QoS manager adjusts the QoS parameter values to be sent to the admission controller and to the QoD (Quality of Data) manager. This one will adjust the MDE value and then send it to the precision controller which is used to remove the update transaction accessing data considered as outdated ($DF < MDE$).

The feedback loop ensuring the system balancing is based on the principle of observation and self-adaptation. The self-adaptation takes place throughout the system functioning to continuously adjust its workload in the presence of unpredictable user queries. However, the observation consists on considering the system state to determine if it meets the specified QoS parameters, such as the system utilization rate and the transactions' success ratio. The system will adapt its settings to appropriately change in behaviour and thus deal with instability periods.

In [9], the feedback loop was used for the QoS management in large-scale real-time applications in an architecture called DRACON (Decentralized data Replication and Control). Furthermore, in [10], the authors adapted this technique for the QoS management of real-time embedded databases.

---

[1] The database administrator.

3   Transactions Scheduling

Several scheduling policies of transactions have been proposed, of which there are
on-line and off-line policies. An on-line scheduling selects the next process to be
scheduled at any time of an application execution according to information about
the transactions initiated at this time. However, an off-line scheduling is planned
in advance so it does not adapt with the environment changes. Similarly, we
distinguish preemptive policies allowing the interruption of the process execution
from those non-preemptive where the process is executed until it terminates or
is blocked [11,12].

Among the proposed conventional scheduling protocols, we have the *First
In First Out* (*FIFO*), which allows to run each process according to its arrival
date. Thus, the elected process is the first-arrived one. For processes arriving at
the same time, the *FIFO* randomly picks one of them. However, with the *Round
Robin* protocol, a quantum of time Q is imposed for each process during which
it is allowed to run. Any process that consumed its Q must leave the processor
and be inserted at the end of the list. Hence all processes have the same priority.
However, when it comes to real-time applications where transactions have time
constraints, these protocols are no longer useful.

In order to take into account the time constraints of real-time transactions,
new scheduling policies have emerged. *Earliest Deadline First* (*EDF*) [13] is one
of the main scheduling protocols proposed for a real-time context. It consists
on assigning the highest priority to the transaction having the nearest deadline.
Thus, transactions are executed in an ascending order according to their deadlines in order to increase their opportunities to succeed in their executions. In
[14], the authors extended the *EDF* protocol and propose the *GEDF* (*Generalized Earliest Deadline First*) scheduling policy. With the *GEDF*, the priorities
assignment is based on transactions importance and on their deadlines. In [15],
the authors proposed a real-time adaptive co-scheduling algorithm based on the
EDF. It is called *Adaptive Earliest Deadline First Co-Scheduling* (*AEDF-Co*)
and it aims to schedule transactions such that the deadline constraints are satisfied and the QoD is maximized.

With the scheduling protocols mentioned above, several criteria are taken
into account to decide about the next transaction to be executed. Most of these
criteria consider mainly the time constraints of transactions to be satisfied in a
real-time context such as the deadlines. However, the user choice is never involved
when determining the execution time of transactions. Thereby, we propose to
include this criterion in the definition of transactions' priorities aiming to reach
the users satisfaction. In what follows, we describe in more detail our approach.

4   New Scheduling Protocol for RTDBMSs

Our approach is to propose a new transactions' scheduling protocol in a
RTDBMS. It consists on taking into account, not only the time constraints of
transactions, but also the criteria set by the database users. This protocol functioning is based on the use of three parameters for determining the priorities of

the transactions to be executed. The first parameter is the transaction deadlines that must be met in order to be committed. The second parameter represents the arrival time of each transaction to the system. The third parameter serves of an indicator of the criticality level for each transaction and that reflects, in particular, its importance towards the concerned user. Therefore, we consider the highest priority transaction the one having both the earliest deadline, the earliest arrival time and the highest criticality level. Thus, we define as the following the priority of each transaction t_i:

$$Priority(t_i) = 1/DL(t_i) + 1/AT(t_i) + CR(t_i) \qquad (2)$$

where $DL(t_i)$ represents the deadline of the transaction t_i, $AT(t_i)$ is its arrival time and $CR(t_i)$ is its criticality level. This new protocol that we call the *MRTS (Mixed Real Time Scheduling)* protocol follows the principle of preemptive scheduling policies. Thereby, a lower priority transaction may be interrupted in favour of another one having a higher priority. However, the preemption may take place only after checking that it does not cause the missing deadline of the transaction to be suspended. This means that the suspended transaction can be resumed later from its suspension point.

Afterwards, we applied the *MRTS* protocol at the *FCSA* which is proposed to manage the QoS in a RTDBMS. This architecture presented a good performance for the system stabilization face the overload or underutilization periods caused by the unexpected arrival of user transactions. Figure 2 illustrates the architecture containing the new scheduler and that we call the *FC-MRTS* architecture.

The challenge of our approach is to satisfy the RTDBMS users by improving the QoS provided, and this is by increasing the number of transactions that meet their deadlines, especially those qualified as more crucial for users. Moreover, in collaboration with the *FCSA*, we reduce the risk of overloading our system face a significant number of transactions coming simultaneously.

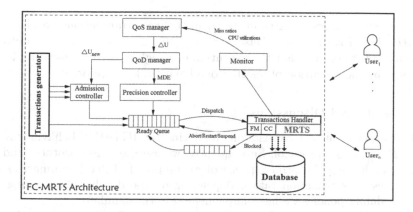

Fig. 2. The MRTS protocol and the FCSA.

We soon plan to introduce a new parameter in our approach reflecting the aggregation links that may exist between transactions. Therefore, if a transaction t_i is a sub-transaction of t_j, we choose to begin by running t_i even if it is of a lower priority. The implementation of such a solution requires an analyzer of transactions' relationships that are usually not obvious enough.

In the following section we discuss the results provided by our approach.

5   Simulations and Results

In this section, we evaluate our approach performance according to the simulation results we obtained.

5.1   Simulation Principle

To evaluate the new scheduling approach, we used a simulator based on the *FCSA* and simulating the RTDBMS behaviour. We performed a set of simulations in which we vary some parameters' values. The system parameters used for generating data and transactions are summarized in Table 1.

Table 1. System parameter settings

Parameter	Signification	Value
Duration (ms)	Simulation duration	10000
Validity (ms)	The data validity duration	[400, 800]
NbOfOperations	The number of a transaction operations	[3, 10]
ReadOpTime	The execution time of a read transaction	[1, 2]
WriteOpTime	The execution time of a write operation	[1, 10]
MDE	The maximum data error	3

We note that for resolving conflicts arising between transactions, we chose the *2PL-HP (High Priority Two Phase Locking)* [16] concurrency control protocol. Its principle is to ensure that the execution of a high priority transaction is not hampered by the execution of another one having a lower priority.

5.2   Results and Discussions

The first experiments set consists on simulating the RTDBMS behaviour using the *EDF* scheduling protocol. Subsequently, we use the *FIFO* protocol and we finish by evaluating the *MRTS* protocol we proposed. Table 2 summarizes the values of users' satisfaction ratios, depending on the database size, using the three scheduling policies we have implemented in the simulator.

The graphical representation of Table 2 is given in Fig. 3 that we analyze in the following.

Table 2. Simulation results of the three schedulers.

		Users' satisfaction ratio(%)		
		EDF	FIFO	MRTS
Number of data	*100*	68	20	73
	200	50	19	64
	300	41	16	62
	400	31	17	59
	500	30	19	51

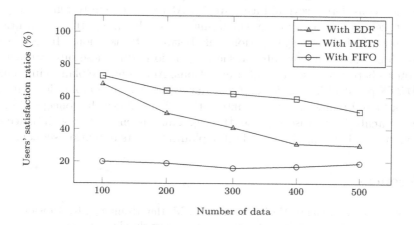

Fig. 3. Users' satisfaction ratio using the three schedulers.

By referring to Fig. 3, we assert that the satisfaction ratios of users using the *MRTS* policy proposed by our approach is always better than those provided by either the *EDF* or the *FIFO* scheduler. This is due to the fact that our approach considers, in addition to the deadlines, the arrival dates of transactions to minimize their waiting time to be executed, and especially their criticality levels defined by users. This is not the case of the other schedulers. Thus, users are more satisfied with *MRTS* that promotes their transactions of highest priorities. Also, we note that the user satisfaction ratio is influenced by the increase of the number of incoming transactions to the system. In this case, the system load will increase and therefore the number of conflicts in accessing data. Thereby, we note a decrease in the ratio values with the three schedulers. Yet, the new scheduler still holds the best results. The *FIFO* scheduler still has the worst results since it does not consider the transaction deadlines, which makes it not very appropriate in a real-time context.

6   Conclusions and Future Works

In this paper, we have implemented a new scheduling policy for improving QoS provided to RTDBMS users. We combined the newly proposed policy with the *FCSA* designed for QoS managing in RTDBMSs. Our approach is mainly based on the integration of the user in the scheduling process by including a user preference setting (queries criticality). The experiments confirmed the contribution of our approach in satisfying the maximum of the critical transactions for users. We promoted the deadlines meeting of these transactions while minimizing the waiting time of users by considering their arrival dates to the system.

We are considering several future works. At first, we plan to refine the parameters taken into account in the scheduling process by integrating an additional parameter which is the aggregation link between transactions. Indeed, some transactions share common sub-transactions. Prioritizing these sub-transactions promotes a better success ratios of transactions. At second, we plan to implement the *MRTS* protocol for spatio-temporal databases [17] in combination with the feedback loop. Finally, it would be interesting to introduce the concept of ontology when analyzing the users queries [18,19]. This will enable to take into account the semantics of the queries for a better planning of its execution sequence.

References

1. Aranha, R.F.M., Ganti, V., Narayanan, S., Muthukrishnan, C.R., Prasad, S.T.S., Ramamritham, K.: Implementation of a real-time database system. Inf. Syst. **21**, 55–74 (1996)
2. Amirijoo, M., Hansson, J., Son, S.H.: Specification and management of QoS in real-time databases supporting imprecise computations. IEEE Trans. Comput. **55**, 304–319 (2006)
3. Lu, C., Stankovic, J.A., Son, S.H., Tao, G.: Feedback control real-time scheduling: framework, modeling, and algorithms. Real-Time Syst. **23**, 85–126 (2002)
4. Katet, M., Bouazizi, E., Sadeg, B.: Prise en Compte des Données Dérivées Temps Réel dans Une Architecture de Contrôle par Rétroaction. Revue Électronique des Technologies de lInformation (e-TI), vol. 11 (2007)
5. Zeddini, B., Duvallet, C., Sadeg, B.: Une Approche Qualité de Service dans Les Systèmes Multimédias Distribués. Revue Électronique des Technologies de lInformation (e-TI), vol. 4 (2007)
6. Hamdi, S., Bouazizi, E., Faiz, S.: QoS management in real-time spatial big data using feedback control scheduling. ISPRS Ann. Photogrammetry Remote Sens. Spat. Inf. Sci. **II–3/W5**, 243–248 (2015)
7. Xiong, M., Ramamritham, K.: Deriving deadlines and periods for real-time update transactions. IEEE Trans. Comput. **53**, 567–583 (2004)
8. Kang, K.-D., Son, S.H., Stankovic, J.A., Abdelzaher, T.F.: A QoS-sensitive approach for timeliness and freshness guarantees in real-time databases. In: 14th Euromicro Conference on Real-Time Systems, pp. 203–212. IEEE Computer Society, Vienna, Austria (2002)
9. Kang, W., Son, S.H., Stankovic, J.A.: DRACON: QoS management for large-scale distributed real-time databases. JSW **4**, 747–757 (2009)

10. Woochul, K., Jaeyong, C.: QoS management for embedded databases in multicore-based embedded systems. Mob. Inf. Syst. **2015**, 14 (2015)
11. Cottet, F., Delacroix, J., Kaiser, C., Mammeri, Z.: Ordonnancement Temps Réel. Hermès (2000)
12. Liu, C.L., Layland, J.W.: Scheduling algorithms for multiprogramming in a hard-real-time environment. J. ACM **20**, 46–61 (1973)
13. Buttazzo, G.C.: Hard Real-time Computing Systems: Predictable Scheduling Algorithms and Applications. Real-Time Systems Series. Springer, New York (2004)
14. Semghouni, S., Amanton, L., Sadeg, B., Berred, A.: On new scheduling policy for the improvement of firm RTDBSs performances. Data Knowl. Eng. **63**, 414–432 (2007)
15. Song, H., Kam-yiu, L., Jiantao, W., Sang, H.S., Aloysius, K.M.: Adaptive co-scheduling for periodic application and update transactions in real-time database systems. J. Syst. Softw. **85**, 1729–1743 (2012)
16. Bernstein, P.A., Hadzilacos, V., Goodman, N.: Concurrency Control and Recovery in Database Systems. Addison-Wesley Longman Publishing Co., Inc., Boston (1987)
17. Jaziri, W., Sassi, N., Damak, D.: Using temporal versioning and integrity constraints for updating geographic databases and maintaining their consistency. J. Database Manag. (JDM) **26**, 30–59 (2015)
18. Jaziri, W., Gargouri, F.: Ontology theory, management and design: an overview and future directions. In: Ontology Theory, Management and Design: Advanced Tools and Models, pp. 27–77 (2010)
19. Sassi, N., Brahmia, B., Jaziri, W., Bouaziz, R.: From temporal databases to ontology versioning: an approach for ontology evolution. In: Ontology Theory, Management and Design: Advanced Tools and Models, pp. 225–246 (2010)

A Comparative Analysis of Algorithms for Mining Frequent Itemsets

Vyacheslav Busarov$^{(\boxtimes)}$, Natalia Grafeeva, and Elena Mikhailova

Saint Petersburg State University, St. Petersburg, Russia
vyacheslavbusarov@gmail.com,
{n.grafeeva, e.mikhaylova}@spbu.ru

Abstract. Finding frequent sets of items was first considered critical to mining association rules in the early 1990s. In the subsequent two decades, there have appeared numerous new methods of finding frequent itemsets, which underlines the importance of this problem. The number of algorithms has increased, thus making it more difficult to select proper one for a particular task and/or a particular type of data. This article analyses and compares the twelve most widely used algorithms for mining association rules. The choice of the most efficient of the twelve algorithms is made not only on the basis of available research data, but also based on empirical evidence. In addition, the article gives a detailed description of some approaches and contains an overview and classification of algorithms.

Keywords: Data mining · Frequent itemsets · Transaction database · Data structure

1 Introduction

The association rule mining task is to discover a set of attributes shared among a large number of objects in a given database. Consider, for example, the order database of a restaurant, where the objects represent guests and the attributes represent guests' orders. An example could be that "80 % of people who order fish also order white wine". There are many other potential application areas for association rule technology, which include customer segmentation, catalog design, and so on.

The use of association rules for data analysis was first proposed in 1993 by Agrawal et al. [1] who used them to analyze what is now referred to as market basket data. The idea behind association rules can be expressed as follows:

Let F be a set of items, and D a database of transactions, where each transaction has a unique identifier and contains a set of items. A set of items is also called an itemset. The support of an itemset f, denoted by $\sigma(f)$, is the number of transaction in which it occurs as a subset. An itemset f is frequent if its support is more than a user-defined minimum support value (MinSupport). Moreover, the value of support can be interpreted at absolute or relative format, it doesn't matter.

The association rule is an expression $A \Rightarrow B$, where A and B are non-overlapping itemsets. The support of rule is given as $\sigma(A \cup B)$, and the confidence as $\sigma(A \cup B)/\sigma(A)$, (i.e., the conditional probability that a transaction contains B, given

© Springer International Publishing Switzerland 2016
G. Arnicans et al. (Eds.): DB&IS 2016, CCIS 615, pp. 136–150, 2016.
DOI: 10.1007/978-3-319-40180-5_10

that it contains A). The rule is confident if its confidence is more than a user-defined minimum confidence (MinConf).

The data mining task is to generate all association rules in the database, which have a support greater than minimum support, i.e., the rules are frequent. The rules must also have confidence greater than minimum confidence, i.e. the rules are confident. The task can be broken into two steps [5]:

- Find all frequent itemsets. Given m items, there can be potentially 2^m frequent itemsets. Efficient methods are needed to traverse this exponential itemset search space to enumerate all the frequent itemsets. This frequent itemset discovery is the main focus of this paper.
- Generate confident rules. This step is relatively straightforward; rules of the form $A \backslash B \Rightarrow B$, where $B \subset A$ are generated for all frequent itemsets A, provided the rules have at least minimum confidence.

We have already stated that finding frequently occurring datasets is an important subtask that helps answer a lot of questions. Since the variety of approaches is wide, practical applications require a reliable method of selecting proper algorithm to fit the task at hand. We look at the most widely used algorithms and some of the state-of-the-art approaches, namely: Apriori [2] 1994, Apriori Hybrid [3] 1994, FP-Growth [4] 2000, Eclat [5] 2000, dEclat [11] 2003, Relim [6] 2005, LCMFreq v.2/v.3 [8, 9] 2004–2005, H-mine [7] 2007, PPV [12] 2010, PrePost [10] 2012, FIN [13] 2014, PrePost+ [14] 2015. It is now our task to determine which of the algorithms above perform better than the others and should thus be chosen for optimal performance.

2   Selection Criteria

Let us consider some of the possible criteria of optimality that we may use.

1. *Asymptotics.* In the theory of algorithms, this parameter is considered important, but it is completely devoid of objectivity in our case. The matter is that all algorithms use heuristics that allow to reduce the search area. Their efficiency directly depends on the characteristics of particular data, its density and evenness of its distribution. In such cases, it is the worst case that is looked at, and the number of operations is determined on its basis, but such a situation occurs only while we are dealing with artificial and specially selected data that is not likely to be encountered in real life tasks. We will not use this criterion for comparing the algorithms.
2. *Simplicity of realization.* This criterion is undoubtedly quite subjective. The history of the IT industry offers telling examples of how this criterion was used as a basis for managerial decisions. In the 1990s, numerous attempts were made to pay remuneration to software engineers based on the complexity of their work. All such attempts, including those that involved counting the number of lines of code and measuring the time it took to write them, proved useless and unsuccessful. In an industry like the IT one, simplicity of realization cannot be a valid criterion because everything depends not only on a programming task itself, but also on a developer's

programming background and implementation skills. In one of the reviewed works [6], however, the author uses simplicity of realization as one of the selection criteria for the purpose of his study. We consider this criterion to be intuitively clear, but we will not take it into consideration in comparing the algorithms.

3. *Volume of memory used.* This is a telling criterion. If measured in real experiments with the same data set, the volume of memory used can provide a reliable basis for choosing the most effective algorithm for a particular case. It is important to note that, for the reasons described above, we will use an empiric evaluation rather than an asymptotic one. Thus, the volume of memory used will not be used as a criterion for evaluating the algorithms in our work.

4. *Execution time.* Since we have real life applications in mind, the empirically measured execution time of an algorithm will always be a criterion of utmost importance to us. It is this parameter that we will look at, first of all, while comparing the twelve most common algorithms.

3   Comparative Analysis of Algorithms

As was noted earlier, identifying frequently occurring itemsets is key to searching for association rules. Algorithms have to deal with large bases of initial data, which complicates analysis. It is, therefore, important how the incoming data stream is stored. All most widely used algorithms switch from attributes to symbols or sequences of symbols of a fixed length, thus limiting the task of storing initial data sets to storing a glossary of transactions. Therefore, the data structure chosen in this or that case is an important element of an algorithm.

The bulk of research devoted to discovering association rules focuses on two categories of algorithms [7]:

- "candidate-generation-and-test"
- "pattern-growth method"

A characteristic example of the first category of algorithms is Apriori (Agrawal and Srikant [2]). This category also includes all the subsequent variations of this algorithm based on the anti-monotony principle (the support of a set of items does not exceed the support of any of its subsets) [1]. Such algorithms generate itemsets of length $(k + 1)$ based on the previous itemsets having length k. Even though the anti-monotony property principle allows us to disregard quite a few variants, such algorithms are not efficient computationally if initial data is extensive (the number of itemsets or the length of sets).

A good example of the second category of algorithms is called FP-Growth (Han et al. [4]). Algorithms of this type perform in a recursive manner by breaking down a data set into several parts and looking for local results that are subsequently combined into an overall result. The algorithms of this type generate fewer candidates than the algorithms described above, which allows to save a considerable amount of memory. However, the productivity of such algorithms largely depends on the homogeneity of initial data.

The comparative analysis of the most widely used algorithms for finding frequent itemsets was made on the basis of studies published at different times. All the experiments described in those studies were run on well-known data sets in this subject area, such as: Pumsb, Mushroom, Connect, Chess and Accidents (FIMI repository – http://fimi.ua.ac.be). The Pumsb data set contains census data, while the Mushroom data set consists of characteristics of mushroom species, Connect и Chess are sources of data on progress in the corresponding games. The Accidents data contains information about traffic and accidents.

In order to obtain an overall view of the compared algorithms, we introduced binary directed relationships between the algorithms to reflect improvement of productivity and decreasing of memory used (based only on the experiments described in the articles mentioned above). According to the authors of the articles, each pair of algorithms was compared under identical conditions pertaining to the hardware characteristics, data, the language of realization, etc. Based on the relationships introduced between the algorithms, we made a chart of execution time (Fig. 1) and a chart of memory requirements (Fig. 2).

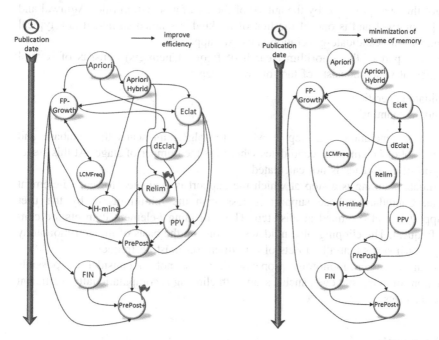

Fig. 1. The relationships between algorithms in terms of execution time.

Fig. 2. The relationships between algorithms in terms of memory requirements.

The introduced binary directed relationships are transitive and, consequently, allow us to perform a comprehensive comparative analysis of all the algorithms. It should be noted that there is a two-direction relationship between some of the algorithms (for example, between FPGrowth and LCMFreq v.2/v.3 in Fig. 1, EClat and dEClat in

Fig. 2), which means that, depending on the type of initial data, such algorithms demonstrate approximately equal results and are considered on a par in this respect. Besides, while it is clearly seen in the first chart which algorithms outperform the others (they are marked with flags), it is impossible in principle to tell the "winner" in the second because there are algorithms that were not compared in terms of the volume of memory used (for example: Apriori, LCMFreqv.2/v.3, PPV) at articles, where they were described. Nevertheless, the second chart is useful in that it gives us some idea of the difference between some of the algorithms in terms of memory requirements.

Just like the authors of the studies we refer to, we consider execution time to be the most important characteristic. The experimental data shown on the charts above (Fig. 1) leads to the conclusion that Relim (Borgelt [6]) and PrePost+ (Deng and Lv [14]) are the most efficient computationally. We will look at them in greater detail alongside with classical algorithms.

4 Apriori

This algorithm was proposed by the author of theory of association rules Agrawal and Srikant [2] in 1994 and is one of the first of its kind. As noted earlier, it is a typical representative of "candidate-generation-and-test" approach.

The main part of the algorithm (search of frequent itemsets) consists of several stages, each of which consists of the following steps:

- candidate generation
- candidate counting

Candidate generation is a step at which the algorithm scans the database and generates a lot of i-element of candidates where i is the number of stage. At this stage, the support of candidates is not calculated.

Candidate counting is a step at which the support is evaluated for each i-element candidate. Candidates whose support is less than the minimum set by the user (MinSupport) are also pruned at this step. The remaining i-element itemsets are considered frequent. The clipping of candidates is done on the basis of the anti-monotony property which states that all subsets of a frequent set must be frequent.

Of course, the anti-monotony property allows us not to consider all possible combinations of items, but even such a scan with clipping is computationally inefficient on large data sets.

5 FP-Growth

This algorithm was published in 2000 by Han et al. [4] and was an important contribution to the field of mining frequent itemsets because it did not involve explicit generation of candidate sets. This approach was later called a pattern-growth method.

As was pointed out earlier, FP-Growth (Frequency-pattern-Growth) algorithm, just like many other algorithms, converts the task of looking for frequent sets to the problem of storing a relevant glossary, which problem is solved by using a prefix tree – an FP-Tree. At the beginning, items in each transaction are sorted in the order of descending support. Naturally, infrequent singleton items have already been discarded. The data structure itself is built in accordance with the following principles:

- The root of the tree v_0 is labeled as null.
- Each node v of the tree T consists: item $f_v \in F$, a set of children nodes $S_v \subseteq T$,
- support $c_v = v(\varphi_v) : \varphi_v = \{f_u | u \in [v_0, v]\}$, where $[v_0, v]$ is the path from the root of the tree v_0 to a node v.
- $V(T,f) = \{ v \in T | f_v = f \}$ – all the nodes of the item f.
- $C(T,f) = \sum_{v \in V(T,f)} c_v$ – overall support of the item f.
- The tree is divided into levels, each of which corresponds to an item, and each item is associated with a single level. At the same time, the next node in the path can be at any level below the current one because all of them are sorted in the descending order of their support of corresponding items.

In the process of the algorithm, a conditional FP-tree (CFP-Tree) is repeatedly formed, such an FP-Tree being built only on transactions with a specified item [4]. Let there be an FP-tree T and item $f \in F$. Conditional FP-Tree $T' = T | f$ will be obtained if we:

1. Leave only the tree nodes on the path from the node v corresponding to the item f upwards to the root $v_0 : T' := \bigcup_{v \in V(T,f)} [v, v_0]$.
2. Increase the value of support c_v of nodes $v \in V(T',f)$ upwards according to the rule $c_u := \sum_{w \in S_u} c_w$ for each $u \in T'$.
3. Delete from T' all the nodes corresponding to the item f, because, by then, all the items lying below the item f will have already been considered.

It should be noted that T' is generated from the tree T without using the transaction database.

The final result is formed by means of a recursive procedure with the following parameters: FP-tree T, itemset φ and a list of frequent itemsets R. As a result, all frequent itemsets containing φ, are added to the R. All the items $f \in F : V(T,f) \neq \emptyset$ are processed sequentially by levels of the tree from the bottom up, and if $C(T,f) \geq MinSupport$, then:

- $R := R \cup \{\varphi \cup \{f\}\}$
- A new tree is generated $T' = T | f$
- The procedure is started again with the parameters: $T', \varphi \cup \{f\}, R$.

For the reader to better understand the above algorithm, we provide a specific example of the FP-Tree construction (Fig. 3).

There is a set of transactions which, after items are replaced by symbols, are transformed into words (corresponding to transactions) stored in the glossary. At the beginning, support is calculated for each singleton itemset, infrequent itemsets are eliminated in accordance with the declared minimal support, and the rest are arranged within each transaction in the order of decreasing support. Let minimal support be equal 3. One pass over the database of transactions in questions results in the formation of the tree represented in Fig. 4.

transactions	dictionary	support	Relim	FP-Growth
- b - - e - g	b e g		b e	e b
- - - d e f -	d e f	a 1	f d e	e d f
- - c - e - -	c e	g 2	c e	e c
- b c d e - -	b c d e	f 3	b d c e	e c d b
- - c d - - -	c d	b 4	d c	c d
- b c - e - -	b c e	d 5	b c e	e c b
- - c - e f -	c e f	c 7	f c e	e c f
a - c d - f -	a c d f	e 8	f d c	c d f
- - - d e - g	d e g		d e	e d
- b c - e - -	b c e		b c e	e c b

Fig. 3. An example of the first stage of algorithms FP-Growth and Relim.

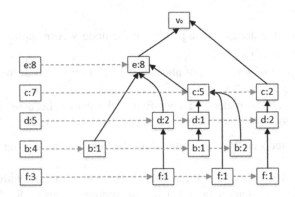

Fig. 4. An example of an initial FP-tree.

Practice has shown that FP-Growth performs worse on test data than many other algorithms in terms of execution time (Fig. 1), but the same experiments show that it makes a more efficient use of memory (Fig. 2). This is attributed to what is a unique feature of an FP-Tree: if the relative density of the data you deal with is the same everywhere, then, beginning with a certain moment, adding new transactions to a tree will not cause the number of nodes to change.

6 Relim

This algorithm was proposed in 2005 in the study by Christian Borgelt [6]. The acronym "Relim" illustrates the underlying principle of the algorithm: "REcursive ELIMination scheme". Relim tries to find all frequent itemsets with a given prefix by lengthening it recursively and renewing support at the same time. The approach utilized here is called a "pattern-growth method". Let us look at the stages of the algorithm:

1. The first iteration on the transaction database allows to calculate support for each item separately and to exclude infrequent items from each transaction (in the example in Fig. 4, MinSupport = 3). To improve the performance, the remaining items of each transaction are arranged in the order of ascending support.

2. In the next iteration, the array of lists is constructed as follows:

   - The head of each list is a particular frequent item. Lists are constructed for each of them.
   - Transactions starting with an item that corresponds to the head of the list are included in each list.
   - Initially, the lengths of the corresponding lists are recorded in the cells of the array, but then the cells are used for storing relative supports of element headers in the context of a given prefix, that is $\{m \times v(\varphi) \mid \varphi \subseteq F$, prefix $\subseteq \varphi$, $v(\varphi)$ - support of φ, m – the number of itemsets with this prefix$\}$. One of the main aspects of the algorithm is this: itemset support $(\varphi \cup f_i)$ is equal to the relative support f_i in the context of prefix φ.

3. Then, the algorithms starts a recursive procedure with the following parameters: (1) the structure described above; (2) current prefix (initially empty), (3) minimal support.

4. Next, the algorithm moves in two directions: a loop for a given data structure and a recursive sequence of procedure calls.

   - In the loop, each element is dealt with separately, and a new data structure of the type described above is built on the basis of the list of transactions of each element header.
   - The support is calculated for the itemset presented as the union $(\varphi \cup f_i)$ of the current prefix φ and the item f_i, and it is then added to the overall result if this itemset is a frequent one.
   - The given element header is added to the current prefix and a recursive procedure with the resulting data structure and updated prefix is run.
   - The lists of the original data structure and the structure obtained in the previous step are combined, the list of the current item header having been removed.
   - Then, the next item in the loop is dealt with.

An example of traversal of data structure is illustrated in Fig. 5.

The white boxes in Fig. 6 stand for the moved elements of the list. The newly generated arrays are shown on the right side of the picture. It is these new arrays that will be used as parameters for recursive calls. To better understand how the procedure works, look at Fig. 6 that illustrates obtaining 2 itemsets (dc and de) having support 3. The first itemset is obtained in step 4.1.1, while the second is obtained in step 4.2.

In addition to high productivity, another advantage of this algorithm is its simplicity. In fact, the task of looking for frequent sets of items is described by one recursive procedure of four lines in length. What is worthy of note is that the use of a single and quite simple data structure makes the realization of Relim on distributed systems much simpler compared to the realization of other algorithms.

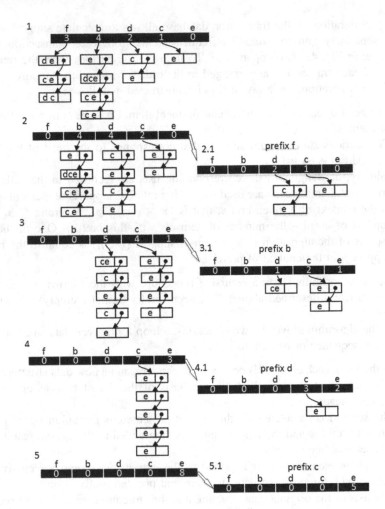

Fig. 5. The cycle of Relim.

7 PrePost+

Proposed by Deng and Lv [14] in 2015, it is the latest algorithm for identifying frequent itemsets. PrePost+ uses three data structures at a time: N-list, PPC-tree and set-enumeration tree, which explains why it requires more memory than FP-Growth does (as you can see in Fig. 3). Although it is an "Apriori-like" algorithm, it has empirically proved superior to many other algorithms in terms of execution time (Fig. 1).

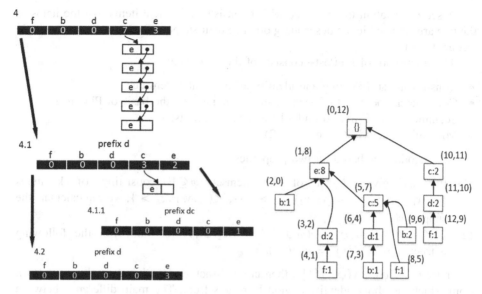

Fig. 6. An example of recursive procedure of Relim.

Fig. 7. An example of a PPC-tree.

A PPC-tree is a tree structure:

1. It consists of one root labeled as "null", and a set of item prefix subtrees as the children of the root.
2. Each node in the item prefix subtree consists of five fields: item-name, count, children-list, pre-order, and post-order.

   - item-name registers which item this node represents.
   - count is the number of transactions presented by the portion of the path reaching this node.
   - children-list contains all the children of the node.
   - pre-order is the pre-order rank of the node.
   - post-order is the post-order rank of the node.

For each node, its pre-order is the sequence number of the node when the tree is scanned by pre-order traversal, and its post-order is the sequence number when the tree is scanned by post-order traversal.

An example of PPC-tree constructed from the data given above (Fig. 3) is shown in Fig. 7.

For each node N in a PPC-tree, we form <(N.pre-order, N.post-order):count> the PP-code of N.

The N-list of a frequent itemset is a sequence of all the PP-codes registering the item from the PPC-tree, with such PP-codes arranged in the ascending order of their pre-order values. The N-list of an itemset is formed, by using special rules, on the basis of the PP-codes corresponding to the items of the itemset.

A set-enumeration tree is a tree which consists of frequent items. All the items in the tree are arranged in the descending order of their support. Each node stores a single frequent item.

The framework of PrePost+ consists of the following:

- Constructing an PPC-tree and identify all frequent 1-itemsets;
- Constructing the N-list of each frequent 1-itemset on the basis of PPC-tree;
- Scanning the PPC-tree to find all frequent 2-itemsets;
- Mine all frequent k-itemsets (k > 2).

The algorithm is based on two properties:

(1) For the given N-list of the itemset $\varphi \subseteq F$ consisting of k items $\{ < (x_1, y_1) : z_1 >, < (x_2, y_2) : z_2 >, \cdots, < (x_k, y_k) : z_k > \}$, we can calculate the support $v(\varphi) = \sum_{i=1}^{k} z_i$.

(2) $\forall \varphi \subseteq F \; \forall f \in F : v(\varphi) = v(\varphi \cup \{f\}) \Rightarrow \forall A \subseteq F : A \cap \varphi = \emptyset, f \notin A$ the following is true: $v(\varphi \cup A \cup \{f\}) = v(\varphi (\cup) A)$.

Indeed, if $v(\varphi) = v(\varphi \cup \{f\})$, then any transaction containing φ, also contains f, from which the above identity obviously results from. The main difference between PrePost and PrePost+ lies in the strategy of pruning candidates for the status of frequent itemsets. PrePost+ uses itemset equivalence as a pruning strategy while PrePost utilizes the single path property of an N-list as a pruning strategy [10]. For the purpose of facilitating the mining process, PrePost+ uses a set-enumeration tree to provide the search space for frequent itemsets.

A more detailed description of the algorithm and its pseudo-code are provided in this article [14].

8 Review of Other Algorithms

Apriori Hybrid [3] (1994). *Category:* "candidate-generation-and-test".
Data structures used: A hash-tree or a hash-table is used for storing generated candidate sets (it simplifies the calculation of support for new sets), and two-dimension number array for storing frequent itemsets. Each of such itemset has a unique identifier. It is these identifiers that are used for indexing.
Some key features: The algorithm applies the same principles as Apriori [2] does, but it does not refer to the initial transaction database in order to calculate the support of each itemset. Instead, the techniques of hashing and intersecting the itemsets are used.

Eclat [5] (2000). *Category:* "candidate-generation-and-test".
Data structures used: What is called Lettuces is partially ordered sets, in which each pair of elements has unique supremum and infinum.
Some key features: All the candidates are stored in a special data structure called Lattices. During the search, the algorithm passes this data structure widthwise and depthwise. One of the main heuristics is that the algorithms strives to partition the itemset into subsets and look at them separately. In the process of looking for frequent itemsets, Eclat tries to identify the equivalence classes, thus pruning the range of possible candidates.

dEclat [11] (2003). *Category:* "candidate-generation-and-test".
Data structures used: Diffset is a data structure that is capable of storing, instersecting and combining itemsets.
Some key features: dEclat is a modification of the previous algorithm by the same authors [5]. The main departure of this algorithm from Eclat is the use of a new data structure that allows to prune a large number of candidates. This algorithm also uses the concept of equivalency based on the values of the function defined for itemsets. One diffset stores equivalence classes, and the other diffset stores the prefixes of candidates of varying length.

LCMFreqv.2/v.3 [8, 9] (2004–2005). Category: "pattern-growth method".
Data structures used: This algorithm uses a combination of common structures – bitmap, prefix tree and array list.
Some key features: A prefix tree contains possible candidates for frequent itemsets, the order of items being clearly fixed. Original transactions are stored in an array list. Bitmap data structure is used for calculating support in a more efficient way. The shared use of these data structures varies from one version of the algorithm to another. V.3 is considered by the authors of the algorithm to be the most efficient of all. It is this version that we used for our experiments. Further in the text, we refer to it as LCMFreq.

H-mine [7] (2007). *Category:* "pattern-growth method".
Data structures used: H-struct is a data structure that stores frequent itemsets together with references to the corresponding transactions in the original transaction database. The method of its construction is similar to that of FP-Growth; however, instead of storing transactions explicitly, the algorithm operates with references to them. It is thanks to this feature that H-mine outperforms many algorithms in the efficiency of memory use (as you can see in Fig. 2).
Some key features: The method of scanning H-struct is basically the same as that in FP-Growth. What differentiates one of the two algorithms from the other is the approach to storing initial data.

PPV [12] (2010). *Category:* "candidate-generation-and-test".
Data structures used: A PPC-tree is the same prefix tree as the one used in the algorithm PrePost+. Node-list is a data structure that consists of PP-codes formed on the basis of a PPC-tree. It is absolutely identical to N-list, the only difference between Node-list and N-list being that Node-list use descendant nodes to represent an itemset while N-list represent an itemset by ancestor nodes.
Some key features: The PPC-tree stores the original transactions. Candidate for the status of frequent itemsets are stored in the Node lists. The problem of calculating the support of candidates boils down to the operation of intersecting Node-lists, which is done in linear time. The operation of intersecting Node-lists is the key feature of PPV algorithm.

PrePost [10] (2012). *Category:* "candidate-generation-and-test".
Data structures used: A PPC-tree is the same prefix tree as the one utilized in PrePost+. The example of constructing a PPC-tree is shown above in Fig. 7. The other data structure used in this algorithm is N-list that consists of PP-codes formed on the basis of a PPC-tree.

Some key features: As noted by the authors themselves, the main difference between PrePost and PrePost+ is the pruning strategies used: PrePost+ adopts itemset equivalence as the pruning strategy whereas PrePost utitizes single path property of N-list for this purpose [10]. To facilitate the mining process, PrePost+ uses a set-enumeration tree to represent the search space for frequent itemsets. The remaining steps of these algorithms are similar to each other.

FIN [13] (2014). *Category:* "candidate-generation-and-test".
Data structures used: A Node-set is a structure that is identical to N-list and Node-list, but it uses twice as little memory since it requires to store either pre-order parameter or post-order one instead of storing both of them at a time. A POC-tree (Pre-Order Coding tree) is a prefix tree that is absolutely identical to a PPC-tree, but it does not storeparameter post-order at all. A set-enumeration tree is a tree that stores all possible items in the ascending order of their support. Each node of the tree stores a single frequent item.
Some key features: In many aspects, this algorithm is similar to Pre-Post, the used data structures being the only difference between the two algorithms.

9   Experiments

The chart of relationship shown in Fig. 2 is meaningful because it is built on the basis of real empirical evidence. The facts that each pair of algorithms was compared on the same databases (1), that the algorithms in each pair were realized in one and the same programming language (2), and that the same computing equipment was utilized in comparing each pair of algorithms (3), provide a valid basis for concluding which of the twelve algorithms is the most efficient. However, the fact that the chart of relationship was built based on separate comparisons is a drawback of this chart as a whole. It is for this reason that we ran our own experiments and compared the performance of all the twelve algorithms in likely conditions in terms of execution time.

For the purposes of this study, we selected the Mushroom data set mentioned almost in all the reviewed publications. The average transaction length is 23, the glossary of attributes consists of 119 elements, and the total number of transactions amounts to 8124. The data contain a description of mushroom species and their characteristics. We used implementations, which exactly corresponded author's algorithms and pseudocodes. All the experiments were run on a computer with the following processor: Intel(R) Core(TM) i7-4700HQ CPU @ 2.40 GHz, 12.0 GB RAM. All the algorithms were realized on Java 8.

The value of minimal support largely affects the number of sets in the result, their length and, of course, execution time. For greater clarity, we measured the value of minimum support (MinSupport) by percentage, with this value defined as the ratio of the number of transactions containing such itemsets to the total number of transactions. As follows from the definition above, this value cannot exceed 100 %. By varying this value, we finally arrived at the results presented in Fig. 8.

As shown by the experiments, the algorithms for finding frequent item sets rank as follows in terms of execution time: Relim (1), PrePost+ (2), FIN (3), PrePost (4), PPV

Minimal support	Relim	PrePost+	FIN	PrePost	PPV	H-mine	dEclat	FP-Growth	LCMFreq	Eclat	Apriori Hybrid	Apriori
70%	397,62	401,15	412,45	415,23	417,56	419,35	426,12	427,89	429,96	434,92	437,53	438,24
75%	372,18	373,32	383,78	385,11	389,98	393,02	403,43	404,14	409,07	416,07	419,23	422,91
80%	341,12	351,87	362,91	365,12	373,32	374,79	379,81	377,02	382,59	391,19	392,98	395,01
85%	332,89	337,6	349,45	348,56	358,76	360,16	373,14	375,51	380,31	390,69	394,11	394,09
90%	327,92	328,12	340,11	345,61	349,08	350,38	359,26	362,97	364,82	375,19	378,85	379,02

Fig. 8. Execution time of experiments (sec).

(5), H-mine (6), dEclat (7), FP-Growth (8), LCMFreq (9), Eclat (10), Apriori Hybrid (11), Apriori (12). We did not run any experiments at low values of minimal support. It is reasonable to conclude that, due to a larger number of itemsets and an increase in their lengths, the execution time would be considerably longer. However, the experiments themselves are quite significant since they reveal a stable relationship between the performanceof the algorithms at different values of minimal support. It should be noted that the language of implementation affects certain nuances without changing the overall picture. What is most important is that the results of our experiments are fully consistent with the relationship chart of execution time which was built on the basis of previous studies of the algorithms.

10 Conclusion and Ideas for Further Research

This study is an innovative comparative analysis of all the widely existing algorithms for mining frequent itemsets. What makes this work valuable is that it provides a rating of the algorithms in terms of execution time. Being the first of its kind, this rating can make it easier for developers to choose a proper algorithm for their practical applications. Our experimental evidence shows that Relim and PrePost+ algorithms outperform others in practical applications, with each of the two belonging to a different category: "candidate-generation-and-test" and "pattern-growth method" respetively. Although the results are almost equal, in these experiments Relim [6] outperformed its "younger" competitor – PrePost+ [14]. It is important to note that the relationship chart of execution time that was built based on the previously published studies (Fig. 1), did not contradict the empirical result obtained in this work, which suggests that these results are quite objective.

We are planning to expand this study to cover any new approaches to finding frequent itemsets if and when such approaches appear. It is also our intention to undertake a more serious comparison of the algorithms both in terms of execution time and memory requirements by running experiments on large databases, such as Kosarak, the latter database notably containing 990,002 transactions and 41,270 possible attributes. It also makes sense in the future to look at the implementation of these algorithms in distributed environments.

References

1. Agrawal, R., Imielinski, T., Swami, A.: Mining associations between sets of items in large databases. In: ACM SIGMOD International Conference on Management of Data, SIGMOD 1993, Washington, DC, pp. 207–216 (1993)
2. Agrawal, R., Srikant, R.: Fast algorithms for mining association rules. In: 20th International Conference on Very Large Databases, VLDB 1994, Santiago, Chile, pp. 487–499 (1994)
3. Agrawal, R., Mannila, H., Srikant, R., Toivonen, H.: Fast discovery of association rules. Adv. Knowl. Discov. Data Min. **12**(1), 307–328 (1996). AAAI MIT Press
4. Han, J., Pei, H., Yin, Y.: Mining frequent patterns without candidate generation. In: ACM SIGMOD International Conference on Management of Data, SIGMOD 2000, Dallas, TX, pp. 1–12 (2000)
5. Zaki, M.J.: Scalable algorithms for association mining. IEEE Trans. Knowl. Data Eng. **12** (3), 372–390 (2000)
6. Borgelt, C.: Keeping things simple: finding frequent item sets by recursive elimination. In: Open Source Data Mining Workshop, OSDM 2005, Chicago, IL, pp. 66–70. ACM Press, New York (2005)
7. Pei, J., Han, J., Lu, H., Nishio, S., Tang, S., Yang, D.: H-mine: fast and space-preserving frequent pattern mining in large databases. IIE Trans. **39**(6), 593–605 (2007)
8. Uno, T., Kiyomi, M., Arimura, H.: LCM ver. 2: efficient mining algorithms for frequent/ closed/maximal itemsets. In: Workshop on Frequent Itemset Mining Implementations, FIMI 2004, Brighton, UK (2004)
9. Uno, T., Kiyomi, M., Arimura, H.: LCM ver.3: collaboration of array, bitmap and prefix tree for frequent itemset mining. In: Open Source Data Mining Workshop, OSDM 2005, Chicago, IL, pp. 77–86. ACM Press, New York (2005)
10. Deng, Z.H., Wang, Z., Jiang, J.: A new algorithm for fast mining frequent itemsets using N-lists. Sci. China Inf. Sci. **55**(9), 2008–2030 (2012)
11. Zaki, M.J., Gouda, K.: Fast vertical mining using diffsets. In: 9th ACM SIGKDD International Conference on Knowledge Discovery and Data Mining, KDD 2003, Washington, DC, pp. 326–335. ACM Press, New York (2003)
12. Deng, Z.H., Wang, Z.: A new fast vertical method for mining frequent patterns. Int. J. Comput. Intell. Syst. **3**(6), 733–744 (2010)
13. Deng, Z.H., Lv, S.L.: Fast mining frequent itemsets using Nodesets. Expert Syst. Appl. **41** (10), 4505–4512 (2014)
14. Deng, Z.H., Lv, S.L.: PrePost+: an efficient N-lists-based algorithm for mining frequent itemsets via Children-Parent Equivalence pruning. Expert Syst. Appl. **42**(10), 5424–5432 (2015)

A WebGIS Application for Cloud Storm Monitoring

Stavros Kolios[1,2(✉)], Dimitrios Loukadakis[2], Chrysostomos Stylios[2],
Andreas Kazantzidis[3], and Aleksandr Petunin[4]

[1] Faculty of Pure and Applied Sciences,
Open University of Cyprus, Nicosia, Cyprus
stavroskolios@yahoo.gr
[2] Laboratory of Knowledge and Intelligent Computing (KIC-LAB),
Technological and Educational Institute of Epirus, Arta, Greece
dimitris.loukadakis@gmail.com, stylios@teiep.gr
[3] Laboratory of Atmospheric Physics, Department of Physics,
University of Patras, Patras, Greece
akaza@upatras.gr
[4] Institute of Mechanics and Machine-Building,
Ural Federal University, Yekaterinburg, Russia
a.a.petunin@urfu.ru

Abstract. Extreme weather phenomena (i.e. heavy precipitation, hail and lightings) frequently cause damages in properties and agricultural production and usually originate from the cloud storms. Automated systems able to provide timely and accurate monitoring and predictions would contribute to prevent the effects of physical disasters and reduce economic losses. Nowadays, meteorological satellites have a significant role in weather monitoring and forecasting, providing accurate and high resolution data. Such data can be analyzed using Geographical Information Systems (GIS) and modern web technologies to develop integrated automated web based monitoring systems. This study describes a WebGIS application focused on monitoring and forecasting cloud tops of storm evolution. The application has developed using modern tools, to exploit their features through an innovative web based monitoring system. There are used open source framework to ensure mobility, stability and portability of the application.

Keywords: Webgis · Cloud monitoring · Satellite images · Remote sensing

1 Introduction

Most of the times, cloud storms have significant and direct impacts on human lives and properties because of the extreme weather phenomena that they produce like heavy rain, hail, strong winds, tornadoes, lightning, and flooding (e.g. [1–5]). The detection and forecasting of cloud storms evolution is extremely difficult and highly complicated not only due to their small scale internal dynamics but also because they are produced under various favorable conditions depending on topography, synoptic weather conditions, atmospheric instability, wind shear and many other factors ([6–12]).

© Springer International Publishing Switzerland 2016
G. Arnicans et al. (Eds.): DB&IS 2016, CCIS 615, pp. 151–163, 2016.
DOI: 10.1007/978-3-319-40180-5_11

Nowadays, the availability of modern geostationary meteorological satellites with their fine time (typically, 15 min) and space (3 km at the sub-satellite point) sampling, comprise a modern alternative approach to face the uncertainness and the restrictions of many numerical models and radars for satellite monitoring/forecasting of cloud storms.

Using satellite images as data inputs along with GIS capabilities and modern web technologies, fully automated applications providing useful information regarding storms evolution, can be developed. Characteristic examples of such integrated systems based on satellite datasets to monitor/forecast cloud storms through web-based applications were developed is the "ForTraCC" system [13], for tracking and forecasting cloud storms over South America using GOES (Geostationary Operational Environmental Satellite) information as well as the fully automated system called "NEFODINA" [14] focused over the Italian peninsula and its surroundings based on multispectral MSG (Meteosat Second Generation) imagery.

This study comprises an analytic description of all the stages regarding the development of a fully automated WebGIS application focused on monitoring and forecasting cloud tops of storms (and/or cloudy areas, which may be evolved to storms and produce extreme weather conditions in the next few hours) using exclusively satellite images. The satellite images come from Meteosat satellite platform. The application domain includes the greater area of the Mediterranean basin.

The system is developed using modern tools and methods like the "MongoDB" database and Google APIs. Innovative characteristics are the provision of monitoring and forecasting capabilities at a 15-min temporal resolution. In addition to this, the forecasting capabilities of the system are extended in four hours, which is one of the most extended statistical forecasts provided through similar systems worldwide (based exclusively on satellite data). Continuous hail and lightning events estimations for such a large geographical area (greater area of the Mediterranean basin) can also be considered as innovative characteristic of the developed system because such information is not provided from other similar systems.

2 Data and Methods

2.1 Data and Parameters

The initial datasets are exclusively satellite images (raw data) coming from the satellite "Meteosat-10", which is the primary European operational meteorological satellite. The images are referred to five channels (multispectral imagery) of the infrared region (Table 1).

The initial multispectral imagery (Table 1) is automatically pre-processed and the images are analyzed in pixel basis to provide operationally monitoring of the cloud tops evolution as well as forecasts up to 4 h ahead (short-range forecasting). All the available information of the five spectral channels (Table 1) is used to calculate the final products of precipitation, hail and lightning as well as their relative interactive maps, using map tiles.

Table 1. Basic characteristics of Meteosat used at the application

Channel (Band)	Spectral region (μm)	Spectral center (μm)
5	5.35–7.15	6.2
6	6.85–7.85	7.3
7	8.3–9.1	8.7
9	9.8–11.8	10.8
10	11–13	12.0

As abovementioned, four types of information are provided and illustrated through this web-based application in pixel basis, either into the monitoring or into the forecasting module:

- Cloud top temperature (°C)
- Precipitation rate (mm/h)
- Hail probability (%)
- Lightnings probability (%)

The typical temporal resolution of the provided information is 15 min because this is the typical temporal resolution for the data flow coming from the Meteosat satellite.

2.2   Monitoring/Forecasting Methodology

For the detection module, a set of criteria has identified (Table 2) to detect all the cloud pixels that either belong to storms or can evolve to storms. Hereinafter, these pixels are referred as convective cloud pixels. The adopted criteria comprise a combination of well-known and recent thresholding methods for the detection of convective cloud patterns based on satellite imagery ([6, 9, 12]).

Table 2. The five citeria used for the detection of the cloud pixels of interest at the Meteosat multispectral imagery.

Criteria
$T_{6.2\mu m} < 240$ K
$(\Delta T_{10.8\mu m}/\Delta T) < -6$ K $(15$ min$)^{-1}$
$(\Delta T_{(6.2\mu m-10.8\mu m)}/\Delta T) > 3$ K $(15$ min$)^{-1}$
$\Delta T_{(6.2\mu m-7.3\mu m)} > -20$ K
$\Delta T_{(12.0\mu m-10.8\mu m)} > -3$ K

The forecasting module produces 15-min forecasts using linear multivariate functions (in the current version of the application). More specifically, there are developed five different analytical functions (for each one of the spectral channels) following the general model of the Eq. 1. Each of them forecasts the relative channel temperature on a pixel basis. The coefficients of the functions are calculated using information from four previous timeslots (typical one hour before). For example, for forecasting one hour after the current time (t_o), the mean values (at pixel basis) of the four previous timeslots are used. The general statistical model to estimate channel temperature values is:

$$Y = A_0 + A_1 X_1 + \cdots + A_n X_n \qquad (1)$$

Where "Y" is the dependent variable (i.e. the pixel temperature of a specific channel at a specific timeslot), A_0 is a specific constant and A_n (with n = 1, 2, 3, ...) is the coefficient of the relative independent variable X_n, where the independent variables are referred at the Table 2. As aforementioned, the linear multiple regression analysis is calculated automatically through the programming environment of the application, in pixel basis where the pixel parameter values of the four latest previous timeslots are considered so that to provide forecast for the next specific timeslot ahead of the current time. Figure 1 illustrates an example regarding the accuracy assessment of the temperature forecast for two basic channels (6.2 μm and 10.8 μm). The Mean Absolute Error (MAE) and the Mean Error (ME) between the measured and forecasted values of 10.933 pixels are calculated. The accuracy assessment of the temperature forecasts for all the spectral channels of Table 1 are concluded to quite satisfactory results.

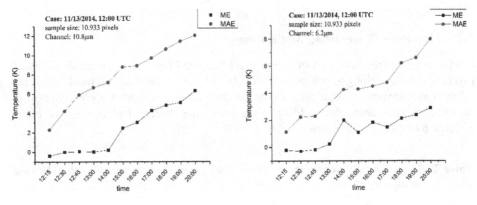

Fig. 1. Accuracy assessment of a sample of temperature pixel values regarding the channels at 6.2 μm and 10.8 μm.

Then, the forecasted temperature values are used to perform pixel based rain, hail and lightning estimations, which are provided as interactive maps through the WebGIS application. There was used the analytic equation proposed by [15] for the rain estimation and the mathematical formula proposed by [16] for the hail estimation. Finally, for the lightning events estimation, a spatiotemporal correlation of lightning data and temperature pixel values in the channel of 10.8 μm, was performed. The analytic equation, after a log-normal fitting to the frequency of occurrence distribution for the temperature values correlated with lightning was used.

2.3   Structure of the System

The WebGIS application comprises a system with interconnected modules as Fig. 2 presents. The system is comprised of a server station used for receiving and storing the continuous flow of the satellite images (in segments) through the satellite antenna. Then, JAVA-written modules automatically implement all the pre-processing steps i.e. georeference, calculation of Brightness Temperatures in pixel basis, errors/missing data checking, unification of different segments of raw images.

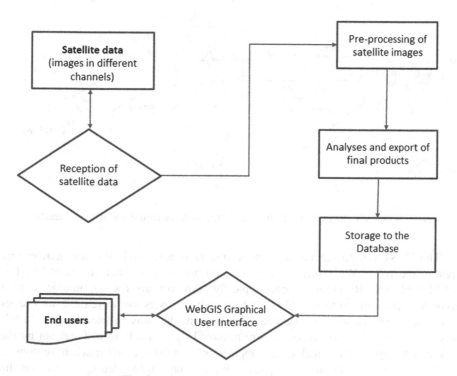

Fig. 2. The overall structure of the system

After the images' pre-processing stage, a threshold (brightness temperature in the 10.8 μm channel smaller than 250 K) is applied to delineate cloud areas, which are favorable for storm developing or represent active storms. At next step, well known mathematical equations are applied to calculate and estimate at pixel basis the values of the precipitation rate, hail and lightning probability. All these data are converted to image files that are visualized as map tiles through the graphical user interface and interactive maps provide by the WebGIS (http://weather.kic.teiep.gr). All the data is also stored in the database of the system for further use.

2.4   Technical Description of the WebGIS

Figure 3 presents the WebGIS general flow of information and the corresponding procedures that are similar to any web based application [17]. The web browser stands for the GUI (Graphical user interface) that interacts with the end users. The web browser transmits any user's requests to the web server and the web server access the corresponding data at the database. The database provides data/information to the web server and the web server shares them through the REST API and the map tiles module.

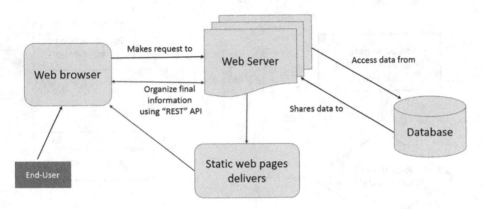

Fig. 3. Schematic diagram presenting the data flow through the applications' modules.

The REST (REpresentational State Transfer) is a general software architectural style of the World Wide Web. Any system conforming to the constraints of REST is called RESTful. RESTful systems typically, but not always, communicate over Hypertext Transfer Protocol (HTTP), which web browsers use to retrieve web pages and to send data to servers. The static web page model, was adopted because in this way, many requests and procedures are managed by user's browsers and not by the server of the application, making the application more stable and quick in responses. And by this way, all the dynamic parts of the interface are loaded apparently via the "REST".

In order to create RESTful services, the SPRING (version 4.2.4) framework with a Controller mechanism has been used, which enrich the features and functionalities on top of the Java EE platform. The SPRING framework is open source, whose core features are used by any Java application and there are many extensions for building web applications (Fig. 4). The Spring Web model-view-controller (MVC) framework is designed around a Dispatcher Servlet that dispatches requests to handlers, with configurable mapping, view resolution, locale, time zone and theme resolution as well as support for uploading files.

The web server (in our case an Apache Tomcat) implements the Java servlets specifications. The server uses Web application ARchive (WAR) package file format, the JAR (Java Archive) and it aggregates any Java class file and associated metadata and resources (text, images, etc.) into one file, which is distributed to application software or libraries on the Java platform.

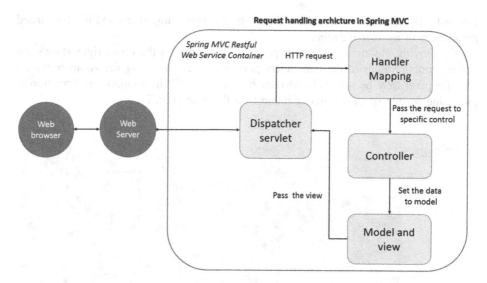

Fig. 4. Schematic diagram showing how the SPRING MVC handles data/information and responses to server and to web browser.

The database is built in the Mongo DataBase framework. The MongoDB is an open source document-oriented database designed with both scalability and developer agility in mind. Instead of storing data in tables as it would be for a relational database, in MongoDB, data are stored in JSON-like documents with dynamic schemas (document-oriented database). The MongoDB GridFS specification is used for storing the raw satellite data, which provides quick process for real-time image generation. This choice was made to handle and organize efficiently the large archive datasets of the WebGIS application but it could also be used for any procedure such as cloud movement tracking and/or weather station data. The MongoDB is characterized by very fast information searching, it has geospatial indexes and features that are essential for the development of the application and its updates (e.g. the capped collections, the aggregation framework etc.). There are also used the Google APIs, a set of application programming interfaces (APIs) developed by Google, which allow communication with Google Services and their integration to other services.

2.5   The Graphical User Interface

The main interface of the application shows an interactive map of the earth globe (google map) focused at the greater Mediterranean basin (Fig. 5).

Above-right of the interface there is a small menu where the user can select the monitoring/forecasting parameter(s) that would like to present (Figs. 5 and 6). Using this menu, interactive map(s) of the parameter(s) for the whole area of interest are appeared. By checking the small check list box above-left of the map, a series of dates ahead of the current date are displayed. These dates are referred to the available forecasts. By choosing one of these dates, the corresponding maps and the forecasted parameters of interest are seen.

By moving the cursor on the map, the stable table on the down-right shows the corresponding values of the parameters pixel by pixel. By using the zoom button, a specific pixel may be selected and then by clicking, all the available information is providing for the specific point of interest on the map (Fig. 7).

Fig. 5. Main interface of the WebGIS application.

Fig. 6. Cloud areas represent potential or active storms. The blue colored areas present rain.

On the left part of the map, there is a series of past dates providing relevant parameters information. By selecting one of these dates, the chosen one data is considered as current date and all the available information for this particular date are presented. For example, Fig. 6 illustrates the clouds storms existed at 30/01/2016 (02:45, UTC) in the greater area of Balearic Islands. The blue colored cloud areas present rain areas.

For the same case of Fig. 6 but zoomed in the eastern area of Spain can be seen at Fig. 7. The left of Fig. 7 presents the cloud storms (along with the estimated data for rainy areas) for the date of 30/01/2016 (02:45, UTC) at the greater area of Balearic Islands. The right part of Fig. 7 illustrates the same area with the forecasting values for three hours later.

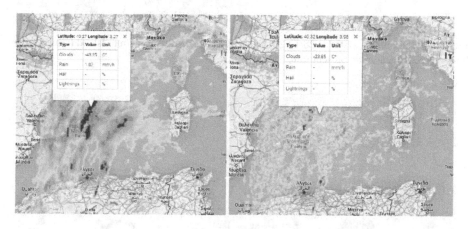

Fig. 7. Cloud storms (along with the estimated rainy areas) at Balearic Islands (left part). On the right, the forecast three hours later (05:45), is displayed.

Figures 8 and 9 illustrate another example for cloud storm monitoring and forecasting.

The same case of Fig. 8 but zoomed in cloud storms areas southern of Crete Island (Greece) is provided at Fig. 9. The blue coloured cloud areas present rain. The right part of Fig. 9 presents the same area with for the forecasting values for three hours later on.

The developed WebGIS includes also capabilities for hail/lightings estimation. Figures 10 and 11 present how this information can seen through the WebGIS interface. The user just has to select the appropriate checkboxes on the right of the interface and the relative information is presented for the time/date which is active in the left of the interface (or the forecast of interest time/date).

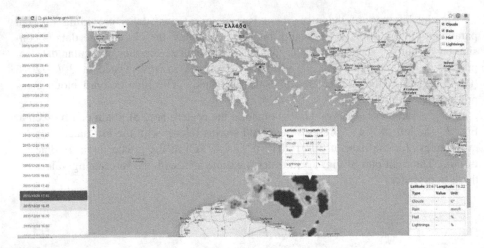

Fig. 8. Cloud storms existed at 28/12/2015 (17:15, UTC) southern of Crete Island (Greece). The blue colored cloud areas present rain.

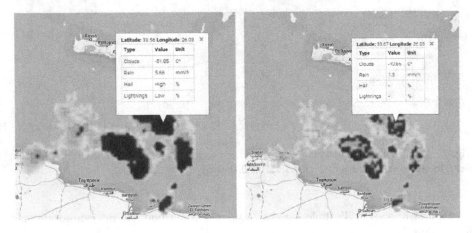

Fig. 9. Cloud storms (along with the estimated rainy areas) southern of Crete Island (Greece) (left part). On the right, the forecast three hours later (Color figure online).

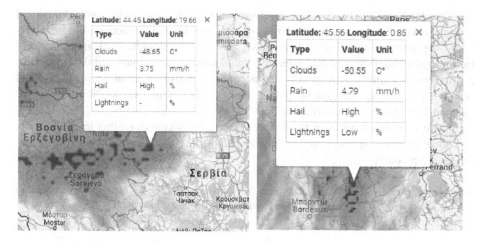

Fig. 10. Hail estimation (red coloured areas) for two cases (a) 03/02/2016, 02:00 UTC (b) 28/01/2016, 08:30 UTC (Color figure online).

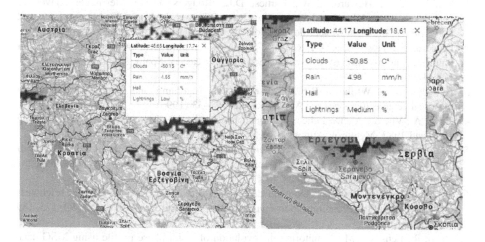

Fig. 11. Lightning estimation (red coloured areas) for two cases (a) 03/02/2016, 02:00 UTC (b) 28/01/2016, 08:30 UTC (Color figure online).

3 Conclusions

A WebGIS application for monitoring and estimating cloud storms has been designed, developed and operated under real-time basis using the map tiles solution for the web mapping. It comprises an automated web interface where basic meteorological data products are visualized at time and space. The data is provided by the Meteosat satellite multispectral imagery. After preprocessing and analysing the data is used for monitoring and short-range forecasting of cloud storms and their potential results (e.g. precipitation, rain and lightnings).

The WebGIS application is developed using modern tools, procedures and frameworks so that to be characterized by easy and immediate access, usage, portability, operational stability and timeless service. It integrates all modern web-technologies, programming and satellite remote sensing techniques and datasets. Thus it provides in a simple way, timely and accurately, useful information for public and private agencies and the general public whose activities and responsibilities can be affected from extreme weather conditions.

There is an on-going work to produce quantitative statistics so that to evaluate the performance of the application and the accuracy of the forecasting/monitoring procedures and products.

Acknowledgements. We thank Laboratory of Atmospheric Physics at the Department of Physics, University of Patras, Greece for providing the Meteosat raw data.

References

1. Maddox, R.A., Howard, K.W., Bartles, D.L., Rodgers, D.M.: Mesoscale convective complexes in middle the latitudes. In: Ray, P.S. (ed.) Mesoscale Meteorology and Forecasting. American Meteorological Society, Boston (1986)
2. Romero, R., Doswell, C.A., Ramis, C.: Mesoscale numerical study of two cases of long-lived quasi-stationary convective systems over eastern Spain. Monthly Weather Rev. **128**, 3731–3751 (2000)
3. Rutledge, S.A., Williams, R.E., Petersen, A.W.: Lightning and electrical structure of mesoscale convective systems. Atmos. Res. **29**, 27–53 (1993)
4. Gaye, A., Viltard, A., De Felice, P.: Squall lines and rainfall over Western Africa during 1986 and 1987. Meteorol. Atmos. Phys. **90**, 215–224 (2005)
5. Correoso, F.J., Hernandez, E., Garcia-Herrera, R., Barriopedro, D., Paredes, D.: A 3-year of cloud-to-ground lightning flash characteristics of mesoscale convective systems over the Western Mediterranean Sea. Atmos. Res. **79**, 89–107 (2006)
6. Bedka, K.M.: Overshooting cloud top detections using MSG SEVIRI Infrared brightness temperatures and their relationship to severe weather over Europe. Atmos. Res. **99**, 175–189 (2011)
7. Drori, R., Lensky, I.M.: Monitoring the evolution of cloud phase profile using MSG data. Atmos. Res. **97**, 577–582 (2010)
8. Lazri, M., Ameur, Z., Ameur, S., Mohia, Y., Brucker, J.M., Testud, J.: Rainfall estimation over a Mediterranean region using a method based on various spectral parameters of SEVIRI-MSG. Adv. Space Res. **52**, 1450–1466 (2013)
9. Merk, D., Zinner, T.: Detection of convective initiation using Meteosat SEVIRI: implementation in and verification with the tracking and nowcasting algorithm Cb-TRAM. Atmos. Measur. Tech. **6**, 1903–1918 (2013)
10. Mikus, P., Mahovic, N.S.: Satellite-based overshooting top detection methods and an analysis of correlated weather conditions. Atmos. Res. **123**, 268–280 (2013)
11. Georgiev, C.G., Santurette, P., Dupont, F., Brunel, P.: Quantitative evaluation of 6.2 μm, 7.3 μm, 8.7 μm Meteosat channels response to tropospheric moisture distribution. In: Joint 2007 EUMETSAT Conference and the 15th Satellite Meteorology and Oceanography Conference of the American Meteorological Society. Amsterdam-Netherlands (2007)

12. Kolios, S., Stylios, C.: Combined use of an instability index and SEVIRI water vapor imagery to detect unstable air masses. In: EUMETSAT Meteorological Satellite Conference, 22–26 September, Geneva, Switzerland (2014)
13. Vila, A.D., Machado, A.L., Laurent, H, Velasco, I.: Forecast and tracking the evolution of cloud clusters (FORTRACC) using satellite infrared imagery: methodology and verification. Weather Forecast. **23**, 233–245 (2008)
14. Puca, S., Biron, D., De Leonimbus, L., Melfi, D., Rosci, P., Zauli, F.: A Neural network algorithm for the nowcasting of severe convective systems. In: CIMSA 2005 – IEEE International Conference on Computing Intelligence for Measurement System Applications. Giardini Naxos, 20–22 July 2005, Italy (2008)
15. Vicente, G.A., Scofield, R.A., Menzel, W.P.: The operational GOES infrared rainfall estimation technique. Bull. Am. Meteorol. Soc. **79**, 1883–1898 (1998)
16. Merino, A., Lopez, L., Sanchez, J.L., Garcia-Ortega, E., Cattani, E., Levizzani, V.: Daytime identification of summer hailstorm cells from MSG data. Natural Hazards Earth Syst. Sci. **14**, 1017–1033 (2014)
17. Kolios, S., Stylios, C., Petunin, A.: A WebGIS platform to monitor environmental conditions in ports and their surroundings in South Eastern Europe. Environ. Monit. Assess. **187**, 574 (2015)

Advanced Systems and Technologies

Self-management of Information Systems

Janis Bicevskis[1]([⊠]), Zane Bicevska[2], and Ivo Oditis[2]

[1] Faculty of Computing, University of Latvia, Riga, Latvia
Janis.Bicevskis@lu.lv
[2] DIVI Grupa Ltd, Riga, Latvia
{Zane.Bicevska,Ivo.Oditis}@di.lv

Abstract. The paper discusses self-management features that are intended to support the usage and maintenance processes in the information system life. Instead of a universal solutions that are evolved by many researchers in the autonomic computing field, this approach, called smart technologies, anticipates self-management features by including autonomic components into information systems directly. The approach is practically applied in several information systems, and the gained results show that the implementation of self-management features requires relatively modest resources. Thereby the approach is suitable even for smaller projects and companies.

Keywords: Autonomic computing · Smart technologies · Business process modeling · Smart technologies chain

1 Introduction

Information technologies provide unprecedented opportunities to automate many processes of human life. Actions which have been the preserve of human beings only a few decades ago can be executed by programmable equipment now. But the mankind's progress has also brought up new challenges. One of them is complexity of computing systems. The authors [1] refer to as "computing systems with complexity approaching boundaries of human ability". The IBM autonomic computing manifesto [2] claims: "It's time to design and build computing systems capable to manage themselves, adjusting to varying circumstances, and preparing their resources to handle most efficiently the workloads we put upon them".

One of the possible solutions of this problem is to entrust at least some of complex IT supervisory processes to the systems themselves. Wikipedia [3] defines self-management in computer science as "the process by which computer systems shall manage their own operation without human intervention". Peter Van Roy [4] defines self-managing systems in the following way: "systems that can maintain useful functionality despite changes in their environment." IBM autonomic computing manifesto defines the self-management by four fundamental self–* features: self–configuration, self–healing, self–optimisation and self–protection. Later [1] the "self–chop" was extended to eight self-management properties. Today the number of identified self-management properties reaches 20 and more [5], and it continues to increase.

© Springer International Publishing Switzerland 2016
G. Arnicans et al. (Eds.): DB&IS 2016, CCIS 615, pp. 167–180, 2016.
DOI: 10.1007/978-3-319-40180-5_12

There are two possible ways trying to implement the self-management in information systems: (1) to construct an "autonomous supervisor" – to develop autonomous information system for supervision of other computer systems, or (2) to implement the properties by adding "independent" components (or add-ins) to the system.

The autonomous supervisor idea is consistent with the nature of the autonomic computing – independent autonomic components solve self–management problems even without knowing about existence of other components (like natural live organisms). Many of these solutions are implemented using "agents" that are able to provide information about specific events to the autonomous supervisors. This is rather a universal solution but unfortunately it is for the time being faced with serious difficulties and has not found wide application.

The implementation of self-management properties by adding "independent" components to information systems is a rather practically oriented approach. A group of specific solutions should be developed, partially built-in into the target systems, to enhance the non-functional features of information systems with self-management properties. Such solutions are called "smart technologies" [6].

This paper introduces four new self–management properties (see Fig. 1):

- self–management properties to support information system operation: (1) run-time verification - control whether internal processes executing is compliance with business measures, (2) environment testing - control of interaction with the external environment;
- self–management properties for maintenance support (3) business model incorporation – built-in business process model lets to change the functionality of the information system by updating the business process descriptions, (4) self-testing – built-in component for internal system process control usable also in a productive mode.

The approach of smart technologies offers to implement self–management properties into the architecture of system. Since each self–management property is simple to implement, the approach becomes rather applicable and it can be used even by rather small team of developers or companies (40–50 employees). This conclusion is based on the authors' subjective experience and research [7]. However, resource assessments by other companies may be radically different from those as the implementation of smart technologies requires deep understanding of software engineering methods and the architecture of the specific software solution.

At the same time it is the weak point of this approach: self–management properties can be developed practically from scratch for every information system. To minimize this impact the authors suggest developing self–management properties as ideas applicable for wide spectra of systems. Thereby this paper discusses smart technologies which bring software development towards the objectives of IBM autonomic computing manifesto.

This paper is a continuation of the research described in [7–9]; therefore some of sentences from these papers are cited to keep completeness of the research. However this paper contains more detailed characteristics of smart technologies, including description of practical context, technical solutions and the results achieved.

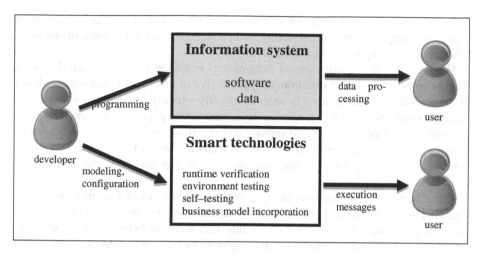

Fig. 1. Smart technologies (overview).

The second section of this paper deals with related research and solutions. The third section describes smart technologies for systems' operation; the fourth section describes smart technologies for systems' maintenance.

2 Related Work

Autonomic computing and smart technologies have a similar goal – to reduce the complexity of system use by delegating some part of user support functions to the information system itself. The autonomic computing manifesto declares a vision of fully independent computer systems that are able to self-management. The manifesto lists four aspects of autonomic computing:

- Self-configuration - automated configuration of components and systems follows high-level policies. Rest of system adjusts automatically and seamlessly.
- Self-optimization - components and systems continually seek opportunities to improve their own performance and efficiency.
- Self-healing - system automatically detects, diagnoses, and repairs localized software and hardware problems.
- Self-protection - system automatically defends against malicious attacks or cascading failures. It uses early warning to anticipate and prevent system wide failures.

In 2003 IBM extended the list to eight characteristic aspects [1], adding system's ability to "know itself" and manage its resources, system's ability to know its environment and the context surrounding its activity, and act accordingly – to adjust, operate in heterogeneous environment accordingly its open standards, as well as anticipate the optimized resources needed while keeping its complexity hidden. The fundamental concept in autonomic computing is the idea of self-regulation and the self-governing operation of the entire information system, thus disburdening users and administrators from complexity of system's use and maintenance. Achievements of autonomic computing movement during its first decade after publication of the manifesto have been explicitly demonstrated in [10].

Later the list of self-management features was extended with new ones, discussed by other authors [5]. The closest from the perspective of this paper is self–diagnosis – a system's ability to analyze itself in order to identify existing problems or to anticipate potential issues. Some of self-properties discussed in this paper are contained by self–diagnosis: testing of execution environment and run-time verification. These properties support system users during its runtime: at first users can verify if the system is running correctly in changing environment and then – whether all of business tasks are accomplished correctly.

Other two of smart technologies are aimed to the support of system maintenance process. The first of them is self–testing: it intends to include testing components into system itself. Practical experience shows that this solution helps to verify software correctness not only during software development likewise it would be done by traditional testing tools but also after the system is taken in production and the maintenance has been begun. However it helps to verify systems correctness in real productive environments.

The last one of smart technologies, business model incorporation, applies to the self–configuring: it is a system's ability to (re)configure itself by (re)setting its internal parameter values to achieve high-level policies or business goals [5]. The existence of this property potentiates implementing the principles of Model Driven Development (MDD). MDD provides business process model integration with computer system, thus allowing updating systems functionality by changing business processes definitions.

As of now, manifesto's targets have been met only to some extent. Paradoxically, to solve the problem — make things simpler for administrators and users of IT — you need to create more complex systems. Continuing efforts on autonomic systems include both, theoretical research and practical implementation [10]. Although many of self–properties are introduced, there is still place for innovative implementations [11]. Many of these are provided as individual compact solutions like smart technologies.

3 Self-management for System Operation

The studies of system life cycle [12] usually focus on the problems of system development. Usually software developer teams are IT professionals and experts therefore they have practically no problems with use of complex technologies. Many of them consider implementing of several non-functional features like self–management properties to be a waste of time and resources.

On the other hand the complex technologies of nowadays lead to complex solutions with awkward usability. Thereby information systems sooner or later are upgraded to improve their usability. The "ordinary" users of information systems become a target audience of self–management properties because they often have difficulties in overcoming of IT complexity. On this account the main focus of the next chapter discussing two smart technology features will be devoted to the support issues for better and easier information system's operation (exploitation) instead of software development phases.

3.1 Runtime Verification

Context. The business process runtime verification is the self-management property which allows verifying whether the business process is executed correctly and in compliance with all of time restrictions. This property is particularly useful when business process is supported by two or more loosely coupled information systems or some of business process steps are not automated at all. There are only few cases when system itself contains component verifying correctness of business process execution. Usually it is assumed that the development time testing ensures correct execution of process in the end user environment. However correctness of business process execution can be verified by checking all of process steps in all linked systems and rather often it cannot be done by users (time restrictions) or an individual system.

Likewise runtime verification is required to prevent conflicts of different processes and systems in collaboration, where one part of the process is done by people, and the other part is supported by software. The software can be designed to support particular processes in different environments at different time frames.

Runtime verification has been well known for years in the area of embedded systems. It is an approach to computing system analysis and execution based on extracting information from a running system and using this information to detect and possibly to react to observed behaviors satisfying or violating certain properties. Such defines mechanisms may be included in the system during its development or they may be included as independent controls from the base process. The independent character of such mechanism allows making later adjustments by adding or disabling the controlling component when a system is developed, and changes are made. These ideas can be applied in business process runtime verification, too.

Solution. The authors [13] propose a solution for business process runtime verification (see Fig. 2) using three objects - verification model/description, agents and controller:

- a verification description contains instructions about the correct execution of the base process;
- an agent is a software module for registering of base process execution events;
- a controller compares the events received from agents with the permissible ("correct") events described in the verification model. As a result, the controller may discover the incorrect behavior of base processes. If inconsistencies are detected, the controller sends messages to the user.

If the base process is already described by a graphical model, the verification process can be created from the base model by indicating those process steps which will be carried out in the runtime verification process.

Results. The developed runtime verification solution was piloted in bank's electronic clearing system (ECS). It identified file processing process delays and some of processing bottlenecks. Some of these delays could be avoided if run–time verification would be introduced sooner.

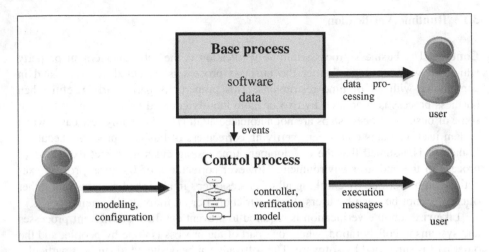

Fig. 2. Smart technology component: runtime verification.

The piloting results leads to 3 main conclusions – (1) solution provides convenient instrument for the tracking of business process execution, (2) the solution is able to detect business process execution defects, and (3) data processing system verification process creates a tiny extra load for the involved information systems infrastructure.

The solution provides a number of interesting possibilities, which bring us closer to the goal defined by ideas of autonomic computing:

- The verification process can be defined without modifying the base process - the base process can have more than one verification process so as to verify all of its various aspects;
- The verification process runs in parallel to a base process and does not interfere with it;
- Process verification can be added dynamically to legacy systems;
- Verification does not depend on modelling language used for process description; it depends only on possibility of verification agents to identity events of the base process.

Likewise, some solution limitations must be taken into account: verification mechanism can detect only those base process steps which leave some modifications in the computer systems "memory".

The implementation of the proposed approach requires programming resources approximately 10000 LOC of C# (implementation of the controller and two agents).

3.2 Environment Testing

Context. The environment testing is the self–management feature for controlling and monitoring of operation (execution) environments to ascertain all involved operation environment fit to the requirements necessary for successful running of the information system. The requirements can relate to operating system, network characteristics, workstation parameters, etc. Discrepancy between the information systems requirements to external environment and the concrete execution environment may occur in several situations:

- Workstation may be incompatible in various means: insufficient memory or processor performance, inadequate network connection and other technical parameters. These are cases when execution environment verification may be done once.
- Workstation may use external resources and availability of these resources may vary during execution time. E.g., some of web services may be unavailable because of lost network connection on server downtime. Obviously in these cases execution environment should be verified continuously.
- Workstation settings do not comply with software requirements: directory structure does not contain all of required subfolders or required permissions, decimal separator must be the symbol ",", data base server must be reachable, etc.
- Developers sometimes assume that software, which works in development environment, will keep working after it is deployed elsewhere, hence encoding some assumptions about the environment into the program. As a result, when the software is installed in other environment, which is different from the development environment, the software may fail or work only partially correct.

Practical use of information systems shows that many incidents and failures are not related to the functionality of the information system itself, but rather are caused by inadequate infrastructure and the execution environment. It means information systems must be accompanied by automatic means for external environmental testing.

The authors do not deal with automated modification of external software environment as it is offered by other researches or tools [14]. It is assumed that one workstation may execute more than one system components simultaneously. Thereby automated external component updates for one system could affect execution of other systems.

Solution. The authors [15] propose a technology, which allows independent environment checks, performed by the software, named – "checker", in order to validate if the execution environment is suitable for normal execution (see Fig. 3). The proposed solution implies gathering these requirements in a "software profile" to be able to validate the execution environment before program's starting. Only if the results of all checks are satisfactory, the program can be considered prepared for work at a given environment, otherwise the session is stopped, giving the user an explanation, why it is not possible to perform work.

A program execution profile is a document achieved when all the requirement descriptions of software are combined together. The profile can be formalized as a separate document and supplemented to typical software deliverables such as code and documentation. The main, but not the only use of the profile is validation of execution environment during program use.

The practical environment testing task is carried out by "checker", which manages environment validation modules- drivers. Each driver is an atomic unit, which enforces

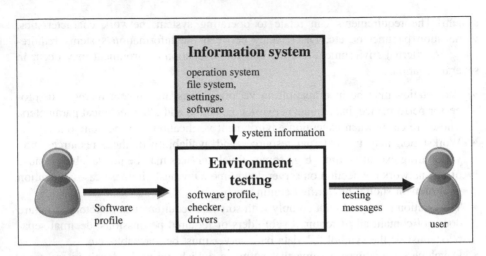

Fig. 3. Smart technology component: environment testing.

validation of a single type of requirement; this is done by reading information from the environment and comparing it to reference values.

To be able to modify the set of checks to be performed without modifying the program code, information about the checks (both the algorithms and reference values) must be stored outside the code of base system – in the software profile. Domain specific language (DSL) was developed for effective software profile definition: it allows describing all external objects and their properties required by the developed software. Software profile is developed by qualified IT professionals as it requires deep knowledge of system execution environment. This concept differs from other approaches used in practice – both from the ones, which validate the environment straightaway after installation or updating, and from the others, which try to "hide" the checks in source code.

Results. The proposed solution since 2009 is used in a number of local information systems in Latvia. An execution environment testing was usually performed when supplying a new version of the information system. The new version was installed only after the current execution environment was checked for its ability to run the new version. Also, receiving alarms from users about the systems malfunctions there was first tested if the execution environment of the concrete workstation meets the environment requirements. In many cases, missing or wrong components of the execution environment were the reason for malfunctions.

The described approach can also be used for other purposes, for instance to monitor the computer systems that are in use in company's internal network and to check the compliance of configurations with standards set by the company. The practical implementation showed that development of the proposed approach requires relatively little programming resources (~ 4000 LOC of C#).

4   Self-management for System Maintenance

This chapter is dedicated to the support, which may be provided to the information systems by self–properties during maintenance of these systems. After the first version of information system is deployed to the operation environment, it is will be updated or modified several times to comply with real user requirements. This leads to regular changes being introduced into the information systems.

In turn, this means that: (1) change requirements must be defined, (2) the software must be updated accordingly, (3) each of software versions must be tested (it should include regression tests and tests for new or updated functionality), (4) software should be deployed to the runtime environment and, if it requires, system data should be migrated. Furthermore usually system execution may be stopped just for rather limited period of time. The authors will discuss two self–properties introduced for support of software maintenance.

4.1   Business Model Incorporation

Context. Business model incorporation is a self-management feature allowing to adapt the functionality of information system without (or with minimal) coding effort, just by changing graphical business process descriptions. One of the implementation options is to apply Model Driven Development (MDD) principles for software development.

Model driven development provides a range of advantages for the system development, maintenance and execution [16, 17]. Same of main advantages are: (1) MDD provides high level of business process abstraction thus providing less error-prone description, meaningful validation and exhaustive testing, (2) MDD bridges the gap between business and IT, (3) MDD captures domain knowledge, (4) MDD results in software being less sensitive to changes in business requirements, (5) MDD provides up-to-date documentation since the models describe the essential issues of the information system's usage. How should the system be developed to gain advantages provided by MDD?

Solution. At the beginning of information system development the business processes (see Fig. 4) should be described as the information system will be designed to support them. A set of business process descriptions are created using DSL, and it serves as business process model. Graphical representations like diagrams can easily be understood and used by domain experts (as a rule, non-IT specialists) for the business process description. After the business process model is created, the information from the diagrams can be transferred to the database of an information system, and it is a task for IT professionals. The business process descriptions are embedded into the information system, and the engine of the information system can interpret information born from the diagrams. Embedded business processes ensure that the information system behaves according to the business process model.

However, the proposed business process incorporation approach differs radically from the model driven architecture (MDA): MDA offers a complete application generation using business process specification described in unified modeling language

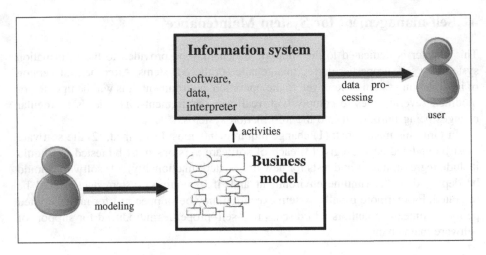

Fig. 4. Smart technology component: business process model incorporation.

(UML). If the business processes are changed, the changes must be implemented in the software specification and then new software should be generated.

The proposed approach of smart technologies provides business process execution engine running according business process definition (domain specific language or DSL is used for process description). It provides the possibility to develop flexible applications with user-friendly interfaces which can be implemented for each of systems independently. Furthermore, business processes can be updated without software modifications, and the functioning of the information system can be updated by changing of business process descriptions, without programming.

Results. The solution described in this chapter leverages use of DSL as language for business process definition providing user – friendly method for process description. Many years there were no sophisticated tools for DSL development available. Even if the DSL could be defined, the supporting tools did not provide user-friendly editor's generation for end users. This problem is solved now: There are tools allowing to define not only the DSL syntax, but also a graphical representation and to generate the corresponding graphical editor. The authors used the platform DIMOD for DSL development [7] which is developed using the tools generation platform GrTP. More detailed GrTP review is given in [18, 19].

As practice shows [20], it is possible to create a special tool for transfer of model's data to executable application relatively quickly. The API of the graphical editor can be used to access the model's repository, to gather the information and to transfer it to applications database. When business model is added to the system database, there is no more need for DSL editor repository in the system's execution environment. In turn, it provides numerous advantages for system performance and usability. Thus guaranties that the application operates according to the model developed in a graphical DSL. And the overall quality of the application – usability, reliability, security, performance etc. – is dependent on the application itself, not on the hypothetical ability of a code generator to create an application in the desired quality.

The smart technology described in the chapter was developed and used in a number of national information systems in Latvian for many years. Therefore, it is problematic to give an accurate assessment of the resources spent for the implementation. However, it should be noted that the information systems (including the appropriate smart technology features) were implemented by teams of 4–6 IT specialists.

4.2   Self-testing

Context. Self-testing is the self-property providing the software with a feature to test itself automatically prior to operation. There is similarity between self-testing and hardware self-checking where computer tests its own readiness for operation when it is just turned on. The purpose of self-testing is to use a built-in support component for automated execution of previously accumulated tests cases not only in test environment, but also in operation (production) environment.

Solution. Self-testing contains two components:

- Test cases of system critical functionality that check the set of functions without which the software could not be used. Identification of critical functionality and designing of tests for it is a part of the requirement analysis and testing process.
- A built-in automated testing mechanism (regression testing) providing automatic execution of tests and comparison of results with benchmark values. These features are typically a part of traditional testing tools.

Implementation mechanism of self-testing approach [21] uses an idea and means of the software instrumentation, which is already known from the 70-ies. Testing operations are put by programmers into certain places of the source code; these points are named as test points. Testing operations allow to track the changing values and to compare them with a benchmark. Thus it is possible to check the correctness of the information system. Unfortunately, this solution is usable only for that software whose development is in the user's influence sphere.

Results. It should be noted that the idea of built-in support for program testing has been offered quite a while ago [22, 23] and it has been implemented in some projects. Regardless the system environment (Development, Test and Production) self-testing functionality can be used in the following system modes (see Fig. 5):

1. Test capture mode - new test cases are captured or existing tests are edited/deleted.
2. Self-testing mode - automated self-testing of software is done by automated execution of stored test cases.
3. Use mode - there are no testing activities – a user simply uses the system. The built-in self-testing mechanism can be used in emergency situations to find out the internal state of the system, which may facilitate the analysis of the causes and consequences of the emergency situation.

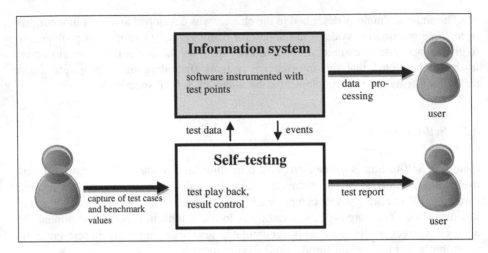

Fig. 5. Smart technology component: self-testing.

4. Demonstration mode. The demonstration mode can be used to demonstrate system's functionality. User can perform system demonstrations using use cases stored in storage files.

The implementation of self-testing feature requires 10000 LOC in C#. Additionally, the source code of the particular system should be instrumented with testing activities like accumulating of test cases and executing of them. These investments are justified when the system is designed and developed for long-term use.

5 Conclusions

There are four smart technology features provided and described in this paper. These extends variety of software self–properties and allow to achieve goals similar to autonomic computing – facilitating the use and maintenance of systems by including support components in them. These technologies provide support for information system execution and maintenance:

- business process run–time verification allows to describe and perceive processes, regardless of systems supporting these processes, to verify correctness of process execution and compliance with time restrictions; this self-property is used during system execution;
- execution environment testing provides verification of system's external environment requirements;
- business model incorporation within system allows to change implemented business processes without (or with minimal) software modifications; this self–property is intended for software maintenance;

- self–testing is intended to use during system maintenance and it allows to verify correctness of systems execution when some software changes or environment updates are done;
- building of smart technologies into information systems requires additional work; the proposed smart technologies have advantages when the information systems are used by many users without profound IT knowledge and the cooperation between the customer and the supplier is long-term;
- according to authors' experience smart technologies can be used even in a small to medium size IT company with 40–50 employees.

The smart technologies which are described in this paper achieve the autonomic computing initiative goals only partially. There may be still a vast variety of smart technologies which would be useful to explore and implement practical systems. For instance, these would include – data quality control, access control, performance monitoring, availability monitoring which are easy enough to implement for a small/medium size organization. Authors are not discussing the research directions where no practical implementations of technologies are achieved yet.

Acknowledgements. This work was supported by the Latvian National research program SOPHIS under grant agreement Nr. 10-4/VPP-4/11.

References

1. Kephart, J., Chess, D.: The vision of autonomic computing. IEEE Comput. Mag. **36**, 41–52 (2003). doi:10.1109/MC.2003.11600552003
2. Horn, P.: Autonomic COMPUTING: IBM's perspective on the state of information technology. IBM (2001). http://libra.msra.cn/Publication/2764258/autonomic-computing-ibm-s-perspective-on-the-state-of-information-technology
3. Wikipedia (2016). https://en.wikipedia.org/wiki/Self-management_(computer_science)
4. Van Roy, P.: Self Management and the Future of Software Design (2007). https://www.info.ucl.ac.be/~pvr/facs06VanRoyFinal.pdf
5. Lalanda, P., McCann, JA., Diaconescu, A.: Autonomic Computing: Principles, Design and Implementation, p. 288 Springer, Heidelberg (2013)
6. Bičevska, Z., Bičevskis, J.: Smart technologies in software life cycle. In: Münch, J., Abrahamsson, P. (eds.) PROFES 2007. LNCS, vol. 4589, pp. 262–272. Springer, Heidelberg (2007)
7. Bicevskis, J., Bicevska, Z.: Business process models and information system usability. Procedia Comput. Sci. **77**(2015), 72–79 (2015)
8. Bicevska, Z., Bicevskis, J., Oditis, I.: Smart technologies for improved software maintenance. Preprints of the Federated Conference on Computer Science and Information Systems, pp. 1549–1554 (2015)
9. Bicevskis, J., Bicevska, Z., Rauhvargers, K., Diebelis, E., Oditis, I., Borzovs, J.: A practitioner's approach to achieve autonomic computing goals baltic. J. Mod. Comput. **3**(4), 273–293 (2015)
10. Kephart, J.: Autonomic computing: the first decade. In: ICAC 2011, pp. 1–2 (2011)

11. Dobson, S., Sterritt, R., Nixon, P., Hinchey, M.: Fulfilling the vision of autonomic computing. IEEE J. **43**, 35–41 (2010). doi:10.1109/MC.2010.14
12. Roger, S.P.: Software Engineering. A Practioner's Approach. The McGraw-Hill Comp., Inc., New York (2010)
13. Oditis, I., Bicevskis, J.: Asynchronous runtime verification of business processes. In: Proceedings of the 7th International Conference on Computational Intelligence, Communication Systems and Networks (CICSyN), Riga, pp. 103–108 (2015)
14. Installshield Installation Information (2016). http://www.sevenforums.com/performance-maintenance/137478-installshield-installation-information.html
15. Rauhvargers, K., Bicevskis, J.: Environment testing enabled software – a step towards execution context awareness. In: Haav, H.-M., Kalja, A. (eds.) Selected Papers from the 8th International Baltic Conference on Databases and Information Systems, vol. 187, pp. 169–179. IOS Press (2009)
16. Haan, J.D.: 15 reasons why you should start using Model Driven Development (2009). http://www.theenterprisearchitect.eu/blog/2009/11/25/15-reasons-why-you-should-start-using-model-driven-development/
17. Haan, J.D.: Opening up the Mendix model specification & tools ecosystem (2015). http://www.theenterprisearchitect.eu/blog/2015/10/30/open-mendix-model-specification-and-tools-ecosystem/
18. Barzdins, J., Zarins, A., Cerans, K., Kalnins, A., Rencis, E., Lace, L., Liepins, R., Sprogis A.: GrTP: transformation based graphical tool building platform (2014). http://sunsite.informatik.rwth-aachen.de/Publications/CEUR-WS/Vol-297/
19. Sprogis, A.: Configuration language for domain specific modeling tools and its implementation. Baltic J. Mod. Comput. **2**(2), 56–74 (2014)
20. Ceriņa-Bērziņa, J., Bičevskis, J., Karnītis, Ģ.: Information systems development based on visual domain specific language BiLingva. In: Szmuc, T., Szpyrka, M., Zendulka, J. (eds.) CEE-SET 2009. LNCS, vol. 7054, pp. 124–135. Springer, Heidelberg (2012)
21. Diebelis, E., Bicevskis, J.: Software self-testing. In: Proceedings of the 10th International Baltic Conference on Databases and Information Systems, Baltic DB&IS 2012, Vilnius, Lithuania, 8–11 July 2012, vol. 249, pp. 249–262. IOS Press (2013)
22. Bichevskii, Y.Y., Borzov, Y.V.: Prioriteti v otladke bolsih programmnih sistem programmirovanie, vol. 3, pp. 31–34 (1982) (in Russian). (Bichevskii, Y.Y., Borzov, Y.V.: Priorities in debugging of large software systems. Program. Comput. Softw. **8**(33), 129–131 (1983))
23. Chengying, M., Yansheng, L., Jinlong, Z.: Regression testing for component-based software via built-in test design. In: Proceedings of the ACM Symposium on Applied Computing, Seoul, Korea, 11–15 March 2007, pp. 1416–1421 (2007)

On the Smart Spaces Approach to Semantic-Driven Design of Service-Oriented Information Systems

Dmitry Korzun[✉]

Department of Computer Science, Petrozavodsk State University (PetrSU),
Lenin Ave. 33, Petrozavodsk 185910, Russia
dkorzun@cs.karelia.ru

Abstract. Smart spaces define a development approach to creation of service-oriented information systems for computing environments of the Internet of Things (IoT). Semantic-driven resource sharing is applied to make fusion of the physical and information worlds. Knowledge from both worlds is selectively encompassed in the smart space to serve for users' needs. In this paper, we consider several principles of the smart spaces approach to semantic-driven design of service-oriented information systems. The developers can apply these principles to achieve such properties as (a) involvement for service construction many surrounding devices and personal mobile devices of the user, (b) use of external Internet services and data sources for enhancing the constructed services, (c) information-driven programming of service construction based on resource sharing. The principles are derived from our experience on the software development for such application domains as collaborative work, e-tourism, and mobile health.

Keywords: Smart spaces · Internet of Things · Edge-Centric computing · Information system · Semantic information broker · Knowledge processor · Ontology · Information-Driven programming

1 Introduction

Smart spaces define a software development approach that enables creating service-oriented information systems for emerging computing environments of the Internet of Things (IoT) [10,15]. Such environments follow the paradigms of ubiquitous and pervasive computing and provide a growing multitude of digital networked devices surrounding their human users. Distributed software system components (from small code pieces to complicated big-data processors) run on various devices (from low-capacity tiny sensors to high-powered supercomputers). This approach supports such emerging vision of human-centered edge-device based computing as edge-centric computing [7].

A smart space is created by interacting agents running on various computing devices. The devices become "smart objects" visible as real participating

© Springer International Publishing Switzerland 2016
G. Arnicans et al. (Eds.): DB&IS 2016, CCIS 615, pp. 181–195, 2016.
DOI: 10.1007/978-3-319-40180-5_13

entities from the physical world. Some agents are associated with web services and other Internet resources, i.e., introducing virtual participants from the information world. Agents interact in the smart space by sharing information. The multi-agent interaction follows similar self-management goals of autonomic computing [3]. The semantic-driven model of information sharing is applied to make fusion of the physical and information worlds [2,13,24,26]. These worlds include many Big Data producers, and such data amounts cannot be straightforwardly duplicated in the smart space. Instead, knowledge about resources from both worlds is selectively encompassed by the agents themselves in the smart space.

As a result, the smart space forms a service-oriented information system to effectively serve for users' needs. Agents, having and making a common view on available resources, construct services and deliver them to the users. The common view is semantic-driven: the smart space contains description of available resources, their semantics, and links to original sources to access the resources. All related processes of the physical and information worlds are virtualized in the smart space by keeping their informational description and semantic relations. Examples and development experience of smart space based information systems can be found in our previous work for such application domains as collaborative work [17,18], e-tourism [25,28], and mobile health [16,22]. In this paper, we continue our concept elaboration study [13,15] on the smart spaces approach to semantic-driven design of service-oriented information systems.

Our research question is to identify some generic principles that can be applied in design of smart space based service-oriented information systems. The identification is based on our experience of smart spaces development. We contribute three principles, where each one is discussed using real application examples. The discussed principles cover such important properties as (a) involvement into the smart space for service construction many surrounding devices of the IoT environment and personal mobile devices of the user, (b) use of external Internet services and data sources for enhancing the services constructed in the smart space, (c) information-driven programming of service constriction using resources shared in the smart space. Although the principles are not strictly formalized to be "universally applied", we expect that they provide systematized understanding of effective options for system design engineering.

The rest of the paper is organized as follows. Section 2 describes the basic properties of smart space based service-oriented information systems deployed in IoT environments. Section 3 briefly introduces the application examples that our study employs for illustration. Section 4 discusses the principle of involvement of surrounding devices into the system. Section 5 discusses the principle of use of external resources for constructing enhanced services. Section 6 discusses the principle of information-driven programming of software agents interacting in the smart space. Finally, Sect. 7 concludes the paper.

2   Smart Space Based Information Systems

We consider IoT environments, which are typically localized by being associated with a physical spatial-restricted place (office, room, home, city square, etc.).

The environment is equipped with variety of devices: sensors, data processors, actuators, consumer electronics, personal mobile devices, multimodal systems, and many other classes of surrounding and embedded devices, including mobile devices of human users (carried, wearable, implantable). Smart spaces follow the vision of ubiquitous computing to support cooperation of all devices in the environment in order to provide its users with convenience, safety, and comfort [2,4]. The support is in the form of a shared information space that describes available resources of the environment.

Information systems that provide digital services of the next generation (so called "smart services") are developed for deploying in such IoT environments. The physical world is digitalized and connected with the information world. This phenomenon refers to the term of smart environment, forming a new challenging subject for software engineering. In particular, various smart technologies has already become of the practical interest to simplify the maintenance and daily use of information systems [3]. Following [13,15,24] let a shared information space (a smart space) be created for a given IoT environment, as Fig. 1 shows. Services are constructed (and delivered to the users) by agents interacting in this environment by sharing and using in the information on available resources.

In this scheme, the IoT environment provides hosting for running the agents and network communication means. IoT objects—real things from the physical world or entities from the information world—can be virtualized by associated agents and keeping digitalized representations (descriptions) as shared information. Therefore, the smart space connects its agents into a distributed system with "central brain support" for multi-agent interaction and knowledge processing on service construction and delivery.

A service is designed in terms of scenarios with knowledge reasoning acts [24,26]. Each scenario defines a control flow initiated from the user side (explicit or implicit detection of user needs) and completed at a point where

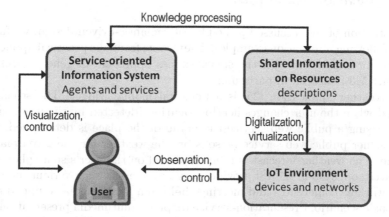

Fig. 1. Smart space based service-oriented information system deployed in IoT environment to create a smart environment providing digital services for users

the user perceives a service (something useful for satisfaction of the needs). The perception can be in form of information delivery (typically, in a visual form) to the user (e.g., recommendation) or the user observes some changes in the physical world (e.g., room lighting becomes lower).

A scenario control flow is event-driven, i.e., assuming the behavior "do something if a certain event occurs". This variant can be extended to the information-driven behavior "do something if certain knowledge becomes available". The reason for the action is appearance of new information in the smart space. An agent can infer the related knowledge from this information clarified with own knowledge the agent has locally (non-shared).

The scope of this paper is limited with the M3 architecture (multi-device, multi-vendor, multi-domain) for smart spaces [2,10,15]. It utilizes the blackboard and publish/subscribe (Pub/Sub) architectural patterns to share information in the environment, rather than have the devices explicitly send messages to one another. The information and its semantics are collected in a smart space using ontological representation models and technologies of the Semantic Web. Operations of an agent with shared information are ontology-oriented, including advanced search and persistent queries. In fact, an M3 smart space forms a knowledge base for interoperable information sharing between agents.

The M3 architecture is implemented in Smart-M3 platform [6,9,21]—open source middleware for development and deployment of smart spaces in various IoT environments. The key architectural component is Semantic Information Broker (SIB) that implements an information hub for agents. Agents are also called knowledge processors (KPs) to distinguish their specifics from the generic term of software agent. Network communication between a KP and SIB follows Smart Space Access Protocol (SSAP) for querying operations from the KP to SIB and returning the result from SIB to the KP. Other dedicated protocols can be used as well to cover the variety of participating devices [6,11].

3   Application Examples

For illustration of the discussed principles of semantic-driven design, we further use the following application examples. They come from our practical application development experience with the smart spaces approach, in general, and with the Smart-M3 platform, in particular.

M3-Weather application [25] is a personal mobile widget. The scenario is activated when the user changes her/his location (detected by the user's smartphone). Using a public web service the name of the place is determined. Similarly, another public web service reasons on the weather for the given place by its name. The weather forecast is then displayed on the user's smartphone.

SmartRoom system [17,18] provides services to assist such variants of collaborative work as conferences and meetings held in a multi-media equipped room. In the basic scenario, Presentation-service displays multimedia presentations and operates with the related content shared the SmartRoom space. Presentation-service is constructed by a single KP running on a local computer connected to the projector. Conference-service dynamically maintains the activity program (i.e., conference section or agenda of talks). The result is visualized by Agenda-service on the second screen in the room. Both public screens can be used to show augmented information, e.g., online discussion of the participants during the talk. Any participating user can also access the services using SmartRoom client running on the personal mobile device (e.g., smartphone or tablet).

Smart assistant for history-oriented tourists [20,28] provides personalized recommendation services for historical e-Tourism. Such a service aims at the automated construction of fragments in the (personal) semantic network around a given point of interests (POI). The network represents the links between the given POI and other objects (other POIs, persons, and events). The semantic network is analyzed and derived knowledge (e.g., the most interesting POIs nearby) are visually presented to the tourist for further analysis.

Personalized mobile Health (m-Health) system [16,19] aims at services for remote health monitoring and situation-aware recommendations. The implemented scenario provides personalized assistance in emergency cases for mobile patients with chronic cardiovascular deceases. An emergency case is detected based on health parameters monitoring or due to an explicit signal from the patient side. The assistance includes notification of all interested parties (physicians and doctors, ambulance staff, attracted people nearby), recommendation for the first aid (instructions), transportation to the hospital (accompany support) as well as preparation starting beforehand the patient's arrival to the hospital.

4   Information Hub

Involvement of many surrounding devices of the IoT environment (as well as many other objects of the physical and information worlds) is one of the essential properties that the smart spaces approach has to take into account in semantic-driven design of service-oriented information systems. Even low-capacity devices act in service construction on the equal basic with more powerful computers. As a result, it opens the information system for data coming from the physical world (embedded and other IoT devices) and from such an overlapped area of the physical and information worlds as human-related and social activity [5] (smartphones and other personal mobile computers, various carried and wearable devices). Many edge IoT devices become responsible for a significant part of system computations, in accordance to the vision of smart objects in IoT [12,27] and of human-centered information systems in edge-centric computing [7].

When developing a smart space based information system, software agents are considered as primary programmable elements, see Fig. 2. They act as knowledge processors that represent participating data producers and consumers: people, equipment, physical objects, and other "original" entities from the physical and information worlds. Some agents produce their share of information and make it available to others in the hub. Similarly, some agents consume information of their interest from the hub. In fact, a hub can be thought as a server that realizes a shared information space for interacting agents. It is an associative

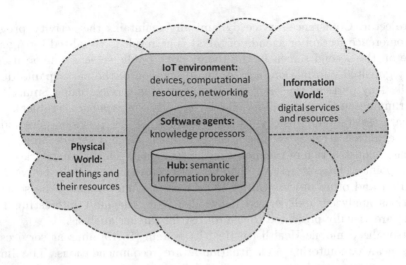

Fig. 2. Information hub keeps semantics of resources accessible in the IoT environment from the physical and information worlds

memory for agents, as it is accepted in space-based computing [23]. Consequently, activity of all agents within their smart space creates fusion of the physical and information worlds. The smart space selectively encompasses related information on ongoing processes and available resources from both worlds.

The key issue is interoperability, which is defined as the ability for agents (written in different programming languages, running on different devices with different operating systems) to communicate and interact with one another (over different networks) in the same smart space. The network-level interoperability is achieved due to IoT technologies, and each agent running on its device has appropriate network communication means to access the hub. To achieve the information-level interoperability, when agents are able to interact understanding the shared information, the methods and models of the Semantic Web are used. Information hub acts as a SIB keeping the content represented with Resource Description Framework (RDF) [8]. The basic data unit is a triple, from which complex semantic structures are formed (RDF graphs). Such RDF-based description enables reasoning over the shared content and inferring new knowledge by means of ontologies. The Web Ontology Language (OWL) is used for creating ontologies. As a result, the smart space becomes a knowledge-based system, where the SIB is an access point for agents to query the shared content.

The above discussion can be summarized in the following principle.

Principle 1 (Information hub). *For a given IoT environment a knowledge base is created in the smart space with ontological representation of involved participants, ongoing process, and available resources.*

The principle leads to the following options for designing a smart space based service-oriented information system.

1. Many surrounding devices of the IoT environment as well as personal mobile devices of the users are involved to participate.
2. Semantic interoperability is achieved due to the ontological representation understandable by the participants.
3. The knowledge base is created as ad-hoc and then maintained cooperatively by the participants themselves.
4. The created knowledge base is localized and customized for a given IoT environment and application needs.

SmartRoom system [17,18] provides an application example to illustrate Principle 1. Many personal mobile devices (smartphones, laptops) and multimedia equipment (public screens, projectors, audio-system) are involved, as schematically shown in Fig. 3. Human participants of the collaborative activity are represented in the smart space as instances of ontological classes (i.e., OWL individuals) with attributes describing each participant and her/his state (name, presented in the room, current speaker, etc.). This description is provided primarily by client KPs running on personal mobile devices of the participants. Similarly, all multi-media devices and objects are represented in the smart space with description of their current state. For instance, the smart space represents all presentations of the participants and which presentation is currently shown on the public screen. Interested KPs can find a link to access the PDF presentation file or even the PDF file of the currently shown slide. Notably that the smart space keeps a link, not the file itself; using a link the file can be downloaded directly from Content-service implemented as a standard web server.

Therefore, each device (or its KP) can observe the current system state by analyzing the content shared in the smart space. When a new device is involved then its representation is published in the smart space. Appropriate semantic relations between representations of the involved devices can be published as well. For instance, the media projector in the room is associated with a current speaker who controls the slide show. When the state of an represented object

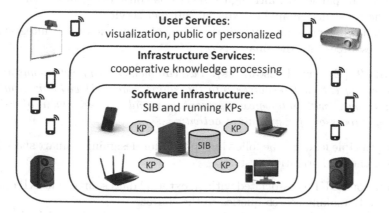

Fig. 3. Personal mobile devices and multi-media equipment in SmartRoom

is changed then the appropriate KP (which detects the change) updates the representation in the smart space. For instance, when human participant leaves the room the fact is observed by a human presence detection system, which can change the status to "not presented in the room".

A similar application example is provided by personalized m-Health system [16,19]. Each patient is equipped with wearable and implantable devices, such as ECG or glucose sensors, insulin pumps, and accelerometers. The smart space represents the latest sensed data as well as derived information that provides semantics. For instance, the fact of glucose level growing can be detected and shared in the smart space. Another illustration is when measurements from two or more medical sensors (even for different health parameters) are becoming correlated, i.e., a new semantic relation is established.

In summary, the discussed principle of information hub supports virtualization of all related processes and resources in the smart space. In addition to the straightforward virtualization, the semantics are also shared to describe relations observed by involved participants in ongoing processes and available resources. The shared content forms a semantic network of represented objects and their relations. The content becomes a dynamic evolving system with properties similar to peer-to-peer systems [14]. In particular, a service construction process is reflected in the smart space (and can observed by interested participants) as some routes in the semantic network.

5   External Resources

In many smart spaces, local resources of the IoT environment are not enough. The rich spectrum of various external resources in the today's Internet can be used for constructing enhanced services. First, this requirement aims at involvement of many data producers (data sources), primarily from the information world. Second, to satisfy the requirement the system design should support construction of composed services, i.e., external resources can also be data processing entities. In particular, although a service is constructed in the smart space locally the construction applies external Internet services.

The following principle enables this kind of semantic-driven integration of external resources into the smart space.

Principle 2 (External resources). *Two complementary mechanisms are used for integrating external resources into the smart space: (i) virtualization of an external resource with ontological representation and (ii) assignment of dedicated knowledge processors for mediation activity.*

The principle leads to the following options for designing a smart space based service-oriented information system.

1. A data source KP is associated with an external data source. The KP operates iteratively making search queries to the source.
2. The ontological representation of a data source and related entities is the way to control the operation of the system with the source.

3. Instead of data duplication the semantics of an external resource are kept to allow a participant to consume target data.
4. Additional KPs can be introduced for improving local processing of external data. For instance, caching voluminous data (e.g., audio or video) or dynamic visualization (e.g., local web pages construction).

First, for Principle 2 let us consider the case of integrating multiple external data sources into a smart space. For illustration we refer to the smart assistant for history-oriented tourists [20, 28]. The today's Internet contains a large corpus of historical data distributed over a multitude of various sources. These data are very fragmented, heterogeneous, and even subjective. Some sources are semantic-oriented, e.g., DBpedia, Freebase, YAGO. Many historical data sources are still in the pure form of web pages with no strict format, i.e., the information cannot be easily extracted by a machine.

For a historian tourist, it is important to selectively encompass certain fragments of information from several data sources and integrate them into a semantic network. Nodes of this network represent historical objects (e.g., person, point of interest). Links represent historical facts (e.g., a person lived in a region). Analysis of this network allows the human to determine the most essential historical objects for further study.

The smart space for such a tourist automates the process of semantic network construction. For each data source a KP is assigned, which can run on a remote host. This KP observes in the smart space which information the tourist needs at the moment and makes queries to the data source. The received data are added to the smart space in accordance to the ontology of historical objects. Since the user context is dynamically changing the data source KPs operate iteratively: context update starts new search queries to the data sources. Notably that this way the smart space creates a kind of "personal DBpedia" for a given tourist or a group of them (when they use the same smart space).

Second, consider the case of composed services for Principle 2. Discussion-service is one of SmartRoom services that allows the participants to discuss by publishing commentaries on conference talks, either already passed ones, currently ongoing, or going to happen. The service can be implemented in the form of integration with an existing Internet service. The basic scenario scheme is depicted in Fig. 4.

The Discussion-service KP fetches from the SmartRoom space the total number of talks, along with their titles/topics and names of presenters. Based on this information, the KP makes queries to the external Internet service—blog comment hosting service Disqus (https://disqus.com/). On the one hand, the discussion is actually formed with the external service, hosting the data on the remote resource. On the other hand, the discussion is virtualized in the smart space due to the ontological representation, based on which a participant can access the discussion using a web mechanism, as described below.

The external service replies with links to widgets, which enable the discussion feature. Then a web-page is created for each talk, embedding the discussion widget into the page. This way, interested SmartRoom participants join a discussion

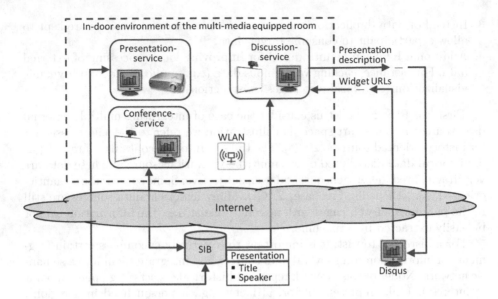

Fig. 4. Discussion activity in SmartRoom is integrated to an external blog service

thread of the talk by browsing its web page (either from SmartRoom clients or by web browsers from any computer with Internet access). The web pages are regularly updated based on the latest activity of participants. For easy navigation, a summary web page is created locally to visually list all available discussion threads. Moreover, this web page provides auto-navigate function, i.e., when the user accesses the service during certain talk, she/he is redirected to the web page with the discussion widget of that talk.

The principle of external resources can be considered as a form of data caching. For instance, in the above example of smart assistant for history-oriented tourists the found historical data (POIs, facts, etc.) from external sources are duplicated in the smart space. This data integration, however, is semantically enriched and personalized, i.e., the smart space keeps not a straight-forwardly collected sum of data from different sources. In the SmartRoom system, caching of media files is used, such as presentation and video files: a dedicated KP implements Content-service as a web server that maintains and shares links to stored files in the smart space. As a result, a participant can access a stored file using the widespread web mechanism (HTTP requests). In the personalized m-Health system, sensed medical data (lengthy time series) from each patient are stored in the specialized healthcare information system. In contrast, the smart space keeps a short window of latest sensed data as well as conclusions derived from the monitoring process.

Another useful option for composing external services is a dedicated KP that implements the service of constructing dynamic web pages. Each page consists of several information fragments. The service makes the pages available by links stored in the smart space. For instance, this option was used in the above example of Discussion-service in the SmartRoom system.

In summary, the principle of external resources opens a smart space for constructing enhanced services. The today's Internet has enough services to solve many everyday problems. However, the puzzle of their combination when solving a given problem is still performed by the users manually because of the high fragmentation of exiting Internet services. Based on the discussed principle, a service-oriented information system provides means for solving such puzzles within the smart space in an automated manner.

6   Information-Driven Programming

Service construction in a smart space can be formulated in terms of flows of information changes [13]. It follows the vision of event-driven and information-driven programming. The events to react are ontologically represented in the smart space. This event-based interaction can be enhanced to information-driven interaction. The reaction is not on a simple event (some values are updated) but on forming a certain informational or knowledge fact, e.g., interaction models of emergent semantics [1] and semantic connections [29].

The following principle enables this kind of programming for implementing service construction as interaction of several cooperative KPs.

Principle 3 (Information-driven programming). *In the service construction, a participating KP implements two basic steps: (i) detection of the specified knowledge formation in the smart space and (ii) reaction for producing new knowledge to share in the smart space.*

The principle leads to the following options for designing a smart space based service-oriented information system.

1. Search query is the basic mechanism for specifying the knowledge and detecting its formation in the smart space.
2. Some variants of detecting the specified knowledge formation can be implemented by subscription operation, including SPARQL-based subscription.
3. One or more reasoning KP can be associated for detecting the specified knowledge formation and reflecting the informational fact in the smart space.

For illustration of Principle 3 let us consider a simple application example of M3-Weather application [25]. There are KP-GPScoords, KP-City, and KP-Weather. Each maintains representation of its own object, respectively: coordinates x, city name s, and weather forecast w. The client KP is denoted KP-GUI and implements a desktop widget on a personal mobile computer of the user. The service is constructed as a distributed pipe-like interaction when objects are affected sequentially $c \rightarrow s \rightarrow w$. Any change of the involved objects defines a simple event to react by an appropriate KP.

The service construction scenario goes as follows. KP-GUI subscribes to weather w in the smart space and visualizes the new information to the user whenever w is updated. KP-GPScoords regularly requests coordinates from device's GPS module. If they show significant change in the user location then KP-GPScoords updates coordinates c in the smart space. KP-City subscribes for coordinates c and for each new c the KP-City queries an appropriate web service (e.g., GeoNames.org) to map the coordinates to the nearest city name. Then KP-City updates the result s in the smart space. KP-Weather, which subscribes for city name s, reacts on new value of s and queries an appropriate web service (e.g., Weather.com) to map the city name to weather forecast w. Then KP-Weather publishes the result w in the smart space, which is finally visualized by KP-GUI to the user.

An example illustrating multiple reactions in service construction scenario comes from our design of SmartRoom system. Whenever the current speaker changes the slide this event is detected by Presentation-service KP as well as by client KPs running on personal mobile computers of the users (if they watch the slide show). This way, the smart space ontologically represents "current slide number" (as well as all other related information), which is subject to the changes during the presentation.

Another useful example of information-driven service construction can be found in the design of personalized m-Health system. On the one hand, the smart space represents the latest sensed medical data of the patient. On the other hand, external services of the healthcare information system can be accessed to analyze personal medical records (e.g., they keep time series data of medical parameters monitoring). Based on an integrated view on these two parts of information, such facts can be detected in the smart space as dangerous increasing of physical activity of the patient or the need to take preventive medicine to reduce the risk of possible complications. This type of integrated analysis can be implemented by a reasoning KP.

An interesting example is provided by the design of smart assistant for history-oriented tourists. In this application, a personalized semantic network of POIs with their relations to historical objects and facts is dynamically formed in the smart space. An important service for a tourist is recommendation of the most interesting POIs for the user in accordance with the recent context. One way is a search query to select some POIs from the semantic network. Although this way is popular in many touristic applications (e.g., selecting all nearby POIs) capturing and applying historical relations for selecting POIs needs more complicated way for recommendation construction. In particular, structural analysis of the whole semantic network can be applied for ranking POIs in accordance with historical relations that the semantic network stores. This analysis can be implemented by a reasoning KP, and the user receives a ranked list of POIs, similarly to a web search engine providing a sorted list of web pages.

In summary, the discussed principle of information-driven programming provides a way to make semantic-driven design of needed interactions to cooperatively construct a service in the smart space. From programming point of view, for each participating KP its input and output interfaces with the smart space should be defined: the output interface design describes the events that the KP initiates, the input interface design describes the reaction that the KP is responsible. The principle supports moving the system design beyond the traditional case when one programmable component (a KP in our smart space terminology) is assigned for constructing one predefined service.

7   Conclusion

This paper addressed the problem of semantic-driven design of smart space based service-oriented information systems for IoT environments. The study is based on our software development experience in such emerging application domains as collaborative work, e-tourism, and m-Health. We discussed the following principles of the smart spaces approach to semantic-driven design: (a) involvement into the smart space for service construction many surrounding devices of the IoT environment and personal mobile devices of the user, (b) use of external Internet services and data sources for enhancing the services constructed in the smart space, (c) information-driven programming with common resources shared in the smart space. Although the discussed principles are not shaped in a formalized definition to be universally applied, the discussion made a step towards a design methodology for smart spaces. We expect that the considered concept definition of principles complemented with several application examples provide systematized understanding of effective options for system design engineering.

Acknowledgments. This research is financially supported by the Ministry of Education and Science of the Russian Federation within project # 1481 of the basic part of state research assignment for 2014–2016. The reported study was partially funded by RFBR according to research project # 14-07-00252.

References

1. Aiello, C., Catarci, T., Ceravolo, P., Damiani, E., Scannapieco, M., Viviani, M.: Emergent semantics in distributed knowledge management. In: Nayak, R., Ichalkaranje, N., Jain, L. (eds.) Evolution of the Web in Artificial Intelligence Environments. SCI, vol. 130, pp. 201–220. Springer, Heidelberg (2008)
2. Balandin, S., Waris, H.: Key properties in the development of smart spaces. In: Stephanidis, C. (ed.) UAHCI 2009, Part II. LNCS, vol. 5615, pp. 3–12. Springer, Heidelberg (2009)
3. Bicevskis, J., Bicevska, Z., Rauhvargers, K., Diebelis, E., Oditis, I., Borzovs, J.: A practitioner's approach to achieve autonomic computing goals. Baltic J. Mod. Comput. 3(4), 273–293 (2015)
4. Cook, D.J., Das, S.K.: How smart are our environments? An updated look at the state of the art. Pervasive Mob. Comput. 3(2), 53–73 (2007)
5. Evers, C., Kniewel, R., Geihs, K., Schmidt, L.: The user in the loop: enabling user participation for self-adaptive applications. Future Gener. Comput. Syst. 34, 110–123 (2014)
6. Galov, I., Lomov, A., Korzun, D.: Design of semantic information broker for localized computing environments in the Internet of Things. In: Proceedings of 17th Conference of Open Innovations Association FRUCT, pp. 36–43. ITMO Univeristy, IEEE, April 2015

7. Garcia Lopez, P., Montresor, A., Epema, D., Datta, A., Higashino, T., Iamnitchi, A., Barcellos, M., Felber, P., Riviere, E.: Edge-centric computing: vision and challenges. SIGCOMM Comput. Commun. Rev. **45**(5), 37–42 (2015)
8. Gutierrez, C., Hurtado, C.A., Mendelzon, A.O., Pérez, J.: Foundations of semantic web databases. J. Comput. Syst. Sci. **77**(3), 520–541 (2011)
9. Honkola, J., Laine, H., Brown, R., Tyrkkö, O.: Smart-M3 information sharing platform. In: Proceedings of IEEE Symposium Computers and Communications (ISCC2010), pp. 1041–1046. IEEE Computer Society, June 2010
10. Kiljander, J., D'Elia, A., Morandi, F., Hyttinen, P., Takalo-Mattila, J., Ylisaukko-oja, A., Soininen, J.P., Cinotti, T.S.: Semantic interoperability architecture for pervasive computing and Internet of Things. IEEE Access **2**, 856–874 (2014)
11. Kiljander, J., Morandi, F., Soininen, J.P.: Knowledge sharing protocol for smart spaces. Int. J. Adv. Comput. Sci. Appl. (IJACSA) **3**, 100–110 (2012)
12. Kortuem, G., Kawsar, F., Sundramoorthy, V., Fitton, D.: Smart objects as building blocks for the Internet of Things. IEEE Internet Comput. **14**(1), 44–51 (2010)
13. Korzun, D.: Service formalism and architectural abstractions for smart space applications. In: Proceedings of 10th Central and Eastern European Software Engineering Conference in Russia (CEE-SECR 2014). ACM, October 2014
14. Korzun, D., Balandin, S.: A peer-to-peer model for virtualization and knowledge sharing in smart spaces. In: Proceedings of 8th International Conference on Mobile Ubiquitous Computing, Systems, Services and Technologies (UBICOMM 2014), pp. 87–92. IARIA XPS Press, August 2014
15. Korzun, D.G., Balandin, S.I., Gurtov, A.V.: Deployment of smart spaces in Internet of Things: overview of the design challenges. In: Balandin, S., Andreev, S., Koucheryavy, Y. (eds.) NEW2AN 2013 and ruSMART 2013. LNCS, vol. 8121, pp. 48–59. Springer, Heidelberg (2013)
16. Korzun, D., Borodin, A., Timofeev, I., Paramonov, I., Balandin, S.: Digital assistance services for emergency situations in personalized mobile healthcare: smart space based approach. In: Proceedings of 2015 International Conference on Biomedical Engineering and Computational Technologies (SIBIRCON/SibMedInfo), pp. 62–67. IEEE, October 2015
17. Korzun, D., Galov, I., Kashevnik, A., Balandin, S.: Virtual shared workspace for smart spaces and M3-based case study. In: Balandin, S., Trifonova, U. (eds.) Proceedings of 15th Conference of Open Innovations Association FRUCT, pp. 60–68. ITMO Univeristy, April 2014
18. Korzun, D.G., Kashevnik, A.M., Balandin, S.I., Smirnov, A.V.: The Smart-M3 platform: experience of smart space application development for Internet of Things. In: Balandin, S., Andreev, S., Koucheryavy, Y. (eds.) NEW2AN/ruSMART 2015. LNCS, vol. 9247, pp. 56–67. Springer, Heidelberg (2015)
19. Korzun, D.G., Nikolaevskiy, I., Gurtov, A.: Service intelligence support for medical sensor networks in personalized mobile health systems. In: Balandin, S., Andreev, S., Koucheryavy, Y. (eds.) NEW2AN/ruSMART 2015. LNCS, vol. 9247, pp. 116–127. Springer, Heidelberg (2015)
20. Kulakov, K., Petrina, O.: Ontological model for storage historical and trip planning information in smart space. In: Proceedings of 17th Conference of Open Innovations Framework Program FRUCT, pp. 96–103. ITMO Univeristy, IEEE, April 2015
21. Morandi, F., Roffia, L., D'Elia, A., Vergari, F., Cinotti, T.S.: RedSib: a Smart-M3 semantic information broker implementation. In: Balandin, S., Ovchinnikov, A. (eds.) Proceedings of 12th Conference of Open Innovations Association FRUCT and Seminar on e-Tourism, pp. 86–98. SUAI, November 2012

22. Nikolaevskiy, I., Korzun, D., Gurtov, A.: Security for medical sensor networks in mobile health systems. In: Proceedings of IEEE International Symposium on a World of Wireless, Mobile and Multimedia Networks (WoWMoM 2014). The 3rd Workshop on the IoT: Smart Objects and Services (IoT-SoS), June 2014

23. Nixon, L.J.B., Simperl, E., Krummenacher, R., Martin-recuerda, F.: Tuplespace-based computing for the semantic web: a survey of the State-of-the-art. Knowl. Eng. Rev. **23**, 181–212 (2008)

24. Ovaska, E., Cinotti, T.S., Toninelli, A.: The design principles and practices of interoperable smart spaces. In: Liu, X., Li, Y. (eds.) Advanced Design Approaches to Emerging Software Systems: Principles, Methodology and Tools, pp. 18–47. IGI Global, Pennsylvania (2012)

25. Samoryadova, A., Galov, I., Borovinskiy, P., Kulakov, K., Korzun, D.: M3-weather: a Smart-M3 world-weather application for mobile users. In: Balandin, S., Ovchinnikov, A. (eds.) Proceedings of 8th Conference of Open Innovations Framework Program FRUCT, pp. 160–166. SUAI, November 2010

26. Smirnov, A., Kashevnik, A., Shilov, N., Oliver, I., Balandin, S., Boldyrev, S.: Anonymous agent coordination in smart spaces: State-of-the-art. In: Balandin, S., Moltchanov, D., Koucheryavy, Y. (eds.) ruSMART 2009. LNCS, vol. 5764, pp. 42–51. Springer, Heidelberg (2009)

27. Tervonen, J., Mikhaylov, K., Pieska, S., Jamsa, J., Heikkila, M.: Cognitive Internet-of-Things solutions enabled by wireless sensor and actuator networks. In: 5th IEEE Conference on Cognitive Infocommunications (CogInfoCom), pp. 97–102. IEEE, November 2014

28. Varfolomeyev, A.G., Ivanovs, A., Korzun, D.G., Petrina, O.B.: Smart spaces approach to development of recommendation services for historical e-tourism. In: Proceedings of 9th International Conference on Mobile Ubiquitous Computing, Systems, Services and Technologies (UBICOMM), pp. 56–61. IARIA XPS Press, July 2015

29. Vlist, B., Niezen, G., Rapp, S., Hu, J., Feijs, L.: Configuring and controlling ubiquitous computing infrastructure with semantic connections: a tangible and an AR approach. Pers. Ubiquit. Comput. **17**(4), 783–799 (2013)

Host Side Caching: Solutions and Opportunities

Shahram Ghandeharizadeh[✉], Jai Menon, Gary Kotzur,
Sujoy Sen, and Gaurav Chawla

Dell Inc., Round Rock, USA
shahram@usc.edu

Abstract. Host side caches use a form of storage faster than disk and less expensive than DRAM to deliver the speed demanded by data intensive applications, e.g., NAND Flash. A host side cache may integrate into an existing application seamlessly using an infrastructure component (such as a storage stack middleware or the operating system) to intercept the application read and write requests for disk pages, populate the flash cache with disk pages, and use the flash to service read and write requests intelligently. This study provides an overview of host side caches, an analysis of its overhead and costs to justify its use, alternative architectures including the use of the emerging Non Volatile Memory (NVM) for the host-side cache, and future research directions. Results from Dell's Fluid Cache demonstrate it enhances the performance of a social networking workload from a factor of 3.6 to 18.

Keywords: Storage · Caching · Non-volatile memory

1 Introduction

Enterprises deploy *host side caches* to enhance the performance of their data intensive applications. These caches utilize a form of storage faster than disk and cheaper than DRAM to stage disk pages referenced by an application, enhancing performance. Example storage used in host-side cache include today's NAND Flash and the emerging STT-RAM [33], Memristor [52] and PCM [6, 15, 31, 46, 55] as future candidates. See Table 1 for details. An example application may be the server component of a database management system (DBMS) such as Oracle and MongoDB that processes queries and updates. A host side cache might be either managed by the application [26, 42] or deployed seamlessly using a storage stack middleware or the operating system (termed the caching software) [8, 12, 28, 30, 34, 38, 48–50].

Figure 1.a shows a traditional architecture of an application consisting of a two level hierarchy, DRAM and disk. With this architecture, the application's read and write operations for disk pages might be serviced by the DRAM (a cache hit) or the disk (a cache miss). The application may manage the writing

S. Ghandeharizadeh—This author was at Dell on a sabbatical leave from USC.

© Springer International Publishing Switzerland 2016
G. Arnicans et al. (Eds.): DB&IS 2016, CCIS 615, pp. 196–210, 2016.
DOI: 10.1007/978-3-319-40180-5_14

of a dirty block from DRAM to disk. For example, a DBMS server may write log records synchronously to implement the ACID property of transactions.

With Fig. 1.b, flash memory is the host side cache and used as an intermediate staging area between DRAM and disk. It is faster than the disk and transparent to the application because the caching software intercepts the disk page read and write requests to service them using the flash. (The caching software may cache at the granularity of a file, however, we assume block based caches in this paper.) The DRAM continues to service the hottest data items. Those data items with a sparser reference pattern reside on the host side cache (flash). Since the host-side cache is realized using technology that is cheaper than DRAM, its size can be much larger than DRAM size. The caching software is aware of both the flash (host side cache) and disk (permanent storage) and may implement different policies to perform the application writes [32]. For example, with a *write-through* policy, the caching software executes writes by writing to both the host side cache and the permanent storage before confirming the write operation to the application. However, with a *write-back* policy, the caching software performs the same write by writing only to the host side cache and confirms the write immediately. Either a background process or a cache eviction policy performs the write to the permanent storage.

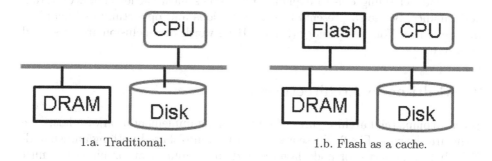

1.a. Traditional.   1.b. Flash as a cache.

Fig. 1. A traditional architecture and its extension with Flash as a host side cache.

A host side cache is different than both a Network Attached Storage (NAS) cache and a Key-Value Store (KVS) cache. We describe each in turn and detail how they differ from a host side cache. A NAS cache is an appliance on the network that acts as an intermediary between a NAS and its applications [14, 53,54]. This appliance is configured with NAND Flash and stages data from a disk based NAS intelligently to expedite application performance. Host side caches reside on the server that processes the application. They provide block level access to implement durable writes required to implement a transaction processing system.

KVS caches manage key-value pairs where a *key* is either a query or a function instance and a *value* is the result of the query or function instance [7,22,27,37,41, 44]. A popular in-memory KVS is memcached in use by social networking sites

Table 1. Alternative data storage technologies [15, 40].

	Memristor	FeRAM	PCM	STT-RAM	DRAM	NAND Flash	Disk
Read time (ns)	<10	20–40	20–70	10–30	10–50	25,000	$2\text{-}8\text{x}10^{6}$
Write time (ns)	20–30	10–65	50–500	13–95	10–50	200,000	$4\text{-}8\text{x}10^{6}$
Retention	>10 years	~10 years	<10 years	Weeks	<100 msec	~10 years	~10 years
Energy/bit $(pJ)^2$	0.1–3	0.01–1	2–100	0.1–1	2–4	$10\text{-}10^{4}$	$10^{6}\text{-}10^{7}$
3D capability	Yes	Yes	Yes	No	No	Yes	N/A

such as Facebook [41]. Similar to a host side cache, a KVS cache may be managed by either the application [7, 41] or deployed transparently [21–23, 27, 37, 44]. The key-value pairs managed by a KVS cache may have different sizes with varying costs [19]. This is in sharp contrast to the fixed size disk pages managed by a host side cache. In order to simplify discussion and without loss of generality, the rest of this paper focuses on host side caches only.

This paper identifies the alternative storage architectures to realize a host side cache, and discusses future research and development directions. The rest of this paper is organized as follows. Section 2 presents several architectures for host side caches. In Sect. 3, we present a case study of the Dell Fluid Cache solution. Section 4 presents analytical models that quantify the cost and benefits of host side caches, enabling one to explain why today's flash caches enhance system performance. Section 5 presents future extensions and opportunities, including the use of Non-Volatile Memory (NVM). Brief words of conclusion are presented in Sect. 6.

2   Multi-node Architectures

Several commercial architectures of a multi-node implementation of a host-side cache are shown in Fig. 2. These are categorized into shared-nothing and shared-disk. In the presence of node failures and disk failures, an architecture must address (1) availability of data on the permanent store and (2) durability of writes with the write back-policy using the host-side cache. Consider each in turn.

With data on the permanent store, a shared-disk architecture may employ multiple host-bus adapters and switches in combination with RAID [43] to enhance availability of data. To realize the same, a shared-nothing architecture may use a variant of techniques used by file systems such as the Google File System [25], database management systems such as Gamma [13, 17, 29], and peer-to-peer data stores such as CAN [47] and Chord [51]. These techniques construct a partition of disk pages, assigning its primary and secondary copies to different nodes. In the presence of failures that render a primary partition unavailable, one of its secondary copies is promoted to be the primary.

To ensure durability of writes with the write-back policy, architectures of Figs. 2, b may pair the host-side caches of multiple nodes with one another [8, 12, 30, 34], designating the content of one as *primary* and the rest as *secondary*.

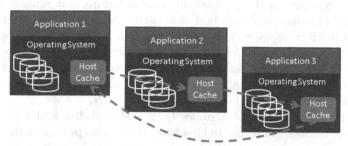

2.a. A shared-nothing architecture.

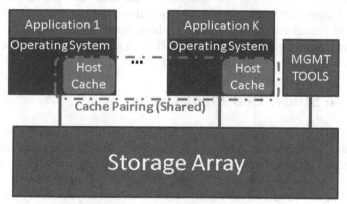

2.b.Shared disk with distributed shared flash.

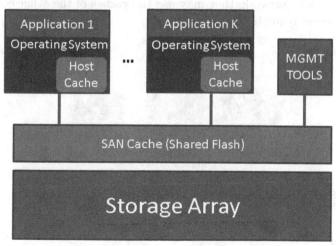

2.c. Shared disk and shared flash.

Fig. 2. Alternative architectures.

A write must be performed to both its primary and secondary host-side cached copies synchronously. In the presence of a node failure that renders the dirty primary copy of a disk page on a host-side cache unavailable, the system writes one of its secondary copies to the SAN [12]. With $N-1$ secondary copies for a disk page and a maximum of δ time units for a dirty disk page to be written to the SAN (in the presence of failures), all N replicas must fail during δ in order for a write to become non-durable. One may minimizie this possibility by either reducing the duration of δ or increasing the value of N.

The architecture of Fig. 2.c assumes both a shared disk and a shared flash (termed SAN Cache). With this architecture, there is no pairing of the host-side caches. Instead, every write to the host-side cache is also written to the SAN Cache synchronously. NetApp presents this architecture in the context of a data center deployment with multiple virtual machines (hosting one or more DBMSs) accessing a hypervisor host that employs a host cache seamlessly [8,34]. The use of SAN Cache is motivated by the observation that a write to the flash requires the same amount of time as a network hop [8].

3   A Case Study: Dell Fluid Cache

Dell Fluid Cache is a leading host-side caching solution for enterprise systems. It implements the architecture of Fig. 2.b using a separate, low-latency private cache network dedicated for processing the asynchronous writes to a pairing of the host-side caches. To tolerate switch failures, the architecture may be extended with a second switch. Fluid Cache does not require every server using the SAN to be configured with a flash cache. As long as these servers are connected to the cache network, they may use the caches of the other contributing servers to process requests [16].

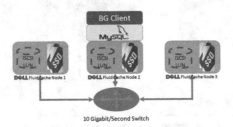

Fig. 3. BG benchmark evaluating MySQL with Fluid Cache.

One may deploy Fluid Cache across multiple nodes in a variety of configurations. Figure 3 shows one deployment evaluated using a social networking benchmark named BG [2]. This deployment consists of three nodes, each with a disk (iSCSI LUN) and one or more SSDs. It shows MySQL version 5.5 server is deployed on one node. The BG Client that generates requests for this server is deployed on the same node.

Fluid Cache can be configured with either the write-back or the write-through policy [32]. In the write-back mode, writes from the application (i.e., MySQL) are written to the SSDs of two different nodes to enhance the durability of data in the presence of node failures. In the write-through mode, writes from the application are written to both the cache and the permanent store (a 145 GB iSCSI LUN in our tests).

When a node is configured with multiple SSD cards, Fluid Cache partitions the disk pages across the available SSDs, harnessing their aggregate bandwidth to service different requests. In our experiments, a node is configured with 2 SSDs each with 16 GB of memory.

With multiple nodes, Fluid Cache may distribute the disk pages across the SSDs of different nodes (termed Proportional) or assign the disk pages to the SSDs local to the node that references these disk pages (termed Client Affinity). In this study, we configure Fluid Cache with the Client Affinity policy.

Table 2. Two BG workloads considered in this study.

BG social actions	Type	Read-Only	Mix of 90 % Read and 10 % Write
View Profile (VP)	Read	100 %	50 %
List Friends (LF)	Read	0 %	20 %
View Friends Requests (VFR)	Read	0 %	20 %
Invite Friend (IF)	Write	0 %	4 %
Accept Friend Request (AFR)	Write	0 %	2 %
Reject Friend Request (RFR)	Write	0 %	2 %
Thaw Friendship (TF)	Write	0 %	2 %

The BG Client creates a social graph consisting of a fixed number of members. In our experiments, the social graph consists of 10 million members and is approximately 50 GB in size[1]. A thread emulates a member of the social graph issuing one of the interactive social networking action shown in Table 2. Each emulated member is termed a *socialite*. A socialite issues an action shown in Table 2 with a fixed probability. These actions are categorized into either read or write actions. An example read action is for a socialite to view another member's profile. An example write action is for a socialite to invite another member to be friends. Table 2 shows two different workloads: (1) Read-only consisting of the View Profile action, and (2) A mixed workload with 90 % reads and 10 % writes. See [2] for the detail specification of each action.

BG quantifies both the service time and the overall processing capability of a solution for processing a workload. The service time is measured with 1 socialite

---

[1] With Fluid Cache configured using the Client Affinity policy, the total size of host-side cache is 32 GB.

(thread) issuing requests one at a time. It is quantified as the elapsed time from when BG issues an action (that may result in one or two SQL commands to MySQL) to the time the action is processed. We use BG with 16 concurrent socialites (threads) to quantify the number of actions performed by the Fluid Cache deployment. This is the throughput of the system.

A novel feature of BG is its ability to quantify the amount of stale, erroneous, and incorrect (termed unpredictable [3]) data produced by a solution. This is due to the NoSQL movement that may employ weaker forms of consistency to enhance system performance [9]. In all our experiments with the mixed workload, there were no unpredictable data as MySQL implements strong consistency guarantees and the Fluid Cache system preserves the correctness of the application.

For the read-only workload, the choice of write-back or write-through policies has no impact on the observed performance. Fluid Cache enhances the average service time by more than a factor of six with one socialite. It enhances the system throughput by more than a factor of 18 with 16 socialites.

For the mixed read/write workload, Fluid Cache configured with the write-back policy enhances service time by a factor of 3.8. It enhances the throughput of the system by a factor of 6.4 with 16 socialites. The write-back policy outperforms the write-through policy by 14 % for service time and 27 % for throughput.

Read-only workloads are improved more than mixed workloads with Fluid Cache. This is primarily due to the higher SSD bandwidth for reads versus for writes. Using the FIO v2.1.7 microbenchmark, the read bandwidth observed with 1 thread is twice the write bandwidth. With 10 threads, the read bandwidth is approximately 10 times higher.

4   Memory Overhead

An implementation of a host side cache requires in-memory meta-data to identify which disk page reads should be serviced by the host side cache. The meta-data should provide sufficient information to process the disk page read in $O(1)$ look up with no false positives, resulting in one flash I/O. This in-memory meta-data is the foot print of the host side cache. It is the overhead of using a host side cache because the system could have used this memory to process application requests instead of implementing the host side cache. For example, Mercury [8] requires 22 byte of in-memory meta-data for each 4 KB disk page frame in the flash cache. Hence, it consumes 2.75 GB of memory for a 512 GB flash cache formatted as 4 KB pages, a 0.5 % overhead.

This section presents simple analytical models to quantify the cost of this overhead and the benefits of the host side cache. These models can be used to configure a NVM to maximize the impact of host side caches, see Sect. 5.

Let p denote the hit rate observed with a host side cache, e.g., 512 GB of flash cache. And, let q denote the hit rate observed with the memory overhead of a host side cache had the system not implemented the host side cache, e.g., 2.75 GB of DRAM. The following proposition relates these two metrics with the physical

Table 3. List of terms and their definitions.

Term	Definition
r	DRAM hit rate with the host side cache
p	Hit rate of the host side cache
q	Hit rate for the memory footprint of the host side cache
o	Number of bytes per disk page frame to implement the host side cache
n	Number of disk page frames supported by the host side cache
L_{Device}	Latency to retrieve a page from a device such as DRAM, Flash, Disk
B_{Device}	Bandwidth to transfer a page from a device such as DRAM, Flash, Disk

properties of DRAM, host side cache, and disk. These physical properties include latency, bandwidth, and service time (Table 3).

Proposition 1.

a. For a read only workload, the ratio of p to q $\left(\frac{p}{q}\right)$ must be greater than $\frac{L_{DRAM}-L_{Disk}}{L_{Flash}-L_{Disk}}$ in order for the flash host side cache to enhance the observed average latency; where L is the latency to read a disk page from DRAM (L_{DRAM}), flash (L_{Flash}), and disk (L_{Disk}).

b. The expression $\frac{L_{DRAM}-L_{Disk}}{L_{Flash}-L_{Disk}}$ results in a value slightly greater than one. This is because L_{Disk} is significantly larger than the latency of DRAM and flash and the latency of DRAM is faster than that of flash, see Table 1. In the example below, $\frac{L_{DRAM}-L_{Disk}}{L_{Flash}-L_{Disk}}$ equals 1.00628.

c. In order for the overhead of the host side cache to outweigh its benefits, $\frac{p}{q}$ must be a value smaller than one, requiring q to exceed p.

d. q may not exceed p because the pages that fit in the memory foot print of the host side cache are most likely a subset of those that fit in the host side cache itself.

Proof: To prove Proposition 1.a consider the latency with and without a flash cache. With no flash cache, the average latency is:

$$\overline{L_{NoCache}} = ((r+q) \times L_{DRAM}) + ((1-r-q)L_{Disk}) \tag{1}$$

where r is the probability of a page reference observing a hit in DRAM with the host side cache. We add q to r because there is no flash cache and the memory overhead of the host side cache is not incurred.

The average latency with the flash cache is as follows:

$$\overline{L_{Cache}} = (r \times L_{DRAM}) + (p \times L_{Flash}) + ((1-r-p) \times L_{Disk}) \tag{2}$$

Note that the latency associated with flash is incurred for those requests that observe a hit with the host side cache.

One may solve for the break even point when Eq. 1 equals Eq. 2. This point is realized when $\frac{p}{q} = \frac{L_{DRAM} - L_{Disk}}{L_{Flash} - L_{Disk}}$. When $\frac{p}{q}$ is less than $\frac{L_{DRAM} - L_{Disk}}{L_{Flash} - L_{Disk}}$, the host side cache is not beneficial in reducing the average latency and should be abandoned in favor of using its memory foot print to service application disk pages. Otherwise, the host side cache is beneficial and enhances the application performance.   ∎

Example: As an example, assume L_{DRAM} is 30 nanoseconds, L_{Flash} is 25 microseconds (25 thousand nanoseconds), and L_{Disk} is 4 milliseconds (4 million nanoseconds). The ratio $\frac{L_{DRAM} - L_{Disk}}{L_{Flash} - L_{Disk}}$ equals 1.00628, requiring $\frac{p}{q}$ to be greater than this value in order for the host side cache to enhance the average latency. This means if the DRAM hit rate is 60 % (r=0.6) and the hit rate attributed to the memory foot print of the flash cache is 20 % (q=0.2) then the hit rate of the flash cache must exceed 20.13 % (p >0.2013) to provide a superior latency. This is feasible because the flash cache is much larger (512 GB) than its memory foot print (2.75 GB) and contains a super set of the disk pages contained by its memory foot print.   ∎

One may conduct a similar analysis with either the bandwidth provided by the different devices or the page service time (latency plus transfer time of a page from a storage medium using its bandwidth). Note that the observation with one metric (say latency) may not apply to another (say bandwidth). To illustrate assume the bandwidth of disk and flash is 3 and 400 MB/sec, respectively (B_{Disk}=3, B_{Flash}=400). Moreover, assume the bandwidth of DRAM is limited by the system bus at 12 GB/sec. Based on proposition 1, the break even point is realized when $\frac{p}{q} = \frac{B_{DRAM} - B_{Disk}}{B_{Flash} - B_{Disk}}$. Note that this equation violates Proposition 1.b because the value of $\frac{B_{DRAM} - B_{Disk}}{B_{Flash} - B_{Disk}}$ might be many folds higher than one (B_{Disk} is lower than B_{Flash} and B_{DRAM}). Thus, the remaining propositions no longer apply. To demonstrate, consider the hit rates assumed in our example (r=0.6 and p=0.2), the maximum possible value of q (40 %) will not justify the use of a host side flash cache as it reduces the observed bandwidth by 30 %. The value of q must exceed 618.9 % in order for a host side cache to break even. Cache hit rates higher than 100 % are not feasible, causing the host side cache to be too costly.

The value of $\frac{B_{DRAM} - B_{Disk}}{B_{Flash} - B_{Disk}}$ becomes negative when B_{Disk} is greater than B_{Flash}, rendering the host side cache undesirable from a bandwidth perspective. For example, with an enterprise class disk array such as a Dell Compellent SC2020 providing a maximum bandwidth of 6 GB/sec with 256 KByte disk pages and a flash cache providing a bandwidth in the order of a few hundred MB/sec, a host side cache is not beneficial to improve bandwidth.

With service time defined as device latency, i.e., L_{Device}, plus the transfer time of a page from that device, i.e., $\frac{Page\ Size}{B_{Device}}$, a host side cache is once again justified. This is because the disk latency (L_{Disk}) is a significant contributing factor.

As described in Sect. 5, these analytical models can be used to configure a NVM to maximize the impact of a host side cache.

5 Opportunities

The host side cache offers several opportunities for additional research and development. We categorize these into effective monitoring and management tools, extended memory hierarchy, and novel software designs. Below, we describe these in turn.

Effective Monitoring and Management Tools. The host side cache requires effective management tools to enable an administrator to identify bottlenecks and resolve them effectively. Such tools may report on cache hit rates, amount of data served from local and remote caches, and the I/O rates for the cache and the disk (for reads and writes). The tools may enable the administrator to turn certain hints on and off to evaluate their impact on system performance.

It is paramount for the tools to enable an administrator to shut down a host cache instance gracefully. Ideally, the load of this host cache should be divided across the other host cache instances in a manner that avoids formation of hot spots and bottlenecks. Moreover, it might be more effective for the popular content of this host cache to be migrated to the other host caches to minimize the adverse impact on performance attributed to switching from a warm cache to a cold one.

This tool that shuts down a host-side cache works differently for the two architectural alternatives. For example, with the shared-nothing architecture of Fig. 2.a, the tool must migrate permanent (disk) data from one node to the others. With a shared-disk architecture, the tool may migrate the popular data items from the host-side cache to warm-up the destination host caches prior to dismounting virtual disks (LUNs) from a source node. Subsequently, it mounts the LUNs of the source node on to one or more destination nodes [4,18,36].

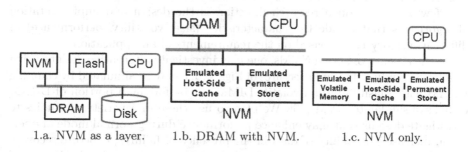

1.a. NVM as a layer.   1.b. DRAM with NVM.   1.c. NVM only.

Fig. 4. Alternative NVM architectures.

Extended Memory Hierarchy. Non-volatile memory (NVM) provides an exciting opportunity to further improve the performance uplift from host side caches. At the time of this writing, NVM is a broad class of technologies including STT-RAM [33], Memristors [52] and Phase-Change Memory [46] (PCM). They are faster than flash and some provide a byte addressable interface similar to DRAM, see Table 1. They are anticipated to provide latencies either comparable to DRAM or an order of magnitude slower depending on their configuration. There is general consensus that NVM will provide higher storage densities than today's DRAM and will be cheaper than DRAM, though more expensive than Flash initially.

Figure 4.a shows an architecture consisting of DRAM, NVM, flash, and Disk. Today's host side caches might be adapted to use this architecture with (a) flash as the host side cache and using NVM as an extension of volatile memory (DRAM), or (b) NVM as the host-side cache and flash as an extension of permanent storage (disk). Another possibility is to implement a host-side cache that manages both NVM and flash separately, caching data across these devices intelligently. A challenge is how to benefit from the byte-addressability features of NVM while the flash (and the application) interface is block-based. Another consideration is the DRAM requirement of such a design, see discussions of Sect. 4. A key research question is what are the tradeoffs associated with these alternative designs? In particular, the additional complexity of the last design must be justified by significant performance benefits when compared with the other two possibilities.

Figure 4.b, c are specific to NVMs that are envisioned to be configurable into N memory banks with varying performance and non-volatility characteristics. In Fig. 4.b, NVM is used as a host-side cache, and it is also separately used to emulate permanent store (disk). The motivation for slowing down a portion of NVM to emulate a memory bank with a higher retention that serves as the permanent store is to enhance the energy efficiency of a solution (battery life, see Table 1) or its physical size by eliminating interfaces and devices. Most likely, extremes such as non-volatility in the order of tens of years (disk) is an overkill for most applications. It might be more appropriate to emulate segments providing storage with latencies several folds slower than DRAM with retention in the order of a few years. An open research direction is the design and implementation of algorithms that decide the characteristics (size, volatility, performance) of different memory banks based on the requirements of an application.

In the absence of actual NVMs, one may investigate alternative designs using simulation studies. The architectures of Figs. 4.b, c may be simulated by coupling DRAM with Flash [39] (non-volatile DIMM) or by forcing a portion of DRAM to behave similar to NVM [45]. We plan to use these in a variety of studies to investigate designs that may enhance performance during normal mode of operation, increase availability of data in the presence of failures, expedite recovery after a failure with the content of the cache intact, and improve both vertical and horizontal scalability of a shared-nothing and a shared-disk deployment.

Novel Software Designs. Today's host side caches are either fully managed by the application or completely transparent. A hybrid may consistent of a transparent cache that accepts hints from an application to enable it to prioritize asynchronous writes to the persistent store. For example, it might be useful for a DBMS to identify its log file to enable the host cache to flush these to the persistent store more aggressively as the likelihood of their repeated reference is very low during the normal mode of operation. This improves the cache hit rate of the host cache to enhance performance. Such a hybrid requires an effective abstraction that is exposed by the caching layer (a storage stack middleware, OS, Hypervisor) and exercised by an application.

A challenge is how to recover from a short lived failure or an intermittent network connectivity event between a host cache instance and its peers without clearing the content of the cache. The time required to warm up a flash cache might be in the order of hours due to its high capacity. As an example, a 300 Gigabyte flash cache required more than 10 hours to reach its maximum hit rate for an enterprise[2] workload [8]. Once warmed up, it is desirable to maintain the content of the cache intact. After a short failure, only a small subset of the cache content might be invalid. An opportunity is to identify the invalid content and discard them either lazily or eagerly. Once a new host cache instance is deployed, a deployment may warm up its cache aggressively using the caches of peer instances [41]. Alternatively, it may monitor the application workload, predict the popular disk pages, and cache them [30].

Data consistency is another challenge faced by host side caches. Consider a scenario where an application modifies a disk page in the host flash cache which is not yet written to the shared disk (a write-back cache policy). If the SAN controller attempts to snapshot or clone a file that contains this dirty disk page, the file will not contain a consistent set of disk pages since the modified disk page will not be visible to the storage controller. Similarly, if the storage controller either reverts to an older snapshot of a file or receives asynchronous mirror changes from a remote primary, it will not be visible to a host flash cache with conflicting data. One may adapt frameworks such as IQ [24] to address this consistency challenge.

In general, today's data stores are not positioned to take advantage of the byte addressability of NVMs [10,11]. Most data stores manage blocks of data with either a row or a column organization of records/documents/key-values [9]. Currently, an object-relational mapping framework such as Hibernate [5, 20,35] might be best suited for NVM as they enable a software designer to identify classes (data structures) that should persist. One may envision adapters for Hibernate to support these data structures effectively using the NVM byte addressability characteristics. However, the time for a complete re-design and re-write of complex applications such as a DBMS is waiting to be seized [11].

6   Conclusions

Host side caches employ a form of storage faster than disk and less expensive than DRAM, e.g., flash or Non-Volatile Memory, to enhance the performance of data intensive applications. They complement disk-based solutions and integrate into an application seamlessly. This paper surveys the different architectures to manage these caches, and opportunities for their future extensions. These

---

[2] A pair of 256 Gigabyte flash caches required about 3 hours to warmup with an OLTP workload [48].

extensions are in the form of monitoring tools, extended memory hierarchy, and novel software designs. The KVS caches [22] (see Sect. 1) are in synergy with host side caches and benefit from similar extensions [1]. We believe both caches will co-exist in future offerings. This is demonstrated by social networking sites such as Facebook that employ both flashcache [38] and memcached [41] to service millions of requests per second.

References

1. Alabdulkarim, Y., Almaymoni, M., Cao, Z., Ghandeharizadeh, S., Nguyen, H., Song, L.: A comparison of flashcache with IQ-twemcached. In: IEEE CloudDM (2016)
2. Barahmand, S., Ghandeharizadeh, S.: BG: a benchmark to evaluate interactive social networking actions. In: CIDR, January 2013
3. Barahmand, S., Ghandeharizadeh, S.: Benchmarking correctness of operations in big data applications. In: MASCOTS (2014)
4. Barker, S., Chi, Y., Moon, H., Hacigümüs, H., Shenoy, P.J.: "Cut me some slack": latency-aware live migration for databases. In: EDBT, pp. 432–443 (2012)
5. Biswas, R., Ort, E.: The Java Persistence API - A Simpler Programming Model for Entity Persistence, May 2006. http://java.sun.com/developer/technicalArticles/J2EE/jpa
6. Breitwisch, M.J.: Phase change memory. In: Interconnect Technology Conference, pp. 219–221 (2008)
7. Bronson, N., Amsden, Z., Cabrera, G., Chakka, P., Dimov, P., Ding, H., Ferris, J., Giardullo, A., Kulkarni, S., Li, H., Marchukov, M., Petrov, D., Puzar, L., Song, Y.J., Venkataramani, V.: TAO: Facebook's distributed data store for the social graph. In: USENIX ATC 2013, pp. 49–60, San Jose, CA (2013)
8. Byan, S., Lentini, J., Madan, A., Pabon, L., Condict, M., Kimmel, J., Kleiman, S., Small, C., Storer, M.: Mercury: host-side flash caching for the data center. In: MSST (2012)
9. Cattell, R.: Scalable SQL and NoSQL data stores. SIGMOD Rec. **39**, 12–27 (2011)
10. Chang, J., Ranganathan, P., Mudge, T., Roberts, D., Shah, M.A., Lim, K.T.: A limits study of benefits from nanostore-based future data-centric system architectures. In: CF, pp. 33–42 (2012)
11. Coburn, J., Caulfield, A.M., Akel, A., Grupp, L.M., Gupta, R.K., Jhala, R., Swanson, S.: NV-Heaps: making persistent objects fast and safe with next-generation, non-volatile memories. In: ASPLOS, pp. 105–118 (2011)
12. Dell: Dell Fluid Cache for Storage Area Networks (2014). http://www.dell.com/learn/us/en/04/solutions/fluid-cache-san
13. DeWitt, D., Ghandeharizadeh, S., Schneider, D., Bricker, A., Hsiao, H., Rasmussen, R.: The gamma database machine project. IEEE Trans. Knowl. Data Eng. **1**(2), 44–62 (1990)
14. EMC: Migrating data from an EMC Celerra or VNS Array to a VNX2 using VNX replicator. EMC White Paper (2014)
15. Fink, M.: Beyond DRAM and Flash, Part 2: New Memory Technology for the Data Deluge, HP Next (2014). http://www8.hp.com/hpnext/posts/beyond-dram-and-flash-part-2-new-memory-technology-data-deluge.vcb6vrbcfe8
16. Gagrani, K., Makransky, K.: Turbocharging Application Response, Dell Power Solutions, Issue 2 (2014). http://www.dell.com/learn/us/en/555/power/ps2q14-20140344-makransky

17. Ghandeharizadeh, S., DeWitt, D.J.: Hybrid-range partitioning strategy: a new declustering strategy for multiprocessor database machines. In: VLDB (1990)
18. Ghandeharizadeh, S., Goodney, A., Sharma, C., Bissell, C., Carino, F., Nannapaneni, N., Wergeles, A., Whitcomb, A.: Taming the storage dragon: the adventures of HoTMaN. In: SIGMOD, pp. 925–930 (2009)
19. Ghandeharizadeh, S., Irani, S., Lam, J., Yap, J.: CAMP: a cost adaptive multiqueue eviction policy for key-value stores. Middleware (2014)
20. Ghandeharizadeh, S., Mutha, A.: An evaluation of the hibernate object-relational mapping for processing interactive social networking actions. In: The 16th International Conference on Information Integration and Web-Based Applications and Services (2014)
21. Ghandeharizadeh, S., Yap, J.: Gumball: a race condition prevention technique for cache augmented SQL database management systems. In: ACM SIGMOD DBSocial Workshop (2012)
22. Ghandeharizadeh, S., Yap, J.: Cache Augmented Database Management Systems. In: ACM SIGMOD DBSocial Workshop, June 2013
23. Ghandeharizadeh, S., Yap, J., Barahmand, S.: COSAR-CQN: an application transparent approach to cache consistency. In: SEDE (2012)
24. Ghandeharizadeh, S., Yap, J., Nguyen, H.: Strong consistency in cache augmented SQL systems. Middleware (2014)
25. Ghemawat, S., Gobioff, H., Leung, S.: The google file system. In: SOSP 2003: Proceedings of nineteenth ACM SIGOPS Symposium on Operating Systems Principles. ACM Press (2003)
26. Graefe, G.: The five-minute rule twenty years later, and how flash memory changes the rules. In: DaMoN, p. 6 (2007)
27. Gupta, P., Zeldovich, N., Madden, S.: A trigger-based middleware cache for ORMs. In: Kon, F., Kermarrec, A.-M. (eds.) Middleware 2011. LNCS, vol. 7049, pp. 329–349. Springer, Heidelberg (2011)
28. Holland, D.A., Angelino, E., Wald, G., Seltzer, M.I.: Flash caching on the storage client. In: USENIXATC (2013)
29. Hsiao, H., DeWitt, D.J.: A performance study of three high availability data replication strategies. Distrib. Parallel Databases 1(1), 53–80 (1993)
30. Kim, H., Koltsidas, I., Ioannou, N., Seshadri, S., Muench, P., Dickey, C., Chiu, L.: Flash-Conscious Cache Population for Enterprise Database Workloads. In: Fifth International Workshop on Accelerating Data Management Systems Using Modern Processor and Storage Architectures (2014)
31. Kim, H., Seshadri, S., Dickey, C.L., Chiu, L.: Evaluating phase change memory for enterprise storage systems: a study of caching and tiering approaches. In: FAST (2014)
32. Koller, R., Marmol, L., Rangaswami, R., Sundararaman, S., Talagala, N., Zhao, M.: Write policies for host-side flash caches. In: FAST (2013)
33. Kultursay, E., Kandemir, M.T., Sivasubramaniam, A., Mutlu, O.: Evaluating STT-RAM as an energy-efficient main memory alternative. In: IEEE ISPASS, pp. 256–267 (2013)
34. Liu, D., Tai, J., Lo, J., Mi, N., Zhu, X.: VFRM: flash resource manager in VMware ESX server. In: IEEE Network Operations and Management Symposium (2014)
35. Marquez, A., Zigman, J.N., Blackburn, S.: Fast portable orthogonally persistent Java. Softw. Pract. Exper. 30(4), 449–479 (2000)
36. Michael, N., Shen, Y.: Downtime-free live migration in a multitenant database. In: TPC Technical Conference (2014)

37. Mitra, L.L.C.: KOSAR (2014). http://kosarsql.com
38. Mituzas, D.: Flashcache at Facebook: From 2010 to 2013 and Beyond (2010). https://www.facebook.com/notes/facebook-engineering/flashcache-at-facebook-from-2010-to-2013-and-beyond/10151725297413920
39. Mogul, J.C., Argollo, E., Shah, M.A., Faraboschi, P.: Operating system support for NVM+DRAM hybrid main memory. In: HotOS (2009)
40. Muller, C., Courtade, L., Turquat, C., Goux, L., Wouters, D.: Reliability of three-dimensional ferroelectric capacitor memory-like arrays simultaneoulsy submitted to x-rays and electrical stresses. In: Non-Volatile Memory Technology Symposium (2006)
41. Nishtala, R., Fugal, H., Grimm, S., Kwiatkowski, M., Lee, H., Li, H.C., McElroy, R., Paleczny, M., Peek, D., Saab, P., Stafford, D., Tung, T., Venkataramani, V.: Scaling Memcache at Facebook. In: NSDI, pp. 385–398. USENIX, Berkeley, CA (2013). https://www.usenix.org/conference/nsdi13/scaling-memcache-facebook
42. Oracle Inc.: Oracle Database Smart Flash Cache (2010)
43. Patterson, D.A., Gibson, G., Katz, R.H.: A case for redundant arrays of inexpensive disks (RAID). In: SIGMOD, pp. 109–116 (1988)
44. Ports, D.R.K., Clements, A.T., Zhang, I., Madden, S., Liskov, B.: Transactional consistency and automatic management in an application data cache. In: OSDI. USENIX, October 2010
45. Rao, D.S., Kumar, S., Keshavamurthy, A., Lantz, P., Reddy, D., Sankaran, R., Jackson, J.: System software for persistent memory. In: Ninth Eurosys Conference, p. 15 (2014)
46. Raoux, S., Burr, G., Breitwisch, M., Rettner, C., Chen, Y., Shelby, R., Salinga, M., Krebs, D., Chen, S.H., Lung, H.L., Lam, C.: Phase-change random access memory: a scalable technology. IBM J. Res. Dev. **52**(4.5), 465–479 (2008)
47. Ratnasamy, S., Francis, P., Handley, M., Karp, R., Schenker, S.: A scalable content-addressable network. In: Proceedings of the ACM Conference on Applications, Technologies, Architectures, and Protocols for Computer Communications, pp. 161–172, August 2001
48. Daniel, S., Jafri, S.: Using NetApp. Flash cache (PAM II) in online transaction processing. NetApp. White Paper (2009)
49. Stearns, W., Overstreet, K.: Bcache: Caching Beyond Just RAM (2010). http://bcache.evilpiepirate.org/, https://lwn.net/Articles/394672/
50. STEC: EnhanceIO SSD Caching Software (2012). https://github.com/stec-inc/EnhanceIO
51. Stoica, I., Morris, R., Karger, D., Kaashoek, M., Balakrishnan, H.: Chord: A Scalable peer-to-peer lookup service for internet applications. In: ACM SIGCOMM, pp. 149–160, San Diego, California, August 2001
52. Strukov, D.B., Snider, G.S., Stewart, D.R., Williams, R.S.: The missing memristor found. Nature **7191**, 80–83 (2008)
53. Tabor, J.: Avere architecture for cloud NAS. Avere White Paper (2014)
54. Trammell, J.: CacheIQ: automatic storage tiering in the age of big data. In: Flash Memory Summit 2012 Proceedings (2012)
55. Vučinić, D., Wang, Q., Guyot, C., Mateescu, R., Blagojević, F., Franca-Neto, L., Moal, D.L., Bunker, T., Xu, J., Swanson, S., Bandić, Z.: DC Express: shortest latency protocol for reading phase change memory over PCI express. In: FAST 2014 (2014)

Conclusions from the Evaluation of Virtual Machine Based High Resolution Display Wall System

Rudolfs Bundulis[✉] and Guntis Arnicans

Faculty of Computing, University of Latvia, Riga, Latvia
{rudolfs.bundulis,guntis.arnicans}@lu.lv

Abstract. There are several approaches to the construction of large scale high resolution display walls depending on the required use case. Some require support of 3D acceleration APIs like OpenGL, some require stereoscopic projection. Others simply require a surface with a very high display resolution. The authors of this paper have developed a virtual machine based high resolution display wall architecture that works for all planar projection use cases and does not require a custom integration. The software that generates the presented content is executed in a virtual machine thus no specific APIs other than those of the virtualized OS are required. Any software that is able to run under a given OS can be run on this display wall architecture without modifications. The authors have performed performance evaluations, virtualization environment comparisons and comparisons among other display wall architectures. All this knowledge along with key conclusions is summarized in this paper.

Keywords: Display wall · Remote visualization · Video streaming · H.264 · Virtualization

1 Introduction

Limitations of the traditional display systems have created a distinct field of research devoted to construction of large scale high resolution display walls. Such display walls are required in environments that require a display surface with characteristics that cannot be achieved by a single physical display - either a very large physical size or a very high resolution or both. The following samples should illustrate this issue more clearly.

There is an ongoing telescope image archive digitalization process at the Baldone observatory in Latvia. The digitalized images have a size of around 500 megapixels. Such images cannot be fully viewed on a single PC system even with multiple monitors. A single display with 4 K resolution would allow viewing only 1.6 % of the whole image. Even if higher level hardware is considered – e.g. a system of 4 GPUs with 6 outputs each at a resolution of 2560 × 1440 - this is still only 88 megapixels in total (17 % of the whole image). Thus a simple conclusion can be derived - a pure hardware solution without any software middleware that would union multiple hardware systems in a unified display surface is simply not enough in case of such a task.

© Springer International Publishing Switzerland 2016
G. Arnicans et al. (Eds.): DB&IS 2016, CCIS 615, pp. 211–225, 2016.
DOI: 10.1007/978-3-319-40180-5_15

The Reality Deck display wall demonstrates a display wall system that combines hardware nodes similar as described above to drive a unified 1.5 gigapixel display surface through a custom software stack. The capabilities of the Reality Deck display wall system can already match the given requirements of displaying a 500 megapixel image. Thus this example demonstrates how software solutions used in the display wall systems can leverage the hardware limitations.

However due to the presence of custom software stack compatibility issues arise. If the presented content is static (e.g. image or video files) it must be stored in a format supported by the software stack. But in many cases the content may be generated in real time by some client software. The possible range of this client software is vast – from web browsers, office applications and games to scientific particle simulators etc. To be able to present itself on the display wall the client software must use specific APIs provided by the display wall software stack. In many cases the display wall systems support the OpenGL 3D rendering API but that still covers only a part of the possible client software scenarios. However the supported OpenGL versions are often not up to date. Moreover it is not known whether the supported versions will ever be increased to the up-to-date ones since the evolution and new functionality of OpenGL imposes restrictions on parallelization. Thus a display wall with a software stack that can handle any generic drawing API available in a given OS would be feasible.

After performing a survey on the field of displays walls [2], the authors of this paper tried to solve this situation with a different approach. The existing solutions can be generalized as a software component that interacts with the client software and exposes the display wall to them through some specific APIs. Instead, why not expose the display wall to the client software through the standard means of the operating system as a large monitor? In such scenario the client software would be able to use all the standard APIs exposed by the operating system and no modifications would be needed. With this though in mind the authors of this paper designed a virtual machine based high resolution display wall architecture and developed a prototype. By using virtualization this display wall system is able to run the client software in a virtualized instance of the operating system that the client software was developed for and provide presentation on the content that would normally appear on the attached physical display device on the display wall instead.

This paper contains the overview of the development process of a prototype for this architecture, summarized conclusions and underlines the pros and cons of such architecture in comparison to the existing ones.

2   Display Wall Architectures and Their Problems

To understand the factors that drive the research related to the construction of display walls one must be introduced the most important limitations in the classical PC display system. For that reason the authors need to introduce a few terms and their explanations.

The terms *display wall* and *video wall* are used interchangeably in this and many of the related works. Both of the terms are used to describe a large sized tiled display surface. Each tile can be either a single physical display or an image from a projector.

The purpose of a display wall is to create a single continuous display surface that is superior when compared to a single display or projector image in terms of maximal size, resolution or both.

Visualization *software* denotes the software that performs either a 2D drawing or 3D rendering work the outcome of which is an image that must be represented to the user. Visualization software is the factor that creates the need for display wall whenever the produced image cannot be presented on a single display. Visualization software is not always a single user space process; a whole operating system can be perceived as visualization software since it provides a way for the processes running inside it to present visual data to the user.

In the ideal case the visualization software should not be aware of the display environment. It should not care whether it needs to use a specific 2D drawing or 3D rendering API when run on display wall and another tradition API when run on a traditional PC system. In the context of this paper *software awareness* is understood as the need for the visualization software to be aware of the display environment and change its behavior accordingly. A display wall is called *software agnostic* if it does not enforce software awareness.

Framebuffer is a continuous array of all pixels that represent the contents of an image. Framebuffer is the source of the image that is presented on a display or projector image. In the same time it is the target of all the rendering and drawing actions performed by visualization software. Framebuffer implementations may differ and this paper will cover researches that have tried very diverse solutions.

In general, the intuitive approach to building of a display wall is quite trivial – one would simply to create a large display surface by physically tiling multiple smaller displays or images of multiple projectors. However that is only a part of the solution. A system that provides the signals for the tiles at one end and exposes the display wall to the visualization software that provides the content at the other end is required. Such systems can be constructed in various ways – from purely hardware based ones that involve simple signal cascading and scaling to complex software systems involving distributed computing.

The traditional display system used in desktop PC systems (Fig. 1) consists of four components:

1. A visualization *software* that creates the visual data to be presented to the user and interacts with the GPU or video adapter by executing 2D drawing and 3D rendering commands;
2. A *GPU* or video adapter – a hardware peripheral in a computer system that exposes and implements standardized 2D drawing and 3D rendering APIs (e.g., OpenGL, DirectX), implements them, renders the results to an internal framebuffer and provides the display signal to a display;
3. A *display* – a device that visualizes the signal from the GPU;
4. A *medium* that connects the GPU and the display.

Each of these four components has some limitations in regards to the resolution of the visualization on the display.

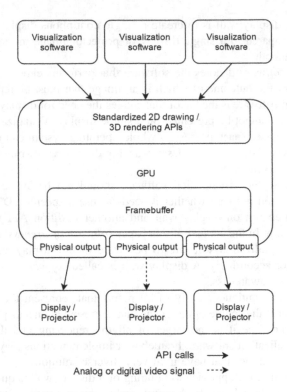

Fig. 1. The traditional PC display system.

The GPU has a limited maximum resolution that is enforced by the memory available on the GPU and the medium used for signal delivery.

The medium which is used to interconnect the GPU and the display has some sort of throughput limitation which limits the amount of data that can pass through in a single unit of time.

The display as well has a limited maximum resolution depending on the limits of the chosen visualization technology (e.g. the maximum resolution of the matrix) and the medium that delivers the signal.

Such dependencies among the components hold back the general progress - e.g. why would a hardware vendor create a GPU with a resolution that no medium could deliver or no display could visualize? Thus due to these limitations domains that actually require large high resolution display surfaces created the need to tackle this issue. That was done by creating hardware peripherals and software systems that leverage and combine the possibilities of the existing GPUs and untraditional mediums.

The mentioned limitations in terms of resolution and size are a direct problem for static content like images. However another aspect that introduced the need for an alternative to the traditional PC display system was the computing power difference between CPUs and GPUs - a mismatch between the speed at which visualization software generates the image and the speed at which the GPU is able to render it. For example graphical representations of fluid flow simulations done by supercomputers could not be visualized by a single GPU at real time. Thus again a scalable rendering solution that could be expanded to meet the needs to visualize the generated graphical data at real time was needed.

Apart from these limitations of the classical PC display system there are multiple other factors that can create a need to have display surface of a size and/or resolution that cannot be achieve with a single physical display:

- Ergonomics – there are tasks for which the response time and error levels decrease as the resolution of the display is increased, thus for such tasks a display with the highest available resolution would be preferable;
- Specific visualization needs:
  - High resolution imaging – there are many imaging sources like satellites, microscopes, X-ray machines etc. that provide very high resolution imaging that cannot be viewed in full size or natural resolution on a single physical display without scaling or cropping;
  - Big data visualization – since the birth of cloud and distributed computing handling big data has become a common problem, and visualization is as well a part of that, a simple example would be a need to fully present a graph with more nodes than pixels in the with the highest possible resolution;
  - Real-time simulations – in many physics and chemistry simulation visualization cases it is very important not to lose image detail due to rendering the output image in a small resolution;
- Collaboration – high resolution display surfaces offer a natural environment for multiple persons to share their visual data and interact with it.

Historically the evolution of GPUs was quite slow. Of course after the appearance of first scientific solutions to the issues mentioned above the hardware manufacturers started to react and evolve the GPUs, displays and transport mediums to target not only the consumer market but also scientific visualization needs.

To address the problems with static content and expand the size of the display surface available on single PC system GPU manufacturers created GPUs with multiple outputs thus allowing connecting multiple displays to a single PC system. Motherboard manufacturers produced motherboards that could host multiple GPUs in such way again multiplying the number of total available outputs. The medium manufacturers improved the underlying transport mechanisms of the digital signal to allow multiple independent digital video streams to be delivered through a single cable thus allowing display chaining and again increasing the maximum number of displays that can be connected to a single PC system. Several multiple GPU interconnection mechanisms like NVIDIA's SLI or ATI's CrossFire were designed to address the gap between the computational and rendering speeds. The display manufacturers developed display technologies that would allow higher resolutions for a single display. However the results of these attempts to exclude the need for further research in this area have not fully succeeded. A pure hardware solution suffers from several efficiency and deployment problems:

- *Scalability* - even though throughout the evolution of the hardware components the number of displays that can be physically connected to a PC system has increased there is still a hard constraint on that number since there is a limit of maximum outputs on a GPU, a limit on maximum GPUs hosted by a motherboard and a limit on the maximum bandwidth of a medium. Of course these limits will grow with time but still they do not meet the needs of scientific visualization software and very high resolution image sources;

- *Component availability and price* – some of the hardware enhancements described above (e.g., 2+ video output count, 4K and 8K displays, 3+ PCIe slots, DisplayPort Multi-Stream Transport) are available only in particular high end hardware (e.g., server motherboards), this will affect both the price (e.g., price per pixel is higher for a 4 K matrix than 4 smaller high definition matrices that can be combined to create an area of the same size and total resolution) and the choice of other hardware components like CPU's components might be limited;
- *Ineffective component use* – high end graphics adapters with multiple outputs are usually built to provide computing power for the most demanding use cases (e.g., games, 3D design and rendering applications, CAD tools) which means that in case if the content being streamed to the display wall does not require intensive computing then the available resources of the graphics adapters are not used to the full extent, this in turn leads to cost ineffectiveness;
- *Power consumption, cooling and noise* – running multiple graphics adapters can be challenging in both providing enough power to the system and cooling it, in the case where the content that is streamed is static or primitive the idle power consumption of the adapter multiplies with each new adapter added to the system, which means it would be more efficient to reduce the number of graphics adapters as much as possible. Complex cooling solutions are needed to prevent such a system from overheating. If the cooling system is not designed carefully enough the noise output can be disturbing to the surrounding users;
- *Wiring* – digital video cables have wire length limitations (HDMI – depending on the category up to 15 m but mostly around 7 m, DisplayPort 1.2 – up to 33 m, older limited at 3 m) which means that the displays themselves must be located in proximity of the system that sends the signal. Apart from the length limitations using a lot of wiring creates chaos and limits the mobility and sustainability of a display wall system;
- *Logical division in case of multiple display walls* – if there is a need to drive multiple independent display walls (e.g., industrial uses as timetables or information boards in an airport, hospital, train station) a single PC system with a single OS that divides its content among subsets of monitors each of which present a single display wall the systems becomes prone to scenarios where errors on the system itself effects all the walls since they are not logically divided.

Thus, the research in the display wall area has not been diminished by the availability of pure hardware level solutions to the general issue. And there is a good reason for that. As stated above even though the hardware peripherals evolve they still impose some kind of maximum limit of displays that can be connected. And removing constraints by providing flexible and scalable solutions is one of the general aims of any research on display walls. A good example to back this statement is the Reality Deck (Papadopoulos 2015) – the world's currently largest display wall that has a resolution of 1.5 gigapixels. Such resolution could not be achieved by pure hardware means - the

Reality Deck is built upon hardware nodes that reach the available hardware constraints and driven by scientifically developed software that composes these nodes in a unified framebuffer.

One of the main characteristics of a display wall system is how the display surface is exposed to the visualization software. To achieve the task of being software agnostic the display wall system must expose itself as similar as possible to the classical PC display system. There have been several approaches to this task:

- Exposing the display wall through a custom implementation of widely used 3D rendering libraries, most often OpenGL;
- Exposing the display wall as additional displays by hooking in the GPU driver. This is a path chosen by many software products that allow devices like phones, tablets and TVs to be used as additional displays for a PC system. However this requires a different implementation on each operating system and leaves rarely used operating systems unsupported. And since the operating system vendors can make arbitrary changes to the device driver architecture such solutions may be rendered dysfunctional after releases of newer versions of the operating system. A good example in this case is Microsoft Windows. Microsoft Windows XP had a driver model name XPDM that supported using mirror drivers. A mirror driver allowed duplicating the output of the primary display driver to a custom memory region. Many software products were developed to support attachment of custom display nodes by using this functionality. However with Windows Vista a new driver model called WDM was introduced that added new requirements for the mirror driver implementations and thus the software products developed for XPDM had to be modified to still be supported;
- Providing a custom library that exposes API for the visualization software to interact with the display wall;

The intuitive assumption that could arise is why not simply provide a fully virtual GPU and hide all the display wall related internals inside the driver? This is a valid approach to the issue but again becomes complicated due to need to handle different operating system architectures. Operating systems expect some amount of hardware presence (interrupt handling, DMA, etc.) from a GPU and implementing this could be problematic due to the existence of closed source operating systems. However, it must be noted that such a solution could fully mimic the traditional PC display system.

Most of the previously developed display wall systems try to solve the mentioned hardware scalability limitations by separating the many functions of the GPU among several components that are usually implemented as separate computer systems (Fig. 2). For more scalability clusters of such systems can be used for each task instead of a single instance. For example, to increase the rendering power the display wall systems either expose custom 2D drawing or 3D rendering APIs or provide their own implementations for the standardized ones like OpenGL. Then the display wall software stack forwards these calls to distributed rendering nodes thus increasing the rendering power. Further down the pipeline the pixel regions produced by the rendering nodes are reassembled and synchronized by one ore many framebuffer components. The framebuffer components then present the complete image on the display wall.

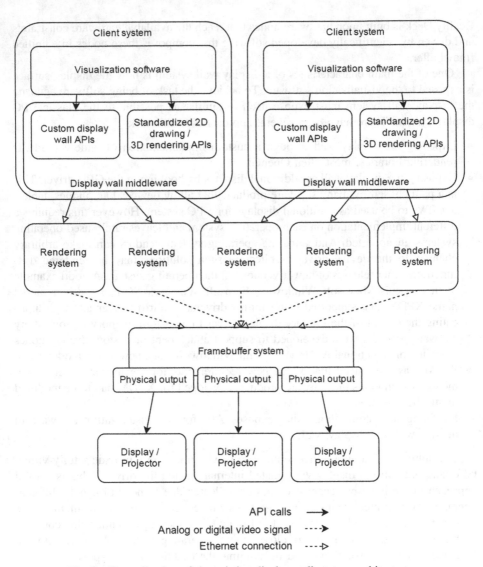

Fig. 2. Generalization of the existing display wall system architectures.

The research in the display wall construction domain has given birth to numerous display wall systems like Reality Deck [1], SAGE [3], SAGE2 [4], DisplayCluster [5], Chromium [6], WireGL [7], Equalizer [8], CGLX [9], XMegaWall [10], SGE [11] and others.

A detailed analysis of all these display wall solutions is outside the scope of this paper. However to provide a ground for comparison with the display wall architecture proposed by the authors of this paper a brief introduction is needed.

In general the existing solutions can be divided in two groups:

- Ones that expose themselves as OpenGL implementations (WireGL, Chromium, Equalizer, CGLX);
- Ones that perform pixel streaming from client workstations, host their own applications that can accept external input and provide custom or standardized frameworks for adoption of existing software (SAGE/SAGE2, DisplayCluster, XMegaWall, Reality Deck). SAGE2 that offers hosting HTML5 content directly on the display wall system is a good example of using standardized technologies other than OpenGL for software adoption on a display wall.

3   Virtual Machine Based High Resolution Display Wall

The authors of this paper propose a new display wall architecture (Fig. 3) [12]. This architecture in contrast to most of the current systems does not host the framebuffer outside of the client system. Instead it uses virtualization for the framebuffer implementation. It implements a virtualized GPU that works on top of one or many physical GPUs. This approach allows removing a hard dependency among physical outputs on the physical GPUs and the size of the display surface available. The proposed architecture does this in a manner which hides this fact from the visualization software and sets no specific requirements on it. Any visualization software running in the client system interacts with a virtualized GPU that works just like a normal GPU and exposes all the standard 2D drawing and 3D rendering APIs. Underneath a custom display wall software stack implements these calls by using the physically available GPUs in the system thus allowing efficient scaling by adding more GPUs to the system in the case if the previously described gap between the computing power and rendering power is encountered. The rendered data is then encoded in a video stream that is transmitted over Ethernet to a display endpoint system where it is decompressed and displayed on a connected display or projector. To satisfy the needs of a multi-client environment the display endpoint can receive different independent streams and display them in a layered mode.

GPU virtualization has become a trending technique in the virtualization technologies. Leading solution providers like VMware and VirtualBox have implemented a way to provide the acceleration features like DirectX and OpenGL support of the host GPU directly to the virtualized guest operating system. Similar technology is provided for XEN by NVIDIA vGPU and RemoteFX for Microsoft Hyper-V. The GPU virtualization technology has successfully helped to utilize a single GPU in a multi operating system environment.

This approach can be applied to solve the issues that exist in the field of display walls. If the virtualization technology provides the guest system with a purely simulated and freely configurable GPU that supports hardware acceleration for 3D APIs like OpenGL and Direct 3D by using the physically available GPUs this can solve both problems – remove the dependencies on the physically available video outputs and increase hardware utilization.

Let's look at the architecture in more detail. The physical host system runs a software stack currently denoted as *Framebuffer Manager* and some kind of virtualization platform which in turn hosts the guest operating system that is running the

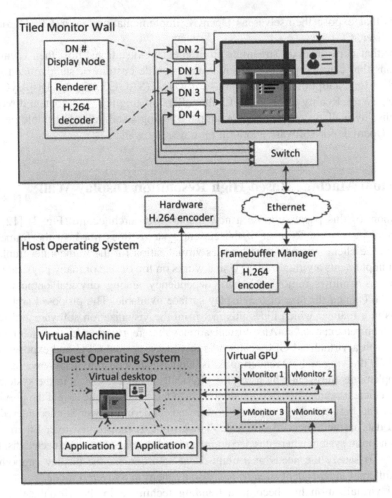

Fig. 3. The proposed display wall architecture.

content that needs to be visualized on the display wall. The virtualization platform simulates a virtual GPU that can be freely configured in terms of virtual monitors and resolutions to exactly match each desired use case depending on the amount of data that needs to be visualized. The virtualization platform interacts with the Framebuffer Manager software stack by providing notifications about drawing operations on the guest operating system and access to the video memory contents of the virtual GPU. The Framebuffer Manager itself performs event-driven management of the frame-buffers and handles (crops/scales) the mapping of image data from the virtual monitors to the display nodes in the monitor wall. After the logical partitioning of the image the Framebuffer Manager uses hardware based video encoding capabilities in the host system to encode the image and provide an encoded video stream to each display node.

The proposed architecture removes direct dependencies between the needed monitor setup and presence of physical GPUs – all of this is taken care by the

Framebuffer Manager and virtualization platform. The Framebuffer Manager takes care of using the present hardware for video encoding and 3D acceleration support for the virtualization platform, while the virtualization platform provides unconstrained display and resolution configuration options.

4 Prototype of Proposed Display Wall

Further on the authors developed a working hardware prototype of the proposed display wall architecture [13]. The prototype consists of a 5 × 5 22" monitor wall and two different host systems:

1. with moderate hardware – Gigabyte Brix Pro mini PC (Intel Core i7 4770R CPU (4 physical cores, 8 virtual cores at 3.2 GHz), Intel Iris 5200 Pro GPU, 12 GB of RAM and Windows 8.1);
2. With high end hardware – 2 Intel Xeon e5-2630 v2 CPUs (12 physical cores, 24 virtual cores at 2.60 GHz), NVIDIA Quadro K4200, Windows 8.1.

Both host servers runs VirtualBox as the virtualization platform with both Windows and Linux guests. VirtualBox was chosen for two reasons. First, it was one of the few open source virtualization systems that had support for a configurable virtualized GPU. Second, it provided the best scalability options in terms of the total resolution while other virtualization systems were capped at lower limits. Table 1 describes these limitations in more detail.

The authors have developed a Framebuffer Manager implementation that runs alongside VirtualBox on the host operating system and collects the image data from the framebuffers of the simulated GPU, encodes them into a H.264 stream with either the Intel Iris 5200 Pro or the NVIDIA Quadro K4200 GPU. After the video has been encoded the Framebuffer Manager streams it over a gigabit Ethernet to the monitor wall.

Table 1. Resolution limitations of virtualization systems.

Vendor	Maximum resolution for a homogenous surface	Comments
NVIDIA vGPU	16 megapixels (8 displays at 2560 × 1600)	The mentioned results are based on the NVIDIA GRID K260Q card, other cards provide lower capabilities
RemoteFX (Microsoft Hyper-V)	10 megapixels (8 displays at 1280 × 1024)	Windows Server 2012 R2 host operating system and Windows 8/8.1 guest operating system
VMWare vSGA	4 megapixels (2 displays at 1920 × 1200)	Only Windows guest operating systems
Oracle VirtualBox	Any configuration that can fit in the video memory of the virtualized	The video memory of the virtualized GPU currently cannot exceed 256 MB

The monitor wall itself consists of 25 22" DELL displays, each of which is driven by a Raspberry Pi model. The Raspberry Pi devices were chosen to implement the role of the display node because of the low cost, efficient power usage and ability to decode a 1920 × 1080 H.264 at acceptable frame rates for live streaming. The Raspberry Pi units are actually poorest in the terms of scalability in this prototype, since they support H.264 decoding only up to the resolution of 1920 × 1080 meaning using displays with higher resolution is not possible in the current prototype. However, since the embedded systems are developing vastly the authors do not see this as a significant issue. At the moment of writing this article NVIDIA has already released Jetson K1 embedded system that is able to drive a 4 K monitor.

Figure 4 demonstrates Xubuntu 14.04 running common applications with static content like Google Maps and SVG based graphs in Firefox and a PDF viewer. All these applications seemed to work without issues regarding the high display area.

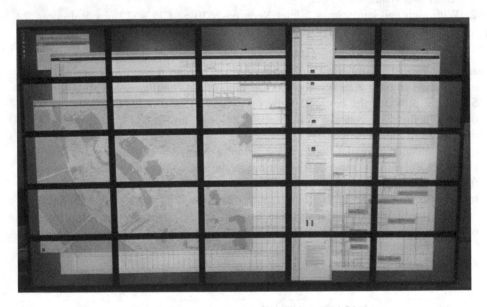

Fig. 4. The prototype running Xubuntu 14.04.

5 Lessons Learned

All the described development and evaluation process lead the authors to several conclusions and observations.

Video compression codec choice for real-time video streaming is not as obvious as it may seem. Currently most of the multimedia content available either in a stored or live format is encoded with the leading inter frame ISO/IEC Moving Pictures Experts Gorup (MPEG) codecs –H.264 and H.265. Both of the codecs are so widespread because they offer better compression ratios compared to older intra frame compression methods like JPEG [14]. Since size is the most important factor for media that is streamed over the Internet the ability to sustain real-time encoding and decoding is not prioritized. However in the case of desktop capture these factors switch places. VNC systems still use JPEG for transmitting the content since JPEG performs better in the scenario where only small regions in the picture change from frame to frame. Both H.264 and H.265 encode the full frame each time, while JPEG can only encode the dirty regions. And in the case where the only thing moving in the picture is the mouse pointer this makes a great difference. However the content of the desktop can swiftly change from static to very dynamic – and with dynamic content JPEG produces bitstreams with much higher size than H.264 or H.265. Thus the lesson here is that for optimum performance the display wall software stack must use a hybrid encoding mode that is able to use the best codec for each type of content and perform real-time switching among them.

The hardware acceleration on GPUs will continue to evolve making the proposed display wall architecture more and more efficient. Currently the hardware accelerated video encoding available in the evaluated GPUs (Intel Quick Sync in Intel Iris Pro 5200 and NVIDIA NVENC in NVIDIA Quadro K4200) is not powerful enough to provide real-time encoding for large scale display wall systems. However according to the roadmaps of both hardware vendors the performance of the GPUs will increase very fast and will eventually exceed the CPU based video encoding technologies.

The main limit of the proposed architecture is the amount of virtual GPU memory in VirtualBox which can be lifted or at least increased to a sufficient amount. The achieved results in terms of the total display resolution from the evaluated prototype may not be impressive if compared to the Reality Deck. But the only limitation here is the hardcoded virtual GPU memory limit in VirtualBox. Thus by removing that and providing sufficient amounts of RAM this display wall architecture can provide higher resolution than a single node used in the Reality deck with only one GPU.

Support for DirectX 10/11/12 presents the biggest challenge in obtaining a fully seamless virtualization. As stated in the introduction the main benefit of the virtualized display wall architecture is the ability to provide all of the needed drawing and rendering APIs through the virtualization environment. VirtualBox supports OpenGL up to the version that is available on the physical GPU present in the host system but is limited at DirectX 9.0. Support for higher DirectX versions like 10/11/12 is a problem for all virtualization environments. But during the development of the prototype the authors have gained enough knowledge about the internals of VirtualBox to be capable of providing support for more recent DirectX versions in the future.

The modern operating systems are still somewhat incapable of handling large amounts of displays with the built in mechanisms. As seen on the prototype, both Linux and Windows are not yet capable to flawlessly support a large number of displays. Windows 7 could not identify more than 16 displays. Even though using a single display with a large resolution is a better choice at least from the ergonomic point of view there may be scenarios where using many separate virtual displays is required. The maximum limits of the operating systems must be determined to understand when the operating system itself becomes the bottleneck.

6 Conclusion

The display wall domain is very heterogeneous. There are several solutions to the same general problem but each targeted at some specific use case or environment. Due to the fast that the use of display walls has not been widespread – in most cases they are used in scientific or research facilities to either view some static high resolution content or display real-time simulations – the requirement to use custom crafted client software to generate the visual content has not been a limiting factor.

However the authors of this paper have proposed a display wall architecture that is able to run and visualize any client software that runs on a given operating system by providing a different display wall software stack architecture that exposes itself through a virtualized instance of the required operating system.

This paper contains evidence gathered from the prototype of this architecture that the given premise is valid – the authors were able to virtualize Linux and Windows operating systems, simulate a virtual GPU with a 52 megapixel resolution and present the contents of this virtualized desktop on a physical display wall. No software incompatibilities were encountered while running both casual desktop software (web browsers and office applications) and domain specific software (CCTV surveillance system).

The issues that were encountered are only relevant to performance – the limitation of the maximum resolution is capped by the limit of the virtual GPU video memory in VirtualBox and the flawlessness of the interaction is limited by the real-time video encoding performance of the physical GPU on the host system. Both of these will decrease with further releases of GPU architectures and VirtualBox.

If compared to the existing solutions the proposed architecture reduces the complexity of the hardware – no distributed systems are involved – just one host system with a set of attached display nodes. This eases deployment and maintenance. The virtualization allows running any client software opposed to the limitations of using OpenGL or custom APIs imposed by other display wall systems.

Acknowledgements. This work was partly supported by the Latvian National research program SOPHIS under grant agreement Nr.10–4/VPP-4/11".

References

1. Papadopoulos, C.: The reality deck–an immersive gigapixel display. IEEE Comput. Graph. Appl. **35**(1), 33–45 (2015)
2. Bundulis, R., Arnicans, G.: Architectural and technological issues in the field of multiple display technologies. In: Databases and Information Systems VII: Selected Papers from the Tenth International Baltic Conference, DB&IS 2012, pp. 317–329 (2012)
3. Renambot, L., Rao, A., Singh, R., Byungil, J., Krishnaprasad, N., Vishwanath, V., Chandrasekhar, V., Schwarz, N., Spale, A., Zhang, C., Goldman, G., Leigh, J., Johnson, A.: SAGE: the scalable adaptive graphics environment. Proc. WACE **9**, 2004–2009 (2004)
4. Renambot, L., Marrinan, T., Aurisano, J., Nishimoto, A., Mateevitsi, V., Bharadwaj, K., Long, L., Johnson, A., Brown, M., Leigh, J.: SAGE2: a collaboration portal for scalable resolution displays. Future Gener. Comput. Syst. **54**(C), 296–305 (2015)

5. Johnson, G., Abram, G., Westing, B., Navr'til, P., Gaither, K.: DisplayCluster: an interactive visualization environment for tiled displays. In: 2012 IEEE International Conference on Cluster Computing (CLUSTER), pp. 239–247 (2012)

6. Humphreys, G., Houston, M., Ng, R., Frank, R., Ahern, S., Kirchner, P., Klosowski, J.: Chromium: a stream-processing framework for interactive rendering on clusters. ACM Trans. Graph. **21**(3), 693–702 (2002)

7. Humphreys, G., Eldridge, M., Buck, I., Stoll, G., Everett, M., Hanrahan, P.: WireGL: a scalable graphics system for clusters. In: Proceedings of the 28th Annual Conference on Computer Graphics and Interactive Techniques, pp. 129–140 (2001)

8. Eilemann, S., Makhinya, M., Pajarola, R.: Equalizer: a scalable parallel rendering framework. IEEE Trans. Vis. Comput. Graph. **15**(3), 436–452 (2009)

9. Doerr, K.-U., Kuester, F.: CGLX: a scalable, high-performance visualization framework for networked display environments. IEEE Trans. Vis. Comput. Graph. **17**(3), 320–332 (2011)

10. Kang, Y.-B., Chae, K.-J.: XMegaWall: a super high-resolution tiled display using a PC cluster. In: Proceedings of Computer Graphics International, pp. 29–36 (2007)

11. Perrine, K.: Parallel graphics and interactivity with the scalable graphics engine. In: Proceedings of the 2001 ACM/IEEE Conference on Supercomputing, p. 5 (2001)

12. Bundulis, R., Arnicans, G.: Concept of virtual machine based high resolution display wall. In: 2014 IEEE 2nd Workshop on Advances in Information, Electronic and Electrical Engineering (AIEEE), pp. 1–6 (2014)

13. Bundulis, R., Arnicans, G.: Virtual machine based high resolution display wall: experiments on proof of concept. In: International Conference on Systems, Computing Sciences and Software Engineering (SCSS 14), Electronic CISSE 2014 Conference Proceedings (2014)

14. Bundulis, R., Arnicans, G.: Use of H.264 real-time video encoding to reduce display wall bandwidth consumption. In: 2015 IEEE 3rd Workshop on Advances in Information, Electronic and Electrical Engineering (AIEEE), pp. 1–6 (2015)

Business Process Modeling and
Performance Measurement

The Enterprise Model Frame for Supporting Security Requirement Elicitation from Business Processes

Marite Kirikova[1]([⊠]), Raimundas Matulevičius[2], and Kurt Sandkuhl[3]

[1] Department of Artificial Intelligence and Systems Engineering,
Riga Technical University, Kalku 1, Riga 1658, Latvia
marite.kirikova@cs.rtu.lv
[2] Institute of Computer Science, University of Tartu, J.Liivi 50409 2,
50409 Tartu, Estonia
rma@ut.ee
[3] Chair Business Information Systems, University of Rostock,
Albert-Einstein-Str. 22, 18059 Rostock, Germany
Kurt.Sandkuhl@uni-rostock.de

Abstract. It is generally accepted that security requirements have to be elicited as early as possible to avoid later rework in the systems development process. One of the reasons for difficulties of early detection of security requirements is the complexity of security requirements identification. In this paper we propose an extension of the method for security requirements elicitation from business processes (SREBP). The extension includes the application of the *enterprise model frame* to capture enterprise views and relationships of the analysed system assets. Although the proposal was used in some practical settings, the main goal of this work is conceptual discussion of the proposal. Our study shows that (*i*) the enterprise model frame covers practically all concepts of the information security related definitions, and that (*ii*) the use of the frame with the SREBP method complies with the common enterprise modeling and enterprise architecture approaches.

Keywords: Security requirements elicitation · Business process models · Enterprise modeling

1 Introduction

Security requirements elicitation is a part of security engineering that plays an important role in high quality system development of [1]. Although the importance of introducing security engineering practices early in the systems development lifecycle has been acknowledged [2, 3], in practice security often is considered at the later stages of the lifecycle or when security problems arise in operations and maintenance. This either causes rework in initial designs and slows down the later stages of system development, or extends system maintenance with unplanned activities. Security is researched in different areas of business and information systems development including enterprise modeling [4]. Nevertheless, new approaches are continuously

G. Arnicans et al. (Eds.): DB&IS 2016, CCIS 615, pp. 229–241, 2016.
DOI: 10.1007/978-3-319-40180-5_16

developed. This may be explained by the fact that, despite a large number of methods created in different research projects, it is still difficult to use them in practice.

In this paper we discuss how *information security solutions (i.e., security countermeasures) could be related to enterprise modeling*. The study was a part of the ITSE project[1], where the main goal was to transfer a method for Security Requirements Elicitation from Business Processes (SREBP) [5, 6] to the practice of small and medium-sized enterprises (SME). The SREBP method was applied to SME with well-defined processes and high awareness of the importance of well-defined security requirements. From discussions with the SME's employees we learned that while in general the SREBP method was quite well appreciated by the enterprise, some underlying enterprise modeling was needed to support and structure the analysis. So the main research question addressed in this paper is the identification of appropriate enterprise modeling support for successful application of the SREBP method in SMEs.

We applied the following research method consisting of four steps:

1. We applied the SREBP method to understand the security requirements within some SMEs. As mentioned above - this resulted in the observation that some enterprise modeling support was needed for putting the SREBP method into the organisational context.
2. We analyzed information security definitions to highlight important enterprise modeling concepts. These concepts were then analysed in Step 4.
3. We surveyed related work on business process related security requirements identification to learn from this research about important enterprise modeling concepts that can support information security requirements elicitation.
4. Taking into account the results from Step 2 and Step 3 we applied a particular *enterprise model frame* and verified it against security and enterprise modeling concepts.

The paper proposes the use of the enterprise model frame to establish the linkage between the security requirements elicitation from business processes and enterprise modeling. The main contributions of this primarily conceptual paper are (1) identification and alignment of concepts from enterprise modeling and information security, which are relevant for security requirements elicitation based on business processes, and (2) discussion of applicability of the use of business process-oriented security elicitation together with enterprise modeling approaches.

The paper is structured as follows: In Sect. 2 we discuss related work on security requirements elicitation focusing on two areas: (*i*) information security related definitions and their linkage to concepts of enterprise modeling and (*ii*) approaches that are based on process modeling for ensuring information security; to learn about already applied concepts of enterprise modeling in the context of business process related security requirements identification. In Sect. 3 we show how the SREBP method can be extended with the enterprise model frame for more informed discussions about

---

[1] ITSE: "Improvement of IT-Security in Enterprises based on Process Analysis and Risk Patterns (ITSE)", involving university partners from: Estonia, Latvia, and Germany, URL: http://hochschulkontor.lv/en/projects/247.

security requirements. The usability of the frame as an extension of the SREBP method is discussed in Sect. 4. In Sect. 5 we briefly present conclusions and future work.

2 Related Work

There exists a number of approaches, methods, and methodologies proposed for security engineering. A comprehensive survey of these is available in [4]. In this section we will focus only on those approaches that utilize business process models or data flow diagrams. We deliberately limit our paper to this type of approaches because (1) our practical experience covers only business process analysis based approach to security requirements elicitation; and (2) business process models and data flow diagrams can represent information assets (on which we focus in this paper) that are to be secured when they are transferred from one process step (or function, or database) to another.

In SubSect. 2.1 we discuss related work concerning Step 2 of the research method presented in the previous section, namely, identification of security relevant enterprise modeling concepts from information security, information, and data definitions. In Sub-Sect. 2.1 we address Step 3 of the research method – identification of relevant enterprise modeling concepts from business process oriented security requirements elicitation approaches. In Sects. 3 and 4 we further focus on Step 4 of the research method.

2.1 Enterprise Modeling Concepts in Information Security Definitions

In this section we will consider some information security related definitions with the purpose of identifying enterprise modeling concepts relevant for supporting information security requirements elicitation. The definitions are extracted from [7] (basic enterprise modeling related concepts are highlighted by italic):

- protection of *information* and *data* so that unauthorized *persons* or *systems* cannot read or modify them and authorized *persons* or *systems* are not denied access to them (ISO/IEC 12207:2008 Systems and software engineering–Software life cycle processes, 4.39);
- the protection of *computer hardware* or *software* from accidental or malicious access, use, modification, destruction, or disclosure. Security also pertains to *personnel, data, communications*, and the physical protection of *computer* installations. (IEEE 1012-2012 IEEE Standard for System and Software Verification and Validation, 3.1);
- all aspects related to defining, achieving, and maintaining confidentiality, integrity, availability, non-repudiation, accountability, authenticity, and reliability of a *system* (ISO/IEC 15288:2008 Systems and software engineering–System life cycle processes, 4.27);
- degree to which a *product* or *system* protects information and *data* so that *persons* or other *products* or *systems* have the degree of *data* access appropriate to their types and levels of authorization (ISO/IEC 25010:2011 Systems and software engineering–Systems and software Quality Requirements and Evaluation (SQuaRE)–System and software quality models, 4.2.6) Security also pertains to

personnel, data, communications, and the physical protection of *computer* installations.

In ISO/IEC/IEEE 24765c:2014 information security is defined as follows: "Preservation of confidentiality, integrity and accessibility of *information* < ... > in addition, other properties such as authenticity, accountability, non-repudiation and reliability can also be involved." The concept "Accessibility" in this context is closely related to the concept "Availability".

In the context of this paper, of the concepts above, the main ones considered are *people, software, hardware,* and *information* and *data.* Concepts of *system* and *product* are also mentioned. However the discussion of these two concepts is beyond the scope the current paper.

Additionally we considered definitions of *information*, utilized by some standards, where *information* is defined based on the notion *knowledge* but not on the notion of *data* that could be expected in information technology related contexts. The information is defined as:

- *knowledge* that is exchangeable amongst users, about things, facts, concepts, and so on, in a universe of discourse (ISO/IEC 10746-2:2009 Information technology – Open Distributed Processing – Reference Model: Foundations, 3.2.6);
- in information processing, *knowledge* concerning objects, such as facts, events, things, processes, or ideas, including concepts, that within a certain context have a particular *meaning* (ISO/IEC 2382-1:1993 Information technology–Vocabulary–Part 1: Fundamental terms, 01.01.01) It is also important to note that although *information* should necessarily have *a representation* form in order to make it communicable, the interpretation of this representation (the *meaning*) is especially relevant.

The emphasis on knowledge requires considering *knowledge* as one more relevant enterprise modeling concept being utilized for information security requirements identification.

Finally, we have considered the definitions of *data*:

- a representation of facts, *concepts*, or instructions in a manner suitable for communication, interpretation, or *processing* by *humans* or by *automatic means* (ISO/IEC/IEEE 24765:2010 Systems and software engineering–Vocabulary);
- collection of *values* assigned to base *measures*, derived measures and/or *indicators* (ISO/IEC 15939:2007 Systems and software engineering–Measurement process, 3.4);
- *representations of information* dealt with by information *systems* and *users* thereof (ISO/IEC 10746-2:2009 Information technology – Open Distributed Processing – Reference Model: Foundations, 3.2.1);
- re-interpretable *representation of information* in a formalized manner suitable for *communication*, interpretation, or processing. (ISO/IEC 25000:2014 Systems and software Engineering–Systems and software product Quality Requirements and Evaluation (SQuaRE) – Guide to SQuaRE, 4.4) (ISO/IEC 2382-1:1993 Information technology–Vocabulary–Part 1: Fundamental terms, 01.01.02).

From the point of view of enterprise modeling, it is important to highlight concepts of *concept*, *value*, and *measure*. However, we will not address them explicitly further in this paper. We will limit our discussion to the fact that in information security the notions *knowledge* and *meaning* are important.

The highlighted concepts will be further elaborated in Sect. 4.

2.2 Process Related Information Security Requirements Identification

There exist approaches for handling security concerns via business processes [4, 8] and to enforce information security by introducing security mechanisms [9–11]. For instance, the UMLsec approach [9] introduces stereotypes to define secure systems from business processes expressed in activity diagrams. Elsewhere security extensions [10, 11] to the BPMN language are proposed to define access control, separation of duties, and similar constraints.

DFD for security risk management. In [12] Spears proposes a holistic method for information systems risk identification. This approach is interesting as it uses data flow diagrams (which are better equipped for information modeling than BPMN) and information systems architecture (see Step 3 below). The method's steps are:

1. Identify core business functions within the organization and their critical business processes.
2. For each business process identify the critical information system.
3. Obtain an updated architecture diagram of the critical information system that includes its supporting infrastructure, and develop a list of IT assets.
4. Obtain updated data flow diagrams (DFD) to identify user groups, subprocesses, external (including subordinate) systems, and information flows through the systems.
5. Identify confidential information from the DFDs.
6. Update the list of IT assets based on information obtained from DFDs.
7. Determine the relative necessity (or importance) of each IT asset to the business process.
8. Develop a risk scenario for each technical asset of high importance, each type of confidential information, and each user group with access to confidential information.
9. Identify threats and vulnerabilities for each IT asset being analysed.
10. Estimate the impact of a security breach to the asset.

Socio-technical systems model for eliciting security requirements. The Socio-Technical Systems model [13, 14] uses the social view, the information view and the authorization view that concerns goals and information units from the social and information views. The business process is designed according to and verified against security policies that are defined on the base of security requirements.

The approach is tool-supported and has two phases. In the first phase security requirements are elicited. This means that one applies some graphical annotations (in terms of graphical icons) on the business process diagram to identify accountability, auditability, authenticity, availability, confidentiality, integrity, non-repudiation, and privacy requirements. Based on these annotations, in the next phase, one implements (i.e., generate security policies, define and update processes, and verify business processes) particular security requirements to the security policies (i.e., resulting in controls and countermeasures). If socio-technical systems or security policies change, the approach is repeated from the second step; if security requirements change, the approach is iterated from the first step.

SREBP. The SREBP method [6] has no tool support and has less graphical annotations than the approach proposed in [13, 14]. However the approach is based on a generic threat model, takes into consideration assumed attacker capabilities, and suggests security risk oriented patterns to identify threats, to mitigate these threats by introducing security requirements (and their potential controls as constraints on the business process diagram). The SREBP method consists of two stages. In the first stage one has to identify business assets and determine security objectives. In the second, one applies the security risk-oriented patterns to:

1. Identify pattern occurrences in the business process model.
2. Extract the security model based on the pattern occurrences.
3. Derive (textual) security requirements from the graphical security model.

During the first step, one performs activities [15] to identify patterns in the analysed business process model. Once the pattern occurrences are determined, one can extract the security model depending on the security risk-oriented pattern used. Typically this model is expressed using the (security extensions of the) UML modeling language. For instance, when applying a pattern for *securing data from unauthorised access*, one would need to created a UML class diagram describing the role-based access control model; an application of pattern for *securing data that flow between business entities* would result in the UML activity diagram describing the secure communication establishment. This also means that depending on the chosen contextual area (and its associated patterns) different activities for security model extraction could be performed. Once the security model is derived, one can document security requirements textually.

Using the SREBP method, the business process model is a primary source for deriving information security requirements, like the data flow diagram in the DFD based security risk management [12]. Also [13, 14, 16] demonstrate that in practical settings, the business processes are a convenient abstraction level for discussing information security issues. Therefore, taking into consideration that the business process based security requirements elicitation considers information flows in business processes, the extension of the SREBP method was developed for relating information flows in a business process model to the specific enterprise model element structures - the enterprise model frames. The proposed extension is described in the next section.

3   Relating Business Processes to Enterprise Model Frames

In the SREBP approach [17], five security risk oriented patterns are defined to derive security requirements from business processes. These patterns are based on the domain model for Information Systems Security Risk Management (ISSRM) [18] that supports the definitions of asset-related concepts, risk-related concepts and risk treatment-related concepts. The patterns are used within five contextual areas (i.e., one pattern in each area), such as access control, communication channel, input interfaces, network infrastructure, and data store. Pattern application is performed in three steps:

1. *Pattern occurrence identification* in the business process diagram. Pattern identification potentially could be performed using hierarchical level matching, business perspective matching, structural similarity and semantic similarity methods [15]. Once the pattern occurrences are found in the business process model, the second step – security model extraction – is performed. In our experiments we did the first step manually and found that it is a time consuming activity. Some effort was made to support this process with the XML based pattern identification, however, the discussion of the identification algorithms is beyond the scope of this paper.
2. *Security model extraction* is performed following activities, which differ from pattern to pattern. For instance, to create a security model within the access control contextual area, one needs to (*i*) identify resource, (*ii*) identify roles, (*iii*) assign users, (*iv*) identify secured operations, and (*v*) assign permissions.
3. *Security requirements derivation* from the security model. Typically, the security requirements are expressed as conditions that need to be fulfilled by implementing security controls (i.e., countermeasures).

Although *security model extraction* (i.e., Step 2) differs for each pattern, the information object (i.e., business asset, how it is addressed in the SREBP approach) is always identified in the BPMN diagram when applying each pattern. To enhance the clarity with respect to this asset, we propose to relate the BPMN model to the enterprise model frame, used for the analysis of information circulation in viable systems [19].

Our proposal is illustrated in Fig. 1. Here, the enterprise architecture frame (see the right-side of the figure) is used to distinguish between various types of information processors (e.g., human, software, and hardware). It also helps to separate the levels of

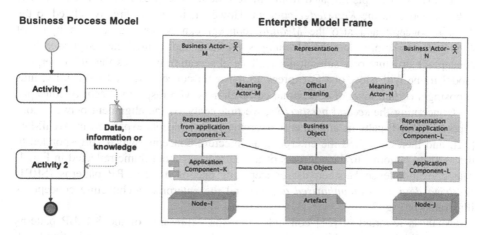

Fig. 1. Security Requirements Elicitations Supported by Enterprise Model Frame (represented in ArchiMate language [20])

security required (e.g., Business, level, Application level, and Technology level). Additionally, the frame illustrates what relationships in the enterprise model should be activated for the particular information object during its transfer from one activity to another. For instance, when sending an e-mail message, only business actors, information representation, e-mail system (application) and hardware are involved; while when transferring the paper format data that is stored in the database, all elements reflected in the frame might be involved. The enterprise model frame helps visualising concerns important in the 2nd step of the SREBP method.

The frame also could extend the data flow based approach with reference to business level elements. The compliance of the frame to ArchiMate language and enterprise modeling methods suggests an opportunity to extend the SREBP patterns to security goals and other concepts if these are present in the enterprise architecture or enterprise model. In Sect. 4 we will discuss the usefulness of the frame with respect to the SREBP method in more detail.

4 Usefulness of the Enterprise Model Frame for SREBP

The analysis of usefulness of the enterprise model frame application to extend the SREBP method is performed taking into account the following perspectives:

1. Compliance with enterprise modeling concepts with respect to security definitions (also see Sect. 2.1).
2. Compliance with enterprise architecture elements directly related to the SREBP patterns.
3. Compliance with enterprise modeling approaches.
4. Practical application.

From the first perspective, Table 1 compares the enterprise modeling related concepts revealed in security oriented definitions with the elements presented in the enterprise model frame. As illustrated in Table 1, the enterprise model frame covers practically all concepts revealed from different definitions presented in Sect. 2.1. The only exception is the *Knowledge* concept. However, it can be directly related to the *Actor*'s concept; and also the *Meaning* concept can be used as a "synonym" of *Knowledge* (assuming that the *Meaning* is explicitly represented, i.e., expressed by a conceptual structure or a sub-ontology). This observation indicates that the enterprise model frame will allow one to express all the main concepts related to security, thus exposing a certain level of conceptual completeness with respect to security concepts.

Concerning the *second perspective*, we first analyzed the alignment between concepts of SREBP patterns and enterprise architecture concepts expressed in ArchiMate [22], and then compared the enterprise architecture concepts, which corresponded to the security patterns, to the concepts of the enterprise model frame reflected in Fig. 1. The correspondence between the concepts of one of the SREBP patterns (SRP1: *Securing data from unauthorised access*) and the enterprise architecture concepts is illustrated in Fig. 2.

Analysis reported in [21] concludes that the expression of the SREBP patterns concerns all enterprise architecture layers – including Business level, Application level, and Technology level – represented in ArchiMate.

Table 1. Security definition concepts versus concepts of the enterprise model frame

Concepts of security definitions	Concepts of enterprise model frame
Information	*Indirectly*: business object, data object
Data	Data object
Persons, personnel	Business actor
Systems (also *product* - not discussed here)	Node, but may be combinations of Node, Application and business actor
Computer hardware, computer	Node
Software	Application, also combination of applications
Communication	The paths between business actors, applications, and nodes
Knowledge (from information definitions)	Can be related to the business actor
Representation of information (from data definitions)	Representation
Meaning (from data definitions)	Meaning

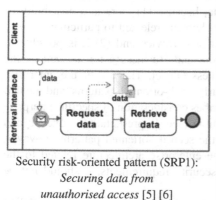

Security risk-oriented pattern (SRP1):
*Securing data from
unauthorised access* [5] [6]

ArchiMate Business Layer Concepts	BPMN Concepts
Business Process	Business Process Diagram, Pools
Function	Task
Business Object	Data Object
Business Event	Event
Business Role	Lane
ArchiMate Application Layer Concepts	**BPMN Concepts**
Data Object	Data Object
ArchiMate Technology Layer Concepts	**BPMN Concepts**
Artefact	Data Object

Fig. 2. Correspondence between Concepts of the SRP1 expressed in BPMN and ArchiMate Concepts

The correspondence between the concepts of the enterprise model frame and the concepts of enterprise architecture identified in [21] is reflected in Table 2. Table 2 indicates that the enterprise architecture concepts identified in [21] are similar to the concepts used in the enterprise model frame. However, the concepts of the enterprise model frame are more *actor-oriented* than the concepts identified by [21]. In the enterprise model frame there is Business actor instead of Business role, which is identified in [21]; and the concepts of the enterprise model frame require considering *representation* and *meaning* of business objects. The use of the enterprise model frame

Table 2. Correspondence between the concepts in the enterprise model frame and the enterprise architecture concepts related to SREBP security patterns as identified in [21]

Concepts of enterprise model frame	ArchiMate concepts related to the SREBP patterns
Business actor	Derived: business role
Representation	Derived: business object
Meaning	–
Application component	–
Data object	Data object
Artefact	Artefact
Node	Device

also requires more thorough analysis at the Application level of the enterprise architecture to compare to what is intended in [21], because in the proposed enterprise model frame particular Application components are also concerned, while in the [21] only Data objects were identified at the Application level of the enterprise architecture.

There are also concepts that are not considered in the enterprise model frame, but were identified in [21] at the Business level of an enterprise's architecture, namely, the enterprise model frame does not consider such concepts as Business event, Business process and Function identified in [21]. Thus we can see that the use of the enterprise model frame in the SREBP method is possible, because (1) there is a possibility of relating the frame to the enterprise architecture elements relevant to particular security risk-oriented patterns via Data object, Artefact, and Device and (2) it is possible to establish the relationship between Business role and Business actor as well as the relationship between Representation and Business Object. The use of the enterprise model frame makes the analysis of the security risk-oriented patterns and related security requirements more concrete as it prescribes consideration of specific enterprise actors and their understanding of the situation (see Representation concept and Meaning concept in the frame) addressed by the security-oriented pattern. However, the frame does not include all concepts of the enterprise architecture related to the security patterns, therefore, for elicitation of security requirements, it should not be separated from the SREBP patterns.

Regarding the compliance of the enterprise model frame with enterprise modeling methods (*the third perspective*), we can see that the enterprise model frame (Fig. 1) consists of concepts adopted from the ArchiMate language [20], which is used in enterprise modeling and enterprise architecture management. In addition the frame is well aligned with contemporary enterprise modeling methods (e.g., 4EM [22]). From the point of view of enterprise modeling, the SREBP method extended with the enterprise model frame is conceptually richer than the approach used in [12], since the Business level, Application level, and Technology level of the enterprise architecture are taken into consideration instead of just information systems architecture addressed in [12]. The frame also allows representing knowledge issues (via meaning), which by definition (Sect. 2.1) are important in information security requirements identification, but are currently scarcely addressed in other business process oriented security requirements elicitation approaches.

From the fourth (practical usage) perspective, the application of the enterprise model frame together with the SREBP method is rather comprehensible and easy. However, tool support is needed to show how relationships in the frame are highlighted depending on chosen security requirements. We have used the frame, also indirectly, as the reference knowledge structure for developing recommendations for practitioners for the application of the SREBP patterns. The recommendations were presented in the excel sheets, which, for each pattern, showed the steps of the user, corresponding security checking activities (representation issues included), and security related steps by the IT system, and the comments representing the meaning of the activities. The detailed description of the excel sheets is beyond the scope of this paper.

The above discussion shows that working with the enterprise model frame is useful and it does not contradict holistic [12] and socio-technical-systems model based [13, 14] methods of information security requirements elicitation. The enterprise model frame might potentially be helpful in supporting these methods; however, further research is needed to understand whether is applicable for security requirements elicitation outside the context of the SREBP method.

5 Conclusions and Future Work

In this paper we analysed how to enrich security requirements elicitation from business process models using enterprise modeling. We considered theoretical concepts of information security definitions and current business process-oriented security requirements elicitation approaches. Our study resulted in the application of the *enterprise model frame*, which is based on the ArchiMate modeling language and complies with the common enterprise modeling methods. In the paper we primarily focused on the conceptual basis of the frame, and its usefulness as an extension the SREBP method, which is of one the business process based security requirements identification approaches. We have concluded that the use of the enterprise model frame as an extension to the SREBP method is useful as it can bridge the security requirements elicitation from business processes and enterprise modeling for the following reasons:

- the enterprise model frame covers practically all concepts identified in theoretical analysis of information security related definitions;
- the enterprise model frame complies with common enterprise modeling and enterprise architecture representation approaches.

At the current state the enterprise model frame is theoretically validated against the basic concepts of information security and some contemporary approaches. Only a small number of practical experiments on a limited scope of process models have been performed. However, the experiments of the enterprise frame application for the SME's procedures, represented in business process modeling language BPMN 2.0, indicated that it is rather easy to use the frame by analysts having knowledge of enterprise modeling, business process modeling, and security requirements engineering. Future work involves further experiments with the enterprise model frame for security requirements elicitation and tool development to further simplify security requirements identification; as well as the comparison of the enterprise architecture frame extended

SREBP method with a larger scope of security-oriented approaches (also beyond the ones directly utilizing business processes) in order to enrich the method, if applicable, with useful new features.

Acknowledgements. The research reflected in this paper was supported in part by the Baltic-German University Liaison Office, German Academic Exchange Service (DAAD), with funds from the Foreign Office of the Federal Republic Germany and in part by the Latvian Council of Science grant for project No 342/2012, and in part by the Latvian National research program SOPHIS under grant agreement Nr.10-4/VPP-4/11.

References

1. Firesmith, D.: Engineering safety and security related requirements for software intensive systems. In: ICSE 2007 Companion, p. 169. IEEE (2007)
2. Jürjens, J.: Secure Systems Development with UML. Springer, Heidelberg (2005)
3. Sindre, G., Opdahl, A.L.: Eliciting security requirements with misuse cases. Requirements Eng. **10**(1), 34–44 (2005)
4. Muñante, D., Chiprianov, V., Gallon, L., Aniorté, P.: A review of security requirements engineering methods with respect to risk analysis and model-driven engineering. In: Teufel, S., Min, T.A., You, I., Weippl, E. (eds.) CD-ARES 2014. LNCS, vol. 8708, pp. 79–93. Springer, Heidelberg (2014)
5. Ahmed, N., Deriving security requirements from business process models. Ph.D. thesis. University of Tartu (2014)
6. Ahmed, N., Matulevičius, R.: Presentation and validation of method for security requirements elicitation from business processes. In: Nurcan, S., Pimenidis, E. (eds.) CAiSE Forum 2014. LNBIP, vol. 204, pp. 20–35. Springer, Heidelberg (2015)
7. Software and Systems Engineering Vocabulary (2015). http://pascal.computer.org/sev_display/index.action
8. Leitner, M., Miller, M., Rinderle-Ma, St.: An analysis and evaluation of security aspects in business process model and notation. In: Proceedings of the Eighth International Conference on Availability, Reliability and Security (ARES), pp. 262–267 (2013)
9. Jürjens, J.: Developing secure systems with UMLsec from business process to implementation. Verlässliche IT-Systeme 2001, DuD-Fachbeiträge, pp. 151–161 (2001)
10. Brucker, A., Hang, I., Lückemeyer, G., Ruparel, R.: SecureBPMN: modeling and enforcing access requirements in business processes. In: Proceedings of the 17th ACM Symposium on Access Control Models and Technologies (SACMAT 2012), pp. 123–126 (2012)
11. Rodriguez, A., Fernandez, M, E., Piattini, M.: A BPMN extension for the modeling of security requirements in business processes. IEICE-TIS(4), pp. 745–752 (2007)
12. Spears, J.L.: A holistic risk analysis method for identifying information security risks. In: Dowland, P., Furnell, S., Thuraisingham, B., Wang, X.S. (eds.) Security Management, Integrity, and Internal Control in Information Systems. IFIP, vol. 193, pp. 185–202. Springer US, New York (2006)
13. Salnitri, M., Dalpiaz, F., Giorgini, P.: Modeling and verifying security policies in business processes. In: Bider, I., Gaaloul, K., Krogstie, J., Nurcan, S., Proper, H.A., Schmidt, R., Soffer, P. (eds.) BPMDS 2014 and EMMSAD 2014. LNBIP, vol. 175, pp. 200–214. Springer, Heidelberg (2014)
14. Salnitri, M., Paja, E., Giorgini, P.: Preserving compliance with security requirements in socio-technical systems. In: Cleary, F., Felici, M. (eds.) CSP Forum 2014. CCIS, vol. 470, pp. 49–62. Springer, Heidelberg (2014)

15. Ahmed, N., Matulevičius, R.: A taxonomy for assessing security in business process Modeling. In: Proceeding of RCIS, pp. 1–10. IEEE (2013)
16. Weske, M.: Business Process Management: Concepts, Languages, Architectures. Springer, Heidelberg (2012)
17. Ahmed, N., Matulevičius, R.: Securing business processes using security risk-oriented patterns. Comput. Stand. Interfaces **36**(4), 723–733 (2014)
18. Dubois, E., Heymans, P., Mayer, N., Matulevičius, R.: A systematic approach to define the domain of information system security risk management. In: Nurcan, S., Salinesi, C., Souveyet, C., Ralyté, J. (eds.) Intentional Perspectives on Information Systems Engineering, pp. 289–306. Springer, Heidelberg (2010)
19. Kirikova, M., Pudane, M.: Viable systems model based information flows. In: Catania, B., Cerquitelli, T., Chiusano, S., Guerrini, G., Kämpf, M., Kemper, A., Novikov, B., Palpanas, T., Pokorny, J., Vakali, A. (eds.) New Trends in Databases and Information Systems. AISC, vol. 241, pp. 97–104. Springer, Heidelberg (2014)
20. ArchiMate 2.1 Specification, Open Group (2013). http://pubs.opengroup.org/architecture/ archimate2-doc/
21. Cjaputa K.: Business process based introduction of security aspects in enterprise architecture. Master thesis, RTU (2016)
22. Sandkuhl, K., Stirna, J., Persson, A., Wißotzki, M.: Enterprise Modeling Tackling Business Challenges with the 4EM Method. Springer, Heidelberg (2014)

Knowledge Management Performance Measurement: A Generic Framework

Latifa Oufkir[✉], Mounia Fredj, and Ismail Kassou

ENSIAS, Mohammed V University in Rabat, Rabat, Morocco
{latifa.oufkir,m.fredj,i.kassou}@um5s.net.ma

Abstract. This theoretical article aims to propose a generic framework for measuring performance of Knowledge Management (KM) projects based on critical literature review. The proposed framework fills the existing gap on KM performance measurement in two points: (i) it provides a generic tool that is able to assess all kinds of KM project as well as the overall organization KM, (ii) it assesses KM projects according to KM objectives in a generic manner. Our framework (GKMPM) relies on a process reference model that provides a KM common understanding in a process based view. It is based on a goal-oriented measurement approach and considers that KM performance dimensions are stakeholder's objectives. The framework application follows a procedural approach that begins with the KM project modelling, followed by the objectives prioritization. The next step consists of collecting and analysing data for pre-designed measures, and produces a set of key performance indicators (KPIs) related to the KM project processes and in accordance with its objectives.

Keywords: Knowledge management project · Performance measurement · Reference model

1 Introduction

Knowledge is well recognized as an important asset for every organization that seeks to sustain in the constantly evolving global competition. Managing knowledge becomes an indisputable necessity [1]. Undoubtedly, a high performing knowledge management (KM) is a key step towards the achievement of the organizational performance. Knowledge Management performance measurement is thus important [2].

Developing an enterprise KM performance measurement system is challenging in many aspects:

- Firstly, researchers and practitioners struggle with establishing Knowledge Management fundamentals. This is due partially to the multidisciplinary and broadness of research spectrum within the KM field [3]. Designing a KM model which is the starting point is still subject to controversy. Consequently, no consensus is reached on what to measure in KM.
- Secondly, KM performance understanding is still confusing since most works dealing with KM performance presumes that either it is linked only to financial performance or it can be intrinsically measured by abstraction of the KM objectives

© Springer International Publishing Switzerland 2016
G. Arnicans et al. (Eds.): DB&IS 2016, CCIS 615, pp. 242–254, 2016.
DOI: 10.1007/978-3-319-40180-5_17

[4]. A literature review [2] shows that KM performance is a multidimensional concept. It is a goal-oriented and it covers other organizational attributes than financial performance such as innovation, decision making, collaboration and learning and competency development.

- Thirdly, existing performance measurement frameworks assess either a specific KM project or the overall organizational KM. literature review shows a lack of generic approach applicable to both cases.
- Fourthly, performance measurement design is more constraining when dealing with a diversity of KM contexts since it requires either multiple performance measurement systems or a generic one that can be applied to all KM projects [5].
- Fifthly, assessing KM would involve multiple measures that require validity and dependency check. They also need to be clearly linked to KM objectives [4].

To address these issues, this paper proposes a generic approach for measuring KM performance based on three principles: (i) a KM process reference model that provides a common KM understanding and discerns what to measure on KM [6], (ii) a goal-oriented measurement that takes into account the multidimensional aspects of performance and (iii) a procedural approach for the application of performance measurement that results in a scorecard of key performance indicators (KPI) clearly linked to KM objectives.

The genericity of the presented framework is related to KM projects scope and KM objectives. Our proposed framework is a starting point that can be enriched through its reference model or KM dimensions when new trends arise.

The remainder of the paper is organized as follows: Sect. 2 provides an overview of the theoretical foundations and sheds light on the literature of KM performance measurement frameworks. Section 3 presents our proposed generic approach structured in three components: the KM process reference model, the identification of KM performance dimensions and the performance measurement approach. In the last section we summarize the main study results and make suggestions for future work.

2 Foundations

2.1 Knowledge Management

KM refers to methods, mechanisms and tools designed towards preserving, valuing, creating and sharing both tacit and explicit knowledge with a view to furthering the organization's objectives [7]. Practically, within organisation, implementing a KM project consists of implementing a socio-technical system that is able to ensure KM processes [1]. From a research perspective, a literature review [2, 8] reveals that research on KM is founded on three perspectives:

- Knowledge resources (KR): are defined as the intangible asset of an organisation [2]. According to Davenport [1], it is commonly broken down into three components: Human capital (employee staff, customer and suppliers), knowledge capital (quantity and quality of knowledge possessed by the firm) and intellectual property (product of knowledge creation that generates value).

- KM processes: refer to the way organization handles their knowledge at various stages considering its dynamic nature. KM is viewed as a collection of processes [7].
- KM factors: are enablers identified as having impact on KM initiatives success [2, 9]. The non-exhaustive enablers list includes cultural, structural, human and technological aspects [1, 10].

In fact, considering the performance measurement purpose, the process-based view seems to be the most appropriate approach [2], as it addresses two main issues:

- Dynamic nature of knowledge: knowledge is both input and output in knowledge activities, it is constantly evolving [11]. As such, KM process based view models the knowledge as a dynamic flow that needs to be monitored [12].
- KM enhancement: within the process based view, the KM performance measurement provides, for deficient processes, actionable indicators. Consequently, KM enhancement is greatly facilitated unlike the resource model that fails to respond to this need due to its static view of the firm knowledge.

However, the assumption that the KM performance is due solely to KM processes disregarding other contextual factors is not accurate [7]. This leads us to consider jointly KM processes and KM factors through the input-process-output Model [13]. Thereby, we will be able to capture the correlation that exists between knowledge enablers, knowledge dynamic and KM project goals achievement.

In order to achieve all benefits of the process-based view, rigorous process identification should be performed and a detailed description of KM activities, inputs and outputs should be processed. The KM processes literature is reviewed based on these requirements.

In fact, several KM process models are proposed in the literature. Globally, these models lack a unified definition and delimitation of processes scope, with the exception of core functions (creating, preserving and transferring knowledge).

SECI is probably the most cited process model. According to Nonaka [11], the dynamic nature of knowledge creation goes through four conversion mechanisms: socialization (tacit knowledge to tacit), externalization (tacit to explicit), internalization (explicit converted into tacit) and combination (explicit to explicit). These four modes constitute a spiralling process of interactions between explicit and tacit, individual and organizational knowledge resulting in new forms of knowledge. SECI model introduces as a KM enabler the BA context; a Japanese term meaning "place", which refers to the specific context needed for knowledge sharing.

Grundstein [6] builds a model composed of four generic processes: identification, preservation, recovery and valuation of knowledge. These processes respond to recurrent knowledge problems recognized in an organization. The proposed model provides a detailed description of knowledge flows by breaking down each generic process into more specific sub-processes.

On the basis of the above review, and to come up with the requirements of the performance measurement, we propose that our KM model comprises two process levels as suggested by Grundstein: (i) the first level includes the generic processes of knowledge identification, preservation, recovery and update, (ii) the second level comprises sub processes issued from level one process decomposition. They correspond when available to the knowledge forms conversions as stressed by Nonaka. Our choice is motivated by two reasons:

- Defining generic aggregated processes in accordance with organisation knowledge problems gives a structuration of KM problems and hence of KM solutions.
- Designing sub-processes as the knowledge forms conversions ensures that all KM activities are captured and documented. Moreover, the granularity and the distinct scope of each sub-process facilitate the performance measures design.

2.2 Goal-Oriented Performance Measurement

This section addresses the second issue of our research questions, which is the performance measurement (PM). It is built on the preliminary studies of performance measurement and on the field progress.

Commonly, it is stated that performance is a multidimensional concept that express stakeholder's goals and expectations. Its measurement designates the reporting and control of project progress toward the achievement of predefined goals. Its purpose is to provide a set of key performance indicators clearly linked to the project goals [14]. Performance measures may address project processes, inputs, direct outputs and outcomes [15].

In order to deepen our understanding of the PM concept in the way that guides us through the design of our PM framework, we review the literature on performance measurement frameworks. Studied systems provide as with a common set of guidelines and best practices for the design of performance measurement system. A good PM system may, in fact, fulfil the following requirements:

- The concept of performance is multidimensional [16]. Dimensions express the expectations and objectives of stakeholders.
- A Performance Measurement Framework should meet two requirements: a model structuring the measurement object and a procedural approach that clearly explains the performance measurement application steps [17].
- A process modelling is a prerequisite step for a performance measurement approach. It provides a structuration and a control of processes that prepares the design of performance measures.
- Performance measures are benchmarks of progress towards the achievement of predefined goals. The links between goals and indicators must be clearly defined and valid through the application of suitable analytical methods [5].

2.3 Review of KM Performance Measurement Frameworks

Over the past years, several attempts have been made to measure the performance of enterprise knowledge management projects. Table 1 summarizes the main characteristics of studied KM performance measurement frameworks. We have suggested five levels of analysis that we consider significant to our research scope:

- *Objective:* it depicts the predefined goal for which the KM project is initiated. In fact, most studies refer to a specific goal which is the achievement of organizational

Table 1. Comparison of studied KM performance measurement framework

Frameworks	Objectives		KM Model			KPI			Scope		Stakeholder	
	Specific	Generic	KM processes	KM processes + others	Others	Normative	To develop	Other	Overall KM	KM project	Employee	Others
[5] KPI for KMS		X	X				X		X			X
[18] KMPI	X		X					X				X
[24] Quality in KMS	X				X	X			X			X
[19] KMPM	X			X		X			X			X
[22] USBS	X				X		X		X		X	
[4] Monte Carlo	X		X			X			X			X
[20] CKMEF	X			X				X	X			X
[23] KMC	X			X					X			X
[21] SKS		X			X	X			X			X
[25] MKMP	X			X			X			X		X

and competitive performance [18–20]. Others keep their study generic and applicable to all kinds of goals [5, 21].

- *KM model*: it refers to what is measured in KM. Among studied frameworks, authors assess KM processes [5, 18, 22] KM resources referred as intelectual capital or KM factors [19, 20, 23].
- *KPI*: it presents how PM frameworks deal with KPI design. This criterion specifies whether KPIs are explicitly produced. Otherwise, only guidelines are provided for KPI design.
- *Scope*: specifies whether the KM performance measurement system that assesses the overall KM is able to assess separately each KM project.
- *Stakeholder*: This criterion specifies whether studied works consider all implied stakeholders when defining KM objectives or is narrowed to a target group. In fact KM objectives are perceived differently from one stakeholder to another. Typical stakeholders are: employees, customer, management…

The critical analysis conducted on the studied frameworks shows that there is a lack of "generic" framework that has the following properties:

- Ability to assess both the overall KM of the organization and the diversity of KM projects separately
- Genericity of objectives (versus framework designed for a single objective, at least the achievement of organizational performance)
- Design of measures and KPI (versus producing some guidelines and templates for KPI design)
- Verification of KPI validity with the application of appropriate analytical methods.

Thus, we define the genericity aspect mainly related to the genericity of objectives, the applicability to all kinds of KM projects, and also to the availability and the validity of KPIs at the end of cycle. This perspective needs to be investigated in the absence of such "generic" framework in the literature.

3 Our Performance Measurement Approach

The previous section shows the diversity and heterogeneity of existing KM performance measurement frameworks. Each of the reviewed frameworks depicts concrete perspectives on KM. Meanwhile, none of the studied framework meets the criterion of genericity as suggested previously which underlines the need for a novel performance measurement approach.

The main idea of our framework is to contribute to the genericity requirement through:

- The design of a "KM process reference model" that depicts the common, shared and justified understanding of KM processes, each process associated with its performance measures.
- The identification of KM performance dimensions and their measures through a literature review, yet the list of dimensions is not exhaustive. The idea is to set up the schema as a starting point of an iterative process that may be enriched each time a new need arise.
- The establishment of a consolidated approach for goal-oriented performance measurement that allows KPI design and validity checking.

3.1 Knowledge Management Process Reference Model

In the enterprise reality, a KM project is regarded as an instantiation of one or more KM processes. It responds to some predefined KM objectives and is influenced by the organization context. Figure 1 illustrates this KM project view as a class diagram using Unified Modeling Language notation (UML).

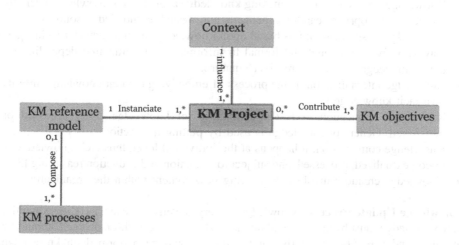

Fig. 1. KM project view, adapted from [27]

Our KM reference model is built mainly on Grundstein model [6] and SECI model [11] and is adapted according to our KM understanding and to the KM processes literature [9, 11, 26]. It is composed of four generic processes:

Knowledge Identification Process. The knowledge identification process responds to the enterprise knowledge location problem. It includes two sub-processes that are:

- Knowledge identification: it relies on knowledge analysis in order to determine knowledge gaps, crucial knowledge and which knowledge needs to be kept, developed or abandoned.
- Knowledge location: once knowledge is identified, it should be located, either by locating experts (owner of tacit knowledge) or locating knowledge sources (for explicit one) [28].

Knowledge Preservation Process. The preservation process relies on memorizing knowledge. The latter includes four sub-processes:

- Knowledge acquisition: knowledge is collected from different sources, either firm repositories or knowledge owners or even external sources like customer, standards, etc. [29].
- Knowledge modelling: once knowledge is captured, it needs to be represented through models in order to make it reusable.
- Knowledge formalization: it refers to the translation of empirical knowledge into structured knowledge that can be processed by a digital system.
- Knowledge storing: it consists of recording and storing codified knowledge for further access [2].

Knowledge Recovery Process. This process deals with how to make knowledge valued. The sub-processes concerned are:

- Knowledge access: consists of making knowledge available for knowledge users by providing appropriate search mechanisms and available knowledge sources.
- Knowledge dissemination: it is based upon knowledge sharing. Transfer techniques may vary from structured and formal to informal and unstructured depending on both knowledge sender and receiver's nature.
- Knowledge internalization: is the process of embodying explicit knowledge into its own tacit knowledge [18].
- Knowledge application: it is the application of knowledge either accessed or transferred. Generally knowledge is used by putting it in action.
- Knowledge combination: it happens at the individual level, thus tacit knowledge is re-contextualized, processed, and subject to induction and deduction reasoning [30].
- Knowledge creation: involves developing new content within the organization.

Knowledge Update Process. Knowledge is very sensitive to changes. Consequently, new knowledge must be incorporated into knowledge sources, those obsolete should be removed and a regular monitoring and update should be performed on knowledge sources [31].

3.2 Measures for KM Process Model

To respond to the performance measurement requirement, the process reference model is enriched by introducing performance measures for all KM sub processes. These measures are grouped in three categories [13, 32]:

- Input metrics: what is used by the process in order to deliver intended result, it may include resources, finance, enablers…
- System metrics: monitor the performance of supporting technology.
- Output metrics: depict what the process aims for in the short term.

We survey the literature on KM performance measures, we retrieve these measures then we categorize them based on our target process reference model as initially they are designed based on their authors adopted KM model.

Table 2 is an excerpt of the classified process measures for "store" sub process.

Table 2. An example of reference model measures

Process	Measures		
	Inputs	Outputs	System
Store	• Strategy for storing knowledge assets [1] • Culture for knowledge store [32] • Management support [32] • Number of knowledge workers [2] • Investment in IT and KMS per year [2]	• Usefulness survey [33] • Amount of the organizational memory (OM) codified and included in the computerized portion of the OM [2] • Working hours per employee spent for inputting knowledge into KMS per month [32] • How often users are contributing to the knowledge resources [2]	• Number of documents and articles uploaded or updated per employee per month [2] • Number of messages or documents stored in the system [34] • Number of active users • Dwell time • Usability survey • Number of registered users • Frequency of updates [34]

3.3 Dimensions of KM Performance

KM performance dimensions are measurable aspects that a KM project aims to achieve. Research conducted on KM goals [35] identified several attributes like financial performance, decision making, innovation, learning and collaboration, competency development and process optimization.

Literature provides a comprehensive content on these dimensions definitions and assessment measures. We present in Table 3 a sample of KM dimensions, at least innovation, its definition and related performance measures.

Table 3. KM dimension definition and measures

Dimension	Definition	Measures
Innovation	Innovation is an abstract human process that produces and implements concrete results(new or modified service, product or process) [36]	Number of new products, R&D, patents, new opportunities, product and service diversification [19]

4   Generic Framework for KM Performance Measurement (GKMPM)

Our GKMPM provides a formal goal oriented approach for KM project performance measurement that requires three prerequisite steps: (i) design the KM process model, (ii) enhance process model with process measures and, (iii) identify all potential KM objectives and related measures from a literature review. Hence, the framework application involves a series of steps that starts with the KM project modelling, followed by prioritization of KM objectives and ends with the performance measurement itself that leads to the interpretation of measurement results.

The framework components as illustrated in Fig. 2 are the following:

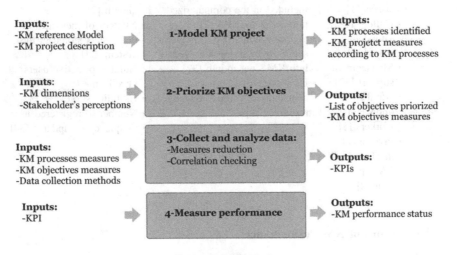

Fig. 2. GKMPM

4.1   Model KM Project

Our Framework aims to assess all kinds of KM projects using the same generic tool. Thus, for a specific KM project, unless project description is provided, we are able to identify all KM processes (derived from the KM reference model) that this project instantiates. Using the transitive property, measures of the instantiated processes are measures of the KM project.

4.2   Prioritize KM Project Objectives

This step foresees to obtain the most relevant KM objectives that the KM project intends to achieve based on stakeholders' perceptions. In fact, surveys about KM objectives importance are submitted to stakeholders who examine all statements (referring to the KM project dimensions previously identified) and prioritize them.

Among decision techniques suitable for prioritizing KM projects objectives, we can use ranking and scoring options like Likert scale (from 1 to 5).

4.3   Collect and Analyse Data

At this point of the framework application, KM assessed project is modelled and instantiated processes obtained and prioritization of KM objectives is provided. Actual step consists of defining how data collection will occur.

To this end, several methods are proposed ranging from survey for qualitative measure to automated data collection system for quantitative ones.

Once data are collected, they need to be processed and analysed as long as we claim that process measures are somehow correlated with goals measures. Accordingly, measures reliability and validity should be verified, correlation between KM processes and KM objectives should be proved as well. To address this point, the following analytical methods are used:

- Exploratory factor analysis: produces a set of distinct non-overlapping variables from the full set of items underlying each construct (sub-process and dimension). This method aims condensing a large number of measures into a smaller set of reliable and valid latent variables.
- Structural equation modelling: by using the Partial Least Square Regression which is a technique that reduces the predictors of a response variable to a smaller set of uncorrelated components and performs least squares regression on these components, we are able to estimate the dependency relationships between KM dimensions and KM processes measures.

4.4   Interpret Measurement Results

After obtaining final measures for KM projects according to both KM processes and KM objectives; the purpose of actual step is to analyze the KM project performance within each KM process and to assess its current performance status. It is also possible to tackle areas of weakness and identify where to operate for enabling substantial performance improvements.

5 Conclusion

This paper presents a generic KM performance measurements approach that emerges from a critical review performed on KM performance measurement frameworks literature. The proposed framework fills the research gap surrounding the KM performance measurement on four points:

- A generic assessment tool applicable to a specific KM project as well as the overall KM in the organization.
- A KM founded understanding of the process based view that results in a KM process reference model.
- A goal-oriented measurement that considers the multidimensional aspect of performance. Dimensions are stakeholder objectives.
- A demonstrated correlation between KM dimensions and KM processes through the application of appropriate analytics methods to the related measures.

The application of the GKMPM implies a series of stages: it starts with the KM project modelling according to the process reference model and is followed by the prioritization of KM objectives. The next stage is to proceed to data collection and analysis, and finally the measurement stage itself that provides an overview of the areas of weakness and the fields of improvement.

This paper provides the theoretical foundations on the framework construction. Nevertheless, the empirical validation of the framework application is not covered. Thus, our future work will address this issue; a multiple case study within a real-world environment will be conducted to obtain insights about reference model validity and framework reliability.

References

1. Davenport, T.H., Long, D.W.De, Beers, M.C.: Successful knowledge management projects. Sloan Manag. Rev. **39**, 43–57 (1998)
2. Wong, K.Y., Tan, L.P., Lee, C.S., Wong, W.P.: Knowledge management performance measurement: measures, approaches, trends and future directions. Inf. Dev. **31**, 239–257 (2013)
3. Lloria, M.B.: A review of the main approaches to knowledge management. Knowl. Manag. Res. Pract. **6**, 77–89 (2008)
4. Kuah, C.T., Wong, K.Y., Wong, W.P.: Monte Carlo data envelopment analysis with genetic algorithm for knowledge management performance measurement. Expert Syst. Appl. **39**, 9348–9358 (2012)
5. Del-Rey-Chamorro, F.M., Roy, R., Wegen, B.Van, Steele, A.: A framework to create key performance indicators for knowledge management solutions. J. Knowl. Manag. **7**, 46–62 (2003)
6. Grundstein, M.: Three postulates that change knowledge management paradigm. In: Hou, H.T. (ed.) New Research on Knowledge Management Models and Methods, pp. 1–22. Intech, Paris (2012)

7. Alavi, M., Leidner, D.E.: Knowledge management and knowledge management systems: conceptual foundations and research issues. MIS Q. **25**, 107–136 (2001)
8. Dudezert, A., Agnes, L.: Performance et Gestion des Connaissances : Contribution à la construction d 'un cadre d'analyse. In: Actes des Journées des IAE, Congrès du cinquantaire., Montpellier, France (2006)
9. Ale, M.A., Toledo, C.M., Chiotti, O., Galli, M.R.: A conceptual model and technological support for organizational knowledge management. Sci. Comput. Program. **95**, 73–92 (2014)
10. Ragab, M.A., Arisha, A.: Knowledge management and measurement: a critical review. J. Knowl. Manag. **17**, 873–901 (2013)
11. Nonaka, I., Toyama, R., Konno, N.: SECI, Ba and leadership: a unified model of dynamic knowledge creation. Long Range Plann. **33**, 5–34 (2000)
12. Lerro, A., Iacobone, F.A., Schiuma, G.: Knowledge assets assessment strategies: organizational value, processes, approaches and evaluation architectures. J. Knowl. Manag. **16**, 563–575 (2012)
13. Lee, H., Choi, B.: Knowledge management enablers, processes, and organizational performance: an integration and empirical examination. J. Manag. Inf. Syst. **20**, 179–228 (2003)
14. Yadav, N., Sagar, M.: Performance measurement and management frameworks: research trends of the last two decades. Bus. Process Manag. J. **19**, 947–971 (2013)
15. Brown, M.G.: Keeping Score: Using the Right Metrics to Drive World Class Performance. AMACOM, New York (1996)
16. Kaplan, R.S., Norton, D.P.: Putting the balanced scorecard to work. In: Schneier, C.E., Shaw, D.G., Beatty, R.W., Baird, L.S. (eds.) Performance Measurement, Management, and Appraisal Sourcebook, pp. 66–74. Human Resource Development Press, Massachusetts (1995)
17. Supply Chain Operations Reference (SCOR ®) model. www.supply-chain.org/scor
18. Chang Lee, K., Lee, S., Kang, I.W.: Measuring knowledge management performance. Inf. Manag. **42**, 469–482 (2005)
19. Chen, M.-Y., Huang, M.-J., Cheng, Y.-C.: Measuring knowledge management performance using a competitive perspective: an empirical study. Expert Syst. Appl. **36**, 8449–8459 (2009)
20. Chen, L., Fong, P.S.W.: Revealing performance heterogeneity through knowledge management maturity evaluation: a capability-based approach. Expert Syst. Appl. **39**, 13523–13539 (2012)
21. Baxter, H., Hower, M.: Improving Organizational KM Through Knowledge Assessment. Technical report, KM World (2014)
22. Chin, K.-S., Lo, K.-C., Leung, J.P.F.: Development of user-satisfaction-based knowledge management performance measurement system with evidential reasoning approach. Expert Syst. Appl. **37**, 366–382 (2010)
23. Zaied, A.N.H.: An integrated knowledge management capabilities framework for assessing organizational performance. int. J. Inf. Technol. Comput. Sci. **4**, 1–10 (2012)
24. Tongchuay, C., Praneetpolgrang, P.: Knowledge quality and quality metrics in knowledge management systems. In: Metro, B. (ed.) The Fifth International Conference on ELearning for Knowledge-Based Society, pp. 21–26. Bangkok Metro, Bangkok (2008)
25. Chua, A., Goh, D.: Measuring knowledge management projects: fitting the mosaic pieces together. In: 2007 40th Annual Hawaii International Conference on System Sciences (HICSS 2007), pp. 192b–192b. IEEE, Waikoloa (2007)

26. King, W.R.: Knowledge management and organizational learning. In: King, R.W. (ed.) Knowledge management and organizational learning, pp. 3–13. Springer US, New York (2009)
27. Rosenthal-Sabroux, C., Grundstein, M.: A global vision of information management. In: CEUR Workshop Proceedings, pp. 55–66. Montpellier (2008)
28. Atwood, C.G.: Knowledge Management Basics. ASTD Press, Alexandria, Virginia (2009)
29. Gasik, S.: A model of project knowledge management. Proj. Manag. J. **42**, 23–44 (2011)
30. Sarrasin, N., Ramangalahy, C.: La gestion cognitive des connaissances dans les organisations. Doc. bibliothèques. **4**, 43–52 (2007)
31. Oztemel, E., Arslankaya, S.: Enterprise knowledge management model: a knowledge tower. Knowl. Inf. Syst. **31**, 171–192 (2012)
32. Hoss, R., Schlussel, A.: How do you measure the knowledge management (KM) maturity of your organization? Metrics that Assess an Organization's KM State. Technical report, USAWC (2009)
33. Robertson, J.: Metrics for knowledge management and content management. http://www.steptwo.com.au
34. Goldoni, V., Oliveira, M.: Knowledge management metrics in software development companies in Brazil. J. Knowl. Manag. **14**, 301–313 (2010)
35. Choy, C.S., Yew, W.K., Lin, B.: Criteria for measuring KM performance outcomes in organisations. Ind. Manag. Data Syst. **106**, 917–936 (2006)
36. Eurostat, O.: Manuel d'Oslo - Principes directeurs pour le recueil et l'interprétation des données sur l'innovation. http://www.uis.unesco.org

Software Testing and Quality Assurance

A Study on Immediate Automatic Usability Evaluation of Web Application User Interfaces

Jevgeni Marenkov[(⊠)], Tarmo Robal, and Ahto Kalja

Tallinn University of Technology, 19086 Tallinn, Estonia
jevgeni.marenkov@gmail.com,
tarmo.robal@ati.ttu.ee, ahto.kalja@ttu.ee

Abstract. More and more web applications are being migrated from desktop platforms to mobile platforms. User experience is extremely different on desktop and portable devices. Changes in user interfaces (UI) could lead to severe violations of usability rules, e.g. changing the text color could lead to decrease of accessibility for users with low vision or cognitive impairments. Manual usability inspection methods are the approaches that help to verify the usability conformance to guidelines. Nevertheless, there are number of difficulties why the aforementioned approaches could not be always applied. The purpose of our research is to develop a conceptual model and corresponding framework including category specific metrics with methodology for immediate automatic usability evaluation of web application user interfaces during design and implementation phase. We address the gap between usability evaluation and development stage of user interface by providing immediate feedback to UI developers.

Keywords: Web usability · Usability guidelines · Web user interface

1 Introduction

The variety of mobile portable devices like smart phones, tablets and smart watches is growing rapidly. As a result, web applications' support for various devices like smart phones and tablets is a crucial requirement for business. With this development, User Interfaces (UI) of web applications should be designed to support various devices with different screen resolution and operating systems. Moreover, user interfaces should be equally compatible with different browsers and their versions. The compatibility is not the only critical requirement to every web application. UI should be understandable, user-friendly, navigable, smooth and easy to use and to learn, having clear structure of information and navigation. All these requirements and characteristics are part of Web usability. Web usability is the ease of use of web application [1]. Usability covers comfort and acceptability of UI to its end- users, efficiency and satisfaction in a context of use and presentation of the content users could understand easily. Usability and user experience as a research branch has been contributed by many researchers from different contemporary research directions. Solutions for automated testing on different platforms [2], enhancement of usability assessing approaches [3, 4], solutions for verifying usability guidelines [5], studies on improvements usability metrics and guidelines [6, 7] are research directions followed by researchers regarding usability

© Springer International Publishing Switzerland 2016
G. Arnicans et al. (Eds.): DB&IS 2016, CCIS 615, pp. 257–271, 2016.
DOI: 10.1007/978-3-319-40180-5_18

analysis. The main outcomes of the aforementioned researches are improved collections of guidelines, enhanced evaluation methodologies for measuring the quality of a UI, and solutions verifying user interface conformance to certain guidelines.

Proper UI design should satisfy multiple requirements and follow usability recommendations and guidelines. There are various studies providing combinations of general usability guidelines and collections of requirements that user interfaces should satisfy [8, 9]. Certainly, most companies have their own adopted guidelines. In common, the evaluation of UI according to the usability guidelines is the responsibility of designers. There are multiple approaches for evaluating the usability starting with the usability inspection methods such as Heuristic Evaluation [10–12], Formal Usability Inspections [13, 14] and The Pluralistic Usability Walkthrough [13, 15]. Another collection of methods for evaluating the usability are the empirical evaluation methods including Card Sorting [16], questionnaires [17], interviews [18] with user-test participants. The main advantage of empirical methods is its direct feedback exposing more severe, more recurring and more global problems [12]. For example, Card Sorting is an effective way to determine how logically and naturally the information and structure of navigation and layout are organised on web sites from the view of its end user. In general, empirical evaluation methods are more effective in finding the problems of workflows and inefficient solutions in UIs. Nevertheless, there are certain obstacles preventing from using such methods widely:

- empirical methods are time and human resource consuming [19];
- it is not always possible to assess every single aspect of UI and increase the coverage of evaluated features due to time, cost and resource constraints [20];
- finding the necessary number of users who belong to a target group is a problem [21];

Certainly, it is impossible to elaborate experimental design without involving large number of quality assurance engineers, designers and applying empirical methods such as questionnaires, interviews with user-test participants. Nevertheless, combining target groups, involving UI testers, and conducting user tests for verifying minor UI changes is not practical in UI designing due to the high time and human resource consumption leading to enormous delays in releases with every UI change. Also, usability inspection methods have been developed to assess usability very fast and at lower cost [21]. Web Content Usability Guidelines (WCAG) [22], UI development recommendations, usability guidelines [8, 9] and organization-specific UI standards and guidelines are the sources and metrics for usability inspection methods. To verify most of the guidelines, there is no need for real users' participation. That is the main reason, why there have been many attempts to develop automatic tools for verifying the UI conformance to predefined guidelines. Nevertheless, these solutions do not consider the opportunity to integrate automatic evaluation to the development process of user interface, structuring their models in a way that could be used only in pre-release testing. Thus, our research addresses the gap between usability evaluation and development stage of user interface by providing feedback immediately to UI developers via a special solution based on defined usability guidelines. Appropriate example could be that even a minor change of user interface (UI) element could lead to severe violations of usability rules; e.g.

changing the text color could lead to the reduce of accessibility for users with low vision or cognitive impairments. The proposed solution would help to highlight this problem of first occurrence.

Assessing UI conformance to guidelines on different platforms has many complexities and as a result is very resource consuming. There is a high variety of platforms and devices with different physical sizes and feature sets. Besides the problem of compatibility, there are also other major factors like platform specific guidelines, information overload and others. Moreover, UI is very fastidious to the modifications, as every change of UI like adding a new element to the screen, updating the element style or color scheme could potentially cause new usability problems. All the aforementioned factors show that modifying the design, structure, color theme or position of UI elements should be done with preliminary analysis of all possible consequences. That begets another critical requirement – assessing the application conformance to usability guidelines every time the developer or designer performs changes of UI.

The purpose of our research is to propose the conceptual model including category specific metrics and the framework containing a language for defining usability guidelines with the tool that is able to automatically evaluate web user interface usability based on predefined guidelines with every single change of user interface.

The rest of the paper is organized as follows. In Sect. 2 we discuss related works of the research area. Section 3 concentrates on the formulation of the problem and its consequences. Section 4 provides an overview of proposed framework, while Sect. 5 provides results after executing sample usability tests on the public governmental portals of Estonia. Finally, Sect. 6 draws conclusions and presents ideas for future research.

2 Related Works

One source of usability guidelines is the Web Content Accessibility Guidelines (WCAG) [22] that provides guidelines to make user interfaces more accessible to people with disabilities. In general, following these guidelines will make UI's more usable for all users. Nevertheless, WCAG as a W3C technical standard does not cover all aspects of usability and does not provide guidelines for all categories of web applications. WCAG provides accessibility guidelines to make content of web applications accessible to a wider range of people with disabilities not covering the other subsets of usability like ease of comprehension, navigation scheme and features, screen based controls and others. That is the reason for designing custom usability guidelines that come together with metrics that assist in accurate guideline evaluation. The goal question metric (GQM) is a de facto standard and the platform for other similar measurement frameworks [6, 7]. There is a set of researches that extend and enhance GQM by introducing area specific usability evaluation metrics [23, 24]. Nebeling et al. present a study about how the visual area of browser is utilised by news web sites at different widescreen resolutions [23]. The main outcome of their study is a number of case sensitive metrics to measure the quality of web page layouts in different viewing contexts. All such results emphasize that there are no common guidelines that could cover all user interfaces from different areas; it is important to introduce and design not just area specific, but project or organization specific UI guidelines.

The development of automatic usability evaluation tools is attractive area of research in the era of multi-platform mobile devices. The main reason for it is that automated solutions tend to detect usability issues quickly and on the early stages of UI development. There is a number of studies that tend to develop a solution that could estimate the usability of UI based on certain rules automatically [5, 25, 26]. Mifsud proposes framework that uses guidelines in predefined format as an input [5]; output is the report presenting UI conformance to the guideline. Particular quality of their research was the practical example of evaluating application UI based on custom guidelines. The solution also provided interface for defining and managing custom evaluation rules. Nevertheless, the definition of custom rules and guidelines was implemented in application specific way; defining custom rules for own purpose is not straightforward. Also, their solution was developed as a web application without possibility to integrate it to any development process.

Evaluation tools that incorporate WCAG guidelines into the framework have been developed to evaluate automatically the web application UI conformance to WCAG [16, 27, 28]. Aizpurua et al. developed framework for evaluation UI conformance to the WCAG guidelines [27]. They also extended their solution by providing the functionality that allows developers to manage their own custom guidelines. The main limitation of their solution is that it was aimed to developers allowing customizing the guidelines only to the people understanding internal structure of user interface. The way, how the guidelines are specified is not fully clear to non-technical team members involved in project.

The next solutions for automatic usability analysis are Multi-Analysis of Guidelines by an Enhanced Tool for Accessibility [29] and OCAWA [30]. Both tools have been developed to evaluate mostly accessibility guidelines of UI. The solutions are based on parsing the HTML and according validation against WCAG recommendations to the HTML syntax excluding any visual analysis of UI. Thus, they cover only accessibility guidelines and there is no opportunity to extend the solution by introducing custom guidelines as they are based only on HTML parsing and not considering executable scripts like the Javascript.

Also, such tools are used with combinations of solutions predicting the usage of UI. Davis and Shipman explored the approach for defining the types of automatically measurable characteristics applicable to a set of pages [31]. They evaluated the usability of UI, and based on collected data developed a model that is effective in providing input data for usability evaluation tools increasing the efficiency of following.

There are researches that analyse the requirements of the automated solutions and modules that they should meet. Dingli and Cassar combined the minimal set of requirements to ideal tool [32]. The most critical of them is to fully automate capture, analysis, and critique activities to be independent of human intervention, employ the Heuristic Evaluation technique for its ability of surfacing the majority of usability problems encountered in a design through the inspection of a set of research-based website usability guidelines compiled in. Their requirements have been used as a platform for the developed model of concept.

3 Problem Formulation

In common, every designed UI contains accordance to various usability guidelines from multiple sources that direct designers to form the interface that would satisfy the requirements of each user. The guidelines are based on the results of researches and experiments from many different fields such as psychology, computer education, user experience and usability. All guidelines are not applicable to all types of applications as each application is unique and needs separate approach. Moreover, in common companies stick to their own developed guidelines filling them maximally into the concept of organization.

Assessing UI conformance to guidelines on different platforms has many complexities and as a result is very resource consuming mainly because of preparing platform specific questionnaires, finding large sample of users and analysis of received feedback. There is a high variety of platforms and devices like smart phones and tablets with different physical sizes and feature sets. Besides the problem of compatibility there are also other major factors like platform specific guidelines, information overload and others. Moreover, UI is very fastidious to the modifications, as every change of UI like adding a new element to the screen, updating the element style or color scheme could potentially cause new usability problems. For example, changing the text color could lead to reduced accessibility for users with low vision or cognitive impairments. All the aforementioned factors show that modifying UI design, structure, color theme or position of elements should be done with preliminary analysis of all possible consequences. That begets another critical requirement – need to assess application conformance to set usability guidelines every time developers or designers change UI structure.

User acceptance tests, manual usability inspection methods such as formal usability inspection and plurastic usability walkthrough are the approaches that help to verify the usability conformance to guidelines and best practices. Nevertheless, there are number of difficulties and problems why the aforementioned approaches could not be always applied. To name a few, these methods are time and human resources [19], it is not always possible to assess every single aspect of UI [20] and finding the necessary numbers of users who belong to target group is a problem [21]. Aforementioned obstacles are motivation for automatic evaluation tools.

The majority of tools for automatic evaluation are Web-based using the URL of web application as an input for automatic UI evaluation. Such approach is unable to evaluate every modification of user interface, and therefore it cannot be fully integrated with user interface design and development process. There are a number of advantages of immediate user interface evaluation after every modification of UI such as the possibility to fix found problems immediately before they could be detected by manual testing. Such approach highlights the usability problem on first occurrence. Appropriate example could be that adding a button or any other element with inappropriate color scheme leads to reduced accessibility for users with low vision or cognitive impairments.

The purpose of our research is to propose the conceptual model including category specific metrics and the framework containing a language for defining usability guidelines. Thus, our research addresses the gap between usability evaluation and development stage of user interface by providing feedback immediately to UI developers via a special solution. The paper delivers framework for immediate usability evaluation.

4   Usability Evaluation Framework

4.1   Usability Evaluation Framework Overview

Before proposing the solution for improving usability and design of web application graphical user interfaces, we designed the concept of the model behind the proposed tool based on category-specific metrics for each category of UI. The model includes core components and separate modules of the framework containing the dependencies between the conceptual components. The main purpose of the model is to formally define the main required components, their dependencies and interactions between components. Figure 1 presents the simplified concept of the model for automatic evaluation of UI conformance to the usability guidelines.

The model is composed of four modules:

- Usability Evaluation Guideline Repositories (UEGR) modules that are responsible for managing the specified guidelines and that contain category-specific metrics for evaluating UI. The model contains category specific guidelines with metrics for each section of UI such as page layout guidelines, navigation guidelines, text appearance guidelines and others. Such categorization simplifies managing the guidelines and provides ability to share common metrics between similar concepts;
- Usability Guideline Definition Language (UGDL) is our custom module responsible for providing the elements and structures for defining custom usability guidelines. We designed UGDL to provide the straightforward way to define custom usability guidelines.
- processing mechanism for transforming the rules defined to the machine understandable way;
- processing Adaptors and usability guideline evaluation mechanism verifies UI conformance to the usability guidelines;

Based on this concept, we have developed a first prototype tool for improving UI of web applications. The input of the tool are the collections of guidelines with the metrics. The tool visually detects elements, sections or groups of elements in UI according to the guidelines, and checks if the values are correspondent to the metrics defined in guidelines or not. Afterwards, it provides report presenting UI conformance to the guideline with detailed test results.

4.2   Usability Guideline Definition Language

Usability Guideline Definition Language is needed to define custom usability guidelines. Usability Guideline Definition Language (UGDL) module (Fig. 1) is responsible for providing the elements and structures for defining them. The main target users of the language are UI designers and product owners being mostly involved in defining

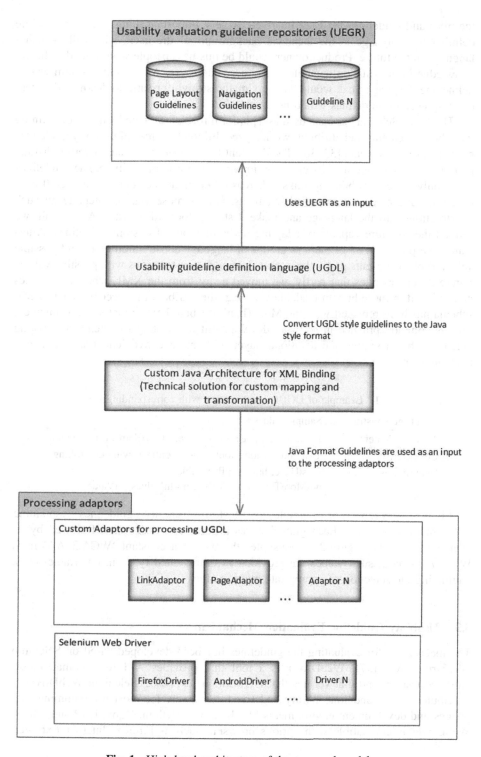

Fig. 1. High level architecture of the proposed model

the rules and guidelines for User Interface. A very important thing to consider, while defining the language is the technical background of the users who will be using language in the future. Product owners could be business people with limited technical knowledge in information technology and user interfaces. As a result, our aim was to define the language that would be maximally straightforward to learn and where defining custom guidelines would be easy.

The XML Schema has been used as a platform for defining the UGDL. For defining the schema elements and attributes we analyzed different sources of usability guidelines published over the years [33–36]. The different formats of guidelines covering distinct parts of user interface and its elements imparted more complexity. Some guidelines cover only the accessibility of links, whereas other guidelines define the overall contrast ratio and color scheme of all elements. Our purpose was to integrate all such constructions into the language and make it straightforward to learn. As a result, we formed the structure capable of adapting different types of custom guidelines. Afterwards, we presented all of them as groups of language construction types and possible value. Table 1 presents partly the various language constructions with possible values. Figure 2 demonstrates that partly, we moved away from the XML schema semantics presenting it in more human readable way; e.g. the attribute minOccures = 1 in XML schema has been replaced with 'notMoreThan' attribute being more self-documented. That is common practice in software development and testing to extend the existing languages by providing the additional layer with more convenient, transparent and readable syntax.

Table 1. Example of UGDL constructions with corresponding values

Language construction	Sample values
Type of element	Link, image, page, everyPage, everyElement, navigation
Characteristic	Title, link, width, font, color, contrast, withoutScrolling
Occurrences	mustHave, has, isFullyVisible notMoreThan, exactValue, notInValues, inValue

Figure 2 presents an example of custom usability guidelines containing self-explained elements. Each guideline has a name and description followed by its evaluation metrics. Figure 2 demonstrates the use of a constant 'WCAG_AA' from WCAG standard; such values are provided as enumerated types in our frameworks, facilitating the selection of appropriate metric value.

4.3 Usability Guideline Evaluation Mechanism

The mechanism for evaluating the guidelines has been developed based on Selenium WebDriver API [37]. WebDriver is a tool that provides API for automated user interface testing, and that verifies that UI works as expected. Selenium WebDriver is distributed as a standalone library and has full support for most programming languages and development environments like Java, C#, Python, Ruby, PHP and others. WebDriver is compatible with various browsers such as Firefox, Internet Explorer,

```
   <guideline name = "4.4 Ensure Link Visual Consistency"
description = "Test that the link color schemes are
consistent with background" >
   <link>
      <color contrastRate = "WCAG_AA" />
   </link>
   </guideline>
   <guideline name = "3.3 Use Bold Text Sparingly"
description = "Test that bold text length is less than
70 characters.">
   <font weight = "BOLD">
      <length notMoreThan = "70"/>
   </font>
   </guideline>
```

Fig. 2. Example of custom usability guideline description

Chrome, Opera, iOS, Android etc. It provides operations and commands for fetching the page, locating UI elements on the fetched pages, clicking the UI elements, filling inputs, moving between windows and others. Thus, it provides the full-stack of instruments and tools that are needed for evaluating various UI guidelines.

Our tool has been developed in Java programming language aiming to be initially tested with Java based web applications. The whole solution is based on fetching the page and its further processing and analysis with the help of Selenium Web Driver API. Such approach reduces the effort needed for processing the page and positioning our solution as an extension to the existing WebDriver library.

Transformation XML Schema based rules into the mechanism readable format is based on Java API for XML Web Service (JAX-WS) providing the functionality for validating the rules according to the schema and generating the mapped Java classes from schema types. Our purpose is to develop the module as independent and all-sufficient as possible. As a result, transformations are done internally by the solution without involving potential users of framework to perform any additional actions.

5   Experiments and Results

To test the proposed usability evaluation framework, nine different web applications of Estonian public service organisations such as hospitals and ministries were selected. These web applications cover different areas such as medicine, real estate, environment and others and they were selected mainly because the target users of such web portals are mainly the inhabitants of Estonia regardless of their social status and experience of using different devices. We selected the web sites of city public hospitals in Estonia covering not only country wide public web sites but emphasizing also the importance of following guidelines in case of local public web sites. As a result, UI of these selected web applications should satisfy common usability guidelines. The next web applications and portals have been selected for experiment: Portal of Ministry of the Environment[1] (20.12.2015) having the same design template as all other ministry web pages of the Republic of Estonia, E-government portal[2] (20.12.2015) providing e-services for all inhabitants in Estonia, Government real estate portal[3] (20.12.2015) providing real estate services, Government info system management portal[4] (21.12.2015) containing repositories for public e-government services, Estonian Research Portal[5] (08.02.2016) providing information about Estonian research activities, and some public healthcare institutions each having absolutely individual design themes: South-Estonian Hospital web site[6] (21.12.2015), Valga hospital web site[7] (21.12.2015), Rakvere hospital web site[8] (21.12.2015), Narva hospital web site[9] (21.12.2015). Selected web pages have absolutely different UI design, target users and categories.

After introducing the language for designing the usability guidelines, we combined a set of predefined guidelines. This set of guidelines containing 117 guidelines from different usability areas such as Navigation, Links, Images, Multimedia and others, has been selected to prepare the test suite covering most usability categories and language structures. For the experiments, the guidelines were grouped into 5 groups based on their category such as page layout guidelines, navigation guidelines and others as follows:

- A: Link Visual Consistency, showing the percentage of inappropriate link visual consistency;
- B: Text visual appearance conformance, showing the amount of lines in bold text used sparingly and the overall amount of bold text lines on page;
- C: Percentage of undescriptive titles;
- D: Maximum size of included graphics in KB and overall size of all graphics;
- E: Navigation visibility, meaning if navigation fits entirely into one screen;

Table 2 shows the results of sample test suite executed on public service portals and web sites highlighting the most critical deviations with bold. Different guidelines contain guideline specific response types. For instance, in case of incorrect link visual consistency the response contains the percentage of elements violating the guidelines and contrast rate of each violated element. Whereas, in case of the guidelines checking if navigation is visible without scrolling, the response contains true or false flag indicating if navigation fits into the page without scrolling or not.

---

[1] http://www.envir.ee.

[2] http://www.eesti.ee.

[3] http://www.rkas.ee.

[4] https://riha.eesti.ee.

[5] https://www.etis.ee.

[6] http://www.leh.ee.

[7] http://www.valgahaigla.ee.

[8] http://www.rh.ee.

[9] http://www.narvahaigla.ee.

Table 2. Results of sample test suite executed on the public sector portal and web sites

Guideline group	A	B	C	D	E
E-government portal	1.5 %	0/5	0 %	109/520	Yes
Government real estate portal	14 %	0/0	0 %	68/245	Yes
Government info system management portal	0 %	1/2	15 %	56/315	Yes
South-Estonian Hospital web site	0 %	2/10	3.5 %	447/543	Yes
Valga hospital web site	9.5 %	0/12	4.5 %	**1951/2500**	Yes
Rakvere hospital web site	**85 %**	1/10	7.5 %	78/392	Yes
Narva hospital web site	39 %	0/66	6 %	154/432	Yes
Portal of Ministry of the Environment	2 %	0/1	1.5 %	95/500	Yes
Estonian research portal	44 %	6/21	2.5 %	16/40	Yes

The majority of guidelines are being satisfied in our test case, nevertheless, there are also critical deviations from usability guidelines. For example, the critical non-conformance to the guidelines has been found in results for Rakvere hospital web site www.rh.ee, in most cases violating (85 % of all links) the next WCAG2 AA color contrast guideline [22]. In this particular case the contrast ratio of most links was 1.91:1, whereas the minimum permitted contrast ration by WCAG 2 AA level is at least 4.5:1. Figure 3 shows the screenshot of the Rakvere hospital web page highlighting the problematic area of user interface. Links have small size and black font on the light blue background. Size of link text, and combination of foreground and background color

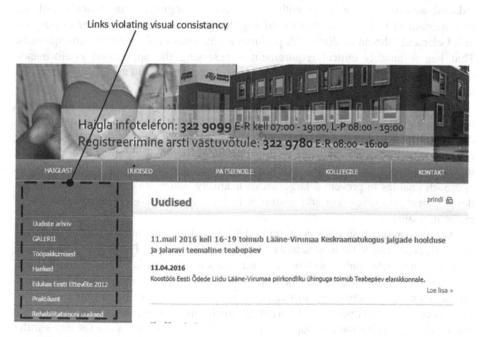

Fig. 3. Rakvere hospital desktop user interface with highlighted problematic area

makes reading of the navigation links complicated. Such violations complicate text reading especially to people with color blindness or other visual impairments, and should be avoided, especially on a hospital website. Estonian research portal also failed that guideline in most cases having 44 % of links violating the correct contrast rate. The results of Valga hospital web site demonstrate a case where the overall size of images is approximately 2 megabytes meaning no optimization has been done to the images used in web application. Such problem increases page load time exceeding the recommended load time of 1 s [38]. Load time becomes especially important while accessing the website over mobile data and on smart devices platforms.

The experiment was based on 117 guidelines combined from different usability areas, and containing different evaluation metrics. The experiment proved that our framework provides enough functionality to define and evaluate most user interfaces based on custom usability guidelines from various categories such as page layout guidelines, navigation guidelines, text appearance guidelines and others. Moreover, framework helps to detect all such violations immediately after modification to UI are introduced. The results of the experiment also showed that portals under study satisfy most of the predefined guidelines having only certain critical deviation from them.

6 Conclusions

User interfaces are very fastidious to the modifications, as every change in UI like adding a new element to the screen, updating the element style or color scheme could potentially cause new usability problems, e.g. changing the text color could lead to reduced accessibility for users with low vision or cognitive impairments. All the aforementioned factors show that modifying design, structure, color theme or position of UI elements should be done with preliminary analysis of all possible consequences. That begets another critical requirement – assessing the application conformance usability guidelines every time the developer or designer perform the changes of the UI structure.

In this paper, we have proposed a conceptual model and a framework containing a language for defining usability guidelines and evaluating UI modifications immediately during UI design and development. Mainly, we address the gap between usability evaluation and development stage of user interfaces by providing immediate feedback to UI developers via a special solution based on defined usability guidelines. This approach enables to prevent a large set of usability issues at once already during UI development and saves cost and resources in later system development phases, especially in testing.

We prepared a test suite containing 117 usability guidelines covering different usability areas and various structures of proposed language. The test suite has been executed on selected public web portals and web sites of Estonia emphasizing the main deviations from the guidelines. The experiment proved that the proposed framework provides enough functionality to evaluate most user interfaces according to the predefined custom guidelines from different categories like Navigation, Color scheme and others already during the development phase. Thus, the solution suits for any similar portal project in the world.

In terms of future research plans, there are several promising directions like finalising the components of framework and performing their extensive testing. Another promising directions is to enhance the framework by providing static repositories with category specific guidelines. Afterwards, the categories guidelines could be integrated as an extra module to our framework.

Acknowledgements. This research was supported by the Estonian Ministry of Research and Education institutional research grant no. IUT33-13.

References

1. Balagtas-Fernandez, F., Hussmann, H.: A methodology and framework to simplify usability analysis of mobile applications. In: 24th IEEE/ACM International Conference on Automated Software Engineering, pp. 520–524. IEEE Press, Washington (2009)
2. Kaasila, J., Ferreira, D., Kostakos, V., Ojala T.: Testdroid: automated remote UI testing on Android. In: Proceedings of the 11th International Conference on Mobile and Ubiquitous Multimedia, Article No. 28. ACM, New York (2012)
3. Wetzlinger, W., Nedbal, D., Auinger, A., Grossauer, C., Holzmann, C., Lettner, F: Mobile usability testing requirements and their implementation in the automation engineering industry. In: Proceedings of the 12th International Conference on Advances in Mobile Computing and Multimedia (MoMM 2014), pp. 62–71. ACM, New York (2014)
4. Porat, T., Schclar, A., Shapira B.: MATE: a mobile analysis tool for usability experts. In: CHI 2013 Extended Abstracts on Human Factors in Computing Systems, pp. 265–270. ACM, New York (2013)
5. Dingli, A.: USEFul: a framework to mainstream web site usability. Int. J. Hum. Comput. Interact. 10–30 (2011)
6. Hussain, A., Ferneley, E.: Usability metric for mobile application: a goal question metric (GQM) approach. In: Proceedings of the 10th International Conference on Information Integration and Web-Based Applications and Services, pp. 567–570. ACM, New York (2008)
7. van Solingen, R., Berghout, E.: The Goal/Question/Metric Method: A Practical Guide for Quality Improvement of Software Development. McGraw Hill, New York (1999)
8. Moreno, A.M., Seffah, A., Capilla, R.: HCI practices for building usable software. Computer **46**, 100–102 (2013)
9. Borges, J.A., Morales, I., Rodríguez, N.J.: Guidelines for designing usable World Wide Web pages. In: Conference Companion on Human Factors in Computing Systems, pp. 277–278. ACM, New York (1996)
10. Nielsen, J., Molich, R.: Heuristic evaluation of user interfaces. In: Proceedings of the SIGCHI Conference on Human Factors in Computing Systems, pp. 249–256. ACM, New York (1990)
11. Yehuda, H., McGinn, J.: Coming to terms: comparing and combining the results of multiple evaluators performing heuristic evaluation. In: CHI 2007 Extended Abstracts on Human Factors in Computing Systems, pp. 1899–1904. ACM, New York (2007)
12. Ekşioğlu, M., Kiris, E., Çapar, B., Selçuk, M.N., Ouzeir, S.: Heuristic evaluation and usability testing: case study. In: Patrick Rau, P.L. (ed.) IDGD 2011. LNCS, vol. 6775, pp. 143–151. Springer, Heidelberg (2011)

13. Hollingsed, T., Novick, D.G.: Usability inspection methods after 15 years of research and practice. In: Proceedings of the 25th Annual ACM International Conference on Design of Communication, pp. 249–255. ACM, New York (2007)

14. Fernandez, A., Abrahao, S., Insfran, E.: Empirical validation of a usability inspection method for model-driven web development. J. Syst. Softw. **86**, 161–186 (2013)

15. Biasm, R.G.: The pluralistic usability walkthrough: coordinated empathies. In: The Pluralistic Usability Walkthrough, pp. 63–76 (1994)

16. Leporini, B., Paterno, F., Scorcia, A.: Flexible tool support for accessibility evaluation. Interact. Comput. **18**, 869–890 (2006)

17. Vuolle, M., Tiainen, M., Kallio, T., Vainio, T., Kulju, M., Wigelius, H.: Developing a questionnaire for measuring mobile business service experience. In: Proceedings of the 10th International Conference on Human Computer Interaction with Mobile Devices and Services, pp. 53–62. ACM, New York (2008)

18. Kantner, L., Sova, D.H., Rosenbaum, S.: Alternative methods for field usability research. In: Proceedings of the 21st Annual International Conference on Documentation, pp. 68–72. ACM, New York (2003)

19. Bak, J.O., Nguyen, K., Risgaard, P, Stage, J.: Obstacles to usability evaluation in practice: a survey of software development organizations. In: Proceedings of the 5th Nordic Conference on Human-Computer Interaction: Building Bridges, pp. 23–32. ACM, New York (2008)

20. Ivory, M.Y., Hearst, M.: The state of the art in automating usability evaluation of user interfaces. ACM Comput. Surv. **33**, 470–516 (2001)

21. Lecerof, A., Paterno, F.: Automatic support for usability evaluation. IEEE Trans. Softw. Eng. **24**, 863–888 (1998)

22. Web Content Accessibility Guidelines. http://www.w3.org/WAI/intro/wcag

23. Nebeling, M., Matulic, F., Norrie, M.C.: Metrics for the evaluation of news site content layout in large-screen contexts. In: Proceedings of the SIGCHI Conference on Human Factors in Computing Systems, pp. 1511–1520. ACM, New York (2011)

24. Nebeling, M., Speicher, M., Norrie, M.: W3touch: metrics-based web page adaptation for touch. In: Proceedings of the SIGCHI Conference on Human Factors in Computing Systems, pp. 2311–2320. ACM, New York (2013)

25. Ivory, M.Y., Mankoff, J., Le, A.: Using automated tools to improve web site usage by users with diverse abilities. Inf. Technol. Soc. **1**(3), 195–236 (2003)

26. Montero, F., González, P., Lozano, M., Vanderdonckt, J.: Quality modles for automated evaluation of websites usability and accessibility. In: Proceedings of the International COST294 Workshop on User Interface Quality Model, pp. 37–43. Interact, Rome (2005)

27. Aizpurua, A., Arrue, M., Vigo, M., Abascal, J.: Transition of accessibility evaluation tools to new standards. In: Proceedings of the 2009 International Cross-Disciplinary Conference on Web Accessibility, pp. 36–44. ACM, New York (2009)

28. Leporini, B., Paternó, F., Scorcia, A.: An environment for defining and handling guidelines for the web. In: Miesenberger, K., Klaus, J., Zagler, W.L., Karshmer, A.I. (eds.) ICCHP 2006. LNCS, vol. 4061, pp. 176–183. Springer, Heidelberg (2006)

29. Leporini, B., Paterno, F., Scorcia, A.: Flexible tool support for accessibility evaluation. Interact. Comput. **18**(5), 869–890 (2006)

30. Automatic Web Accessibility Tool. http://www.ocawa.com/accueilEn.htm

31. Davis, P.A., Shipman, F.M.: Learning usability assessment models for web sites. In: Proceedings of the 16th International Conference on Intelligent User Interfaces, pp. 195–204. ACM, New York (2011)

32. Dingli, A., Cassar, S.: An intelligent framework for website usability. In: Advances in Human-Computer Interaction, Article No. 5 (2014)

33. Nielsen, J., Loranger, H.: Prioritizing Web Usability. New Riders, Berkeley (2006)

34. Dix, A., Finlay, J.F., Abowd, G.D., Beale, R.: Human-Computer Interaction. Pearson Education, Edinburgh (2004)
35. Nielsen, J.: User interface directions for the web: user interface directions for the web. Mag. Commun. ACM **42**(1), 65–72 (1999)
36. Nielsen, J., Tahir, M.: Homepage Usability: 50 Websites Deconstructed. New Riders Publishing, Edinburgh (2001)
37. Selenium WebDriver. http://www.seleniumhq.org/projects/webdriver
38. Nielsen, J.: Usability Engineering. Morgan Kaufmann, London (1993)

Model-Based Testing of Real-Time Distributed Systems

Jüri Vain(✉), Evelin Halling, Gert Kanter, Aivo Anier, and Deepak Pal

Department of Computer Science, Tallinn University of Technology,
Tallinn, Estonia
{juri.vain,evelin.halling,gert.kanter}@ttu.ee
http://cs.ttu.ee/

Abstract. Modern financial systems have grown to the scale of global geographic distribution and latency requirements are measured in nanoseconds. *Low-latency systems* where reaction time is primary success factor and design consideration, are serious challenge to existing integration and system level testing techniques. While existing tools support prescribed input profiles they seldom provide enough reactivity to run the tests with simultaneous and interdependent input profiles at remote frontends. Additional complexities emerge due to severe timing constraints the tests have to meet when test navigation decision time ranges near the message propagation time. Sufficient timing conditions for remote online testing have been proven by Larsen et al. and implemented in Δ-testing method recently. We extend the Δ-testing by deploying testers on fully distributed test architecture. This approach reduces the test reaction time by almost a factor of two. We validate the method on a distributed time-sensitive global financial system case study.

Keywords: Model-based testing · Distributed systems · Low-latency systems

1 Introduction

Modern large scale cyber-physical and financial trading systems have grown to the size of global geographic distribution and their latency requirements are measured in microseconds or even nanoseconds. Such applications where latency is one of the primary design considerations are called *low-latency systems* and where it is of critical importance–to *time critical systems*. A typical example of low-latency system in financial trading domain is multibank online trading (MBOT) where a client can submit a request for quote (RFQ) that is forwarded to multiple participating banks and responded with possibly different prices. When several banks respond with the same price, the fastest bank often wins the client's business. Thus, the latency is the main measure of success in MBOT and often the hardest design concern in both - low-latency and time critical systems.

J. Vain—Department of Computer Science, Tallinn University of Technology, Akadeemia tee 15A, 19086 Tallinn, Estonia; E-mail: juri.vain@ttu.ee.

© Springer International Publishing Switzerland 2016
G. Arnicans et al. (Eds.): DB&IS 2016, CCIS 615, pp. 272–286, 2016.
DOI: 10.1007/978-3-319-40180-5_19

Since large scale systems are mostly distributed systems (by distributed systems we mean the systems where computations are performed on multiple networked computers that communicate and coordinate their actions by passing messages), their latency dynamics is influenced by many technical and non-technical factors. Just to name a few, client profile look up time (few milliseconds) may depend on the load profile, messaging middleware and the networking stacks of operating systems. Similarly, due to cache miss, the caching time can grow from microseconds to about hundred milliseconds [1]. Reaching sufficient feature coverage by integration testing of such systems in the presence of numerous latency factors and their interdependences, is out of the reach of manual testing. Obvious implication is that scalable integration and system level testing presumes complex tools and techniques to assure the quality of the test results. To achieve the confidence and trustability, the test suites need to be either correct by construction or verified against the test goals after they are generated. The need for automated test generation and their correctness assurance have given raise to model based testing (MBT) and the development of several commercial and academic MBT tools. In this paper, we interpret MBT in the standard way, i.e. as conformance testing that compares the expected behaviors described by the system requirements model with the observed behaviors of an actual implementation (implementation under test). For detailed overview of MBT and related tools we refer to [2,3].

2   Related Work

Testing distributed systems has been one of the MBT challenges since the beginning of 90s. An attempt to standardize the test interfaces for distributed testing was made in ISO OSI Conformance Testing Methodology [4]. A general distributed test architecture, containing distributed interfaces, has been presented in Open Distributed Processing (ODP) Basic Reference Model (BRM), which is a generalized version of ISO distributed test architecture. First MBT approaches represented the test configurations as systems that can be modeled by finite state machines (FSM) with several distributed interfaces, called ports. An example of abstract distributed test architecture is proposed in [5]. This architecture suggests the Implementation Under Test (IUT) contains several ports that can be located physically far from each other. The testers are located in these nodes that have direct access to ports. There are also two strongly limiting assumptions: (i) the testers cannot communicate and synchronize with one another unless they communicate through the IUT, and (ii) no global clock is available. Under these assumptions a test generation method was developed in [5] for generating synchronizable test sequences of multi-port finite state machines. However, it was shown in [6] that no method that is based on the concept of synchronizable test sequences can ensure full fault coverage for all the testers. The reason is that for certain testers, given a FSM transition, there may not exist any synchronizable test sequence that can force the machine to traverse this transition. This is generally known as *controllability* and *observability* problem of distributed testers. These problems occur if a tester cannot determine either when to apply a particular input to IUT, or whether a particular output from IUT is generated in response to a specific input [7]. For instance, the controllability problem occurs when the tester at a port p_i is expected to send an input to IUT after IUT has responded to an input from the tester at some other port p_j, without sending an output to p_i. The tester at p_i is unable to decide whether IUT has received that input and so cannot know when to send its input. Similarly, the observability problem occurs when the tester at some port p_i is expected to receive an output from IUT in response to a given input at some port other than p_i and is unable to determine when to start and stop waiting. Such observability problems can introduce fault masking.

In [7], it is proposed to construct test sequences that cause no controllability and observability problems during their application. Unfortunately, offline generation of test sequences is not always applicable. For instance, when the model of IUT is non-deterministic it needs instead of fixed test sequences online testers capable of handling non-deterministic behavior of IUT. But even this is not always possible. An alternative is to construct testers that includes external coordination messages. However, that creates communication overhead and possibly the delay introduced by the sending of each message. Finding an acceptable amount of coordination messages depends on timing constraints and finally amounts to finding a tradeoff between the controllability, observability and the cost of sending external coordination messages.

The need for retaining the timing and latency properties of testers became crucial natively when time critical cyber physical and low-latency systems were tested. Pioneering theoretical results have been published on test timing correctness in [8] where a remote abstract tester was proposed for testing distributed systems in a centralized manner. It was proven that if IUT ports are remotely observable and controllable then 2Δ-condition is sufficient for satisfying timing correctness of the test. Here, Δ denotes an upper bound of message propagation delay between tester and IUT ports. However, this condition makes remote testing problematic when 2Δ is close to timing constraints of IUT, e.g. the length of time interval when the test input has to reach port has definite effect on IUT. If the actual time interval between receiving an IUT output and sending subsequent test stimulus is longer than 2Δ the input may not reach the input port in time and the test goal cannot be reached.

In this paper we focus on distributed online testing of low latency and time-critical systems with distributed testers that can exchange synchronization messages that meet Δ-delay condition. In contrast to the centralized testing approach, our approach reduces the tester reaction time from 2Δ to Δ. The validation of proposed approach is demonstrated on a distributed time-sensitive data acquisition and processing system case study.

3   Preliminaries

3.1   Model-Based Testing

In model-based testing, the formal requirements model of implementation under test describes how the system under test is required to behave. The model, built in a suitable machine interpretable formalism, can be used to automatically generate the test cases, either offline or online, and can also be used as the oracle that checks if the IUT behavior conforms to this model. Offline test generation means that tests are generated before test execution and executed when needed. In the case of online test generation the model is executed in lock step with the IUT. The communication between the model and the IUT involves controllable inputs of the IUT and observable outputs of the IUT.

There are multiple different formalisms used for building conformance testing models. Our choice is Uppaal timed automata (TA) [13] because the formalism naturally supports state transitions and time and there exists a family of tools that support model construction, verification and online model-based testing [16].

3.2 dTron - Extension of TRON for Distributed Testing

dTron[1] [12] extends the functionality of Uppaal Tron [16] by enabling distributed and coordinated execution of tests across a computer network. It relies on Network Time Protocol (NTP) based clock corrections to give a global timestamp (t_1) to events arriving at the IUT adapter. These events are then globally serialized and published to other subscribers using the Spread toolkit [17]. Subscribers can be other IUT adapters, as well as dTron instances. Subscribers that have clocks synchronised with NTP also timestamp the event received message (t_2) to compute and if necessary and possible, compensate for the messaging time overhead $D = t_2 - t_1$. The parameter D is essential in real-time executions to compensate for messaging delays in test verdict that may otherwise lead to false-negative non-conformance results for the test-runs.

3.3 Uppaal Timed Automata

Uppaal Timed Automata [13] (UTA) used for the specification of the requirements are defined as a closed network of extended timed automata that are called *processes*. The processes are combined into a single system by the parallel composition known from the process algebra CCS. An example of a system of two automata comprised of 3 locations and 2 transitions each is given in Fig. 1.

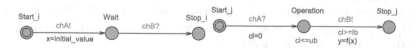

Fig. 1. A sample model: synchronous composition of two Uppaal automata

The nodes of the automata are called *locations* and the directed edges *transitions*. The *state* of an automaton consists of its current location and assignments

---

[1] http://www.cs.ttu.ee/dtron.

to all variables, including clocks. The initial locations of the automata are graphically denoted by an additional circle inside the location.

Synchronous communication between the processes is by hand-shake synchronization links that are called *channels*. A channel relates a pair of edges labeled with symbols for input actions denoted by e.g. chA? and chB? in Fig. 1, and output actions denoted by chA! and chB!, where chA and chB are the names of the channels.

In Fig. 1, there is an example of a model that represents a synchronous remote procedure call. The calling process Process_i and the callee process Process_j both include three locations and two synchronized transitions. Process_i, initially at location Start_i, initiates the call by executing the send action chA! that is synchronized with the receive action chA? in Process_j, that is initially at location Start_j. The location Operation denotes the situation where Process_j computes the output y. Once done, the control is returned to Process_i by the action chB!

The duration of the execution of the result is specified by the interval $[lb, ub]$ where the upper bound ub is given by the *invariant* cl<=ub, and the lower bound lb by the *guard condition* cl>=lb of the transition Operation $\rightarrow$ Stop_j. The *assignment* cl=0 on the transition Start_j $\rightarrow$ Operation ensures that the clock cl is reset when the control reaches the location Operation. The global variables x and y model the input and output arguments of the remote procedure call, and the computation itself is modelled by the function f(x) defined in the declarations section of the Uppaal model.

The inputs and outputs of the test system are modeled using channels labeled in a special way described later. Asynchronous communication between processes is modeled using global variables accessible to all processes.

Formally the Uppaal timed automata are defined as follows. Let Σ denote a finite alphabet of actions $a, b, \ldots$ and C a finite set of real-valued variables p, q, r, denoting clocks. A guard is a conjunctive formula of atomic constraints of the form $p \sim n$ for $p \in C, \sim \in \{\geq, \leq, =, >, <\}$ and $n \in \mathbb{N}^+$. We use $G(C)$ to denote the set of clock guards. A timed automaton A is a tuple $\langle N, l_0, E, I \rangle$ where N is a finite set of locations (graphically denoted by nodes), $l_0 \in N$ is the initial location, $E \in N \times G(C) \times \Sigma \times 2^C \times N$ is the set of edges (an edge is denoted by an arc) and $I : N \rightarrow G(C)$ assigns invariants to locations (here we restrict to constraints in the form: $p \leq n$ or $p < n, n \in \mathbb{N}^+$. Without the loss of generality we assume that guard conditions are in conjunctive form with conjuncts including besides clock constraints also constraints on integer variables. Similarly to clock conditions, the propositions on integer variables k are of the form $k \sim n$ for $n \in \mathbb{N}$, and $\sim \in \{\leq, \geq, =, >, <\}$. For the formal definition of Uppaal TA full semantics we refer the reader to [11,13].

4   Remote Testing

The test purpose most often used in MBT is conformance testing. In conformance testing the IUT is considered as a black-box, i.e., only the inputs and outputs of the system are externally controllable and observable respectively.

The aim of black-box conformance testing according to [10] is to check if the behavior observable on system interface conforms to a given requirements specification. During testing, a tester executes selected test cases on an IUT and emits a test verdict (pass, fail, inconclusive). The verdict shows correctness in the sense of input-output conformance relation (IOCO) between IUT and the specification. The behavior of a IOCO-correct implementation should respect after some observations following restrictions:

(i) the outputs produced by IUT should be the same as allowed in the specification;
(ii) if a quiescent state (a situation where the system can not evolve without an input from the environment [14]) is reached in IUT, this should also be the case in the specification;
(iii) any time an input is possible in the specification, this should also be the case in the implementation.

The set of tests that forms a test suite is structured into test cases, each addressing some specific test purpose. In MBT, the test cases are generated from formal models that specify the expected behavior of the IUT and from the coverage criteria that constrain the behavior defined in IUT model with only those addressed by the test purpose. In our approach Uppaal Timed Automata (UTA) [13] are used as a formalism for modeling IUT behavior. This choice is motivated by the need to test the IUT with timing constraints so that the impact of propagation delays between the IUT and the tester can be taken into account when the test cases are generated and executed against remote real-time systems.

Another important aspect that needs to be addressed in remote testing is functional non-determinism of the IUT behavior with respect to test inputs. For nondeterministic systems only online testing (generating test stimuli on-the-fly) is applicable in contrast to that of deterministic systems where test sequences can be generated offline. Second source of non-determinism in remote testing of real-time systems is communication latency between the tester and the IUT that may lead to interleaving of inputs and outputs. This affects the generation of inputs for the IUT and the observation of outputs that may trigger a wrong test verdict. This problem has been described in [15], where the Δ-testability criterion (Δ describes the communication latency) has been proposed. The Δ-testability criterion ensures that wrong input/output interleaving never occurs.

4.1   Centralized Remote Testing

Let us first consider a centralized tester design case. In the case of centralized tester, all test inputs are generated by a single monolithic tester. This means that the centralized tester will generate an input for the IUT, waits for the result and continues with the next set of inputs and outputs until the test scenario has been finished. This means that the tester has to wait for the duration it takes the signal to be transmitted from the tester to the IUT's ports and the responses

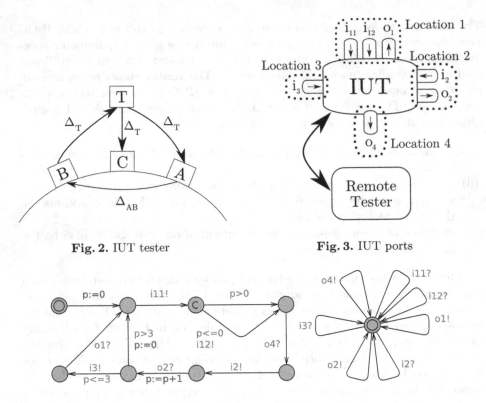

Fig. 2. IUT tester   **Fig. 3.** IUT ports

Fig. 4. Remote tester and IUT models

back from ports to the tester. In the case of IUT being distributed in a way that signal propagation time is nonnegligible, this can lead into a situation where the tester is unable to generate the necessary input for the IUT in time due to message propagation latency. These timing issues can render testing an IUT impossible if the IUT is a distributed real-time system.

To be more concrete, let us consider the test architecture depicted in Figs. 2, 3 and the corresponding model depicted in Fig. 4. In this case the IUT has 4 ports in geographically different locations to interact within the system. In this scenario, the tester sends an input to the port i_{11} and i_{12} at Location 1 and receives a response from port o_4 at Location 4. After receiving the result, the tester sends an input to the port i_2 at Location 2. The IUT responds with an output from port o_2 from Location 2. Next, the tester sends an input to port i_3 at the Location 3 and the IUT will respond from port o_1 at Location 1.

The described IUT is a real-time distributed system, which means that it has strict timing constraints for messaging between ports. More specifically, after sending the first input to ports i_{11} and i_{12} at Location 1 and after receiving the response from port o_4 at Location 4, the tester needs to send the next input to port i_2 at Location 2 in Δ time. But, due to the fact that the tester is not at the same geographical location as the distributed IUT, it is unable to send the next input in time as the time it takes to receive the response and send the next input amounts to 2Δ, which is double the time allotted for the next input signal to arrive.

Consequently, the centralized remote testing approach is not suitable for testing a real-time distributed system if the system has strict timing constraints with nonnegligible signal propagation times between system ports. To overcome this problem, we propose a distributed testing approach, which is described in the next section.

5   Distributed Testing

The shortcoming of the centralized remote testing approach is mitigated with extending the Δ-testing idea by splitting the monolithic remote tester into multiple local testers. These local testers are directly attached to the ports of the IUT. Thus, instead of bidirectional communication between a remote tester and the IUT, only unidirectional synchronization between the local testers is required. The local testers are generated in two steps: at first, a centralized remote tester is generated by applying the reactive planning online-tester synthesis method of [9], and second, a set of synchronizing local testers is derived by partitioning the monolithic tester into a set of location specific tester instances. The partitioning preserves the correctness of testers so that if the monolithic remote tester meets 2Δ requirement then the distributed testers meet (one) Δ-controllability requirement.

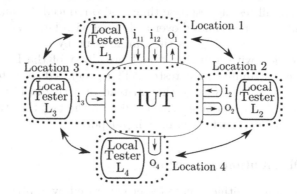

Fig. 5. IUT local tester

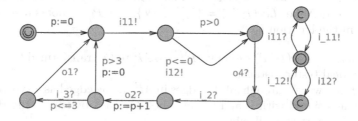

Fig. 6. Local tester at Location 1

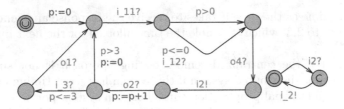

Fig. 7. Local tester at Location 2

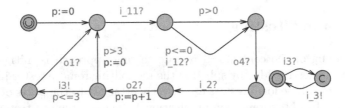

Fig. 8. Local tester at Location 3

We apply the algorithm described in Sect. 5.1 to transform the centralized testing architecture depicted in Fig. 3 into a set of communicating distributed local testers, the architecture of wich is shown in Fig. 5. After applying the algorithm, the message propagation time between the local tester and the IUT port has been eliminated because the tester is attached directly to the port. This means that the overall testing response time is also reduced, because previously the messages had to be transmitted over a channel with latency bidirectionally. The resulting architecture mitigates the timing issue by replacing the bidirectional communication with a unidirectional broadcast of the IUT output signals between the distributed local testers. The generated local tester models are shown in Figs. 6, 7 and 8. Note that instead of active local tester there is only output monitor at Location 4, which is due to the fact that the testers are attached only to the locations which have input ports.

5.1 Tester Distribution Algorithm

Let M^{MT} denote a monolithic remote tester generated by applying the reactive planning online-tester synthesis method [9]. $Loc(IUT)$ denotes a set of geographically different port locations of IUT. The number of locations can be from 1 to n, where $n \in N$ i.e. $Loc(IUT) = \{l_n | n \in N\}$. Let P_{l_n} denotes a set of ports accessible in the location l_n.

1. For each $l, l \in Loc(IUT)$ we copy M^{MT} to M^l to be transformed to a location specific local tester instance.
2. For each M^l we go through all the edges in M^l. If the edge has a synchronizing channel assosiated with it and the channel does not belong to the the set of ports P_{l_n}, we do the following:

- if the channel's action is *send*, we replace it with the co-action *receive*.
- if the channel's action is *receive*, we do nothing.

3. For each M^l we add one more automaton that duplicates the input signals from M^l to IUT, attached to the set of ports P_{l_n} and broadcasts the duplicates to other local testers to synchronize the test runs at their ports.

To verify the correctness of distributed tester generation algorithm we check the bisimulation equivalence relation between the model of monolithic centralized tester and that of distributed tester. For that the models are composed by parallel compositions so that one has a role of words generator on i/o alphabet and the other the role of words acceptor machine. If the i/o language acceptance is established in one direction then the roles of models are reversed. Since the i/o alphabets of remote tester and distributed tester differ due to synchronizing messages of distributed tester the behaviors are compared based on the i/o alphabet observable on IUT ports only. Second adjustment of models to be made for bisimulation analysis is the reduction of message propagation delays to uniform basis either on Δ or 2Δ in both models. The model checking queries to be used for checking bisimulation have structure where the state property under invariance modalities is equivalence of corresponding elements (valuations) of models state vectors in given execution step. For instance when encoding the state elements of model M_i and M_j comparable for bisimulation in arrays S_i and S_j respectively the query to be model checked is $A[](forall(k : dom_k)S_i[k] == S_j[k])$, where $A[]$ expresses invariance on computation tree logic CTL and dom_k is the domain of index variable k on S_i and S_j.

6 Use Case

The need for minimal latency testing is critical when the IUT is a distributed real-time system. A distributed system comprises several subcomponents, which need not necessarily be identical. Two examples of such a system is the interbank trading system and the stock exchange (e.g., NYSE). The components of the interbank trading and stock exchange system are banks and clients that are physically in different places from one another. The interbank trading as well as the stock exchange systems are fully automated and clients use automated trading algorithms (e.g., for hedging or arbitrage). These automated algorithms make the stock market susceptible to flash crashes where the price of the securities can lose billions of dollars in capitalization in a matter of minutes and seconds. Such events have been analyzed post mortem and one of such events, the May 6 flash crash, is discussed in [18]. This is a clear indicator that the processes in this domain can benefit from low-latency distributed online testing to ensure that the processes work according to specifications.

In our sample case, we have an IUT that has four ports. These ports represent clients and banks that are geographically located in different places. In such a situation, the propagation of the input and output signals are not negligible and affect the distributed system process. Each port consists of inputs and outputs,

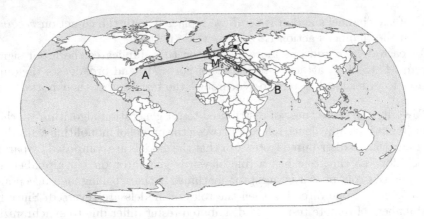

Fig. 9. Geographical layout of IUT ports

but not necessarily both. The inputs of the port represent data input to the subcomponent and output is the output from the subcomponent. In our use case, these represent the quotes (bid and ask prices), orders requests and order confirmations.

In the Fig. 9, we have two banks (A, B), Interbank market (M) and a client (C). The messaging latencies with remote monolithic testing approach are depicted in Fig. 10. The client (C) is connected to each bank information system. In our use case we only have two banks, but in practice there can be any number of banks in the system. The banks are connected to the clients and to the interbank market (M). The client in our use case wants to engage in arbitrage and therefore is waiting for a situation where the banks have different prices for the same financial instrument. The client receives the prices from banks and sends an order request (buy or sell) to the bank they want to buy or sell that financial instrument. The banks receive the interbank prices from the interbank market and they buy and sell the instrument they do not possess themselves from that market for order clearing. The arbitrage opportunity arises when one bank's information system has different prices than the price at the market. The bank forwards the price that it perceives to be the current price to the client, but in reality the price has already changed in the market. This issue is caused completely because of latency and this is also the reason why every participant in the system wants to have as low latency as possible. When the bank receives an order from the client, they fill the order (i.e., send the confirmation of the trade to the client) with the current price that they have received from the interbank market. After they send the confirmation, they send a trade request to the interbank to clear the order (i.e., if the client wants to buy, they must buy from the market). The market advertises the new trade opportunity to the participant banks and the bank, which wants to fill this order will send the trade offer to the market. The client will send opposite order requests to the two banks that have different prices (i.e., buy from one and sell to the other) and if they both get filled, they will make a profit in the amount of the price

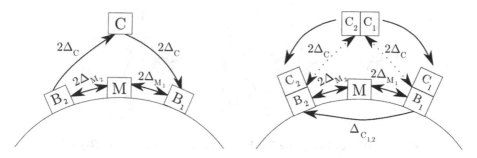

Fig. 10. Remote monolithic testing architecture overview with messaging latencies

Fig. 11. Distributed local testing architecture overview with messaging latencies

difference. The bank which has not up to date prices will suffer this amount as loss. In this use case, we are not considering the bid and ask spreads that are present in the real system. This spread is of course very important in real world system, but is not relevant to demonstrating the Δ-testing methodology.

We start implementing the test architecture, at first, with specifying remote locations and i/o ports of those. After applying the algorithm described in Sect. 5.1, the remote monolithic testing architecture depicted in Fig. 10 is transformed into distributed testing architecture depicted in Fig. 11. The corresponding Uppaal TA model for centralized remote tester represented by Client (C) is depicted in Fig. 12 and IUT in Fig. 13. In this case the banks (B_1 and B_2) and the market (M) all together form the IUT. As can be seen from the Fig. 11, the parallel bidirectional communication channels between the client (C) and banks (B_1 and B_2) have been eliminated as the client tester is split and attached directly to the bank locations. The corresponding local testers at the location of bank B_1 and B_2 are depicted in Figs. 14 and 15. As discussed in Sect. 5, this results in reduced message propagation timing needs and enables testing a real-time distributed system under the timing constraints close to the message propagation time range.

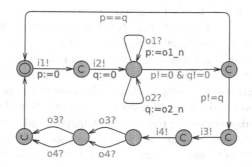

Fig. 12. Client model represents the tester

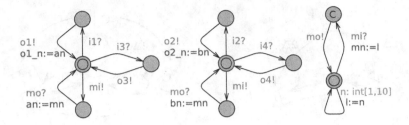

Fig. 13. IUT is formed by Market (M) and banks (B_1 and B_2)

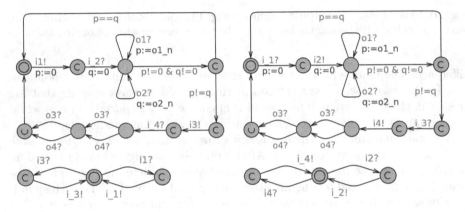

Fig. 14. Local tester at bank B_1 **Fig. 15.** Local tester at bank B_2

7   Conclusion

We extend the Δ-testing method proposed originally for single remote testing by introducing multiple local testers on fully distributed test architecture where testers are attached directly to ports of IUT. Thus, instead of bidirectional communication between a remote tester and IUT only unidirectional synchronization between the local testers is needed in given solution. A constructive algorithm is proposed to generate local testers in two steps: at first, a monolithic remote tester is generated by applying the reactive planning online-tester synthesis method of [9], and second, a set of synchronizing local testers is derived by partitioning the monolithic tester into a set of location specific tester instances. The partitioning preserves the correctness of testers so that if the monolithic remote tester meets 2Δ requirement then the distributed testers meet (one) Δ-controllability requirement. Second contribution of the paper is that distributed testers are generated as Uppal Timed Automata. According to our best knowledge the real time distributed testers have not been constructed automatically in this formalism yet. As for method implementation, the local testers are executed and communicating via distributed test execution environment dTron [10]. We demonstrate that the distributed deployment architecture supported by dTron and its message serialization service allows reducing the total test reaction time by almost a factor of two. The validation of proposed approach is demonstrated on a distributed time-sensitive global financial system case study.

References

1. Brook, A.: Evolution and practice: low-latency distributed applications in finance. Queue - Distrib. Comput. **13**(4), 40–53 (2015). ACM, New York
2. Utting, M., Pretschner, A., Legeard, B.: A taxonomy of model-based testing. Softw. Test. Verif. Reliab. **22**(5), 297–312 (2012). Wiley, Chichester, UK
3. Zander, J., Schieferdecker, I., Mosterman, P.J. (eds.): Model-Based Testing for Embedded Systems. CRC Press, Boca Raton (2011)
4. ISO. Information Technology, Open Systems Interconnection, Conformance Testing Methodology and Framework - Parts 1–5. International Standard IS-9646. ISO, Geneve (1991)
5. Luo, G., Dssouli, R., v. Bochmann, G., Venkataram, P., Ghedamsi, A.: Test generation with respect to distributed interfaces. Comput. Stand. Interfaces **16**(2), 119–132 (1994). Elsevier
6. Sarikaya, B., v. Bochmann, G.: Synchronization and specification issues in protocol testing. IEEE Trans. Commun. 389–395 (1984). IEEE Press, New York
7. Hierons, R.M., Merayo, M.G., Núñez, M.: Implementation relations and test generation for systems with distributed interfaces. Distrib. Comput. **25**(1), 35–62 (2012). Springer
8. David, A., Larsen, K.G., Mikučionis, M., Nguena Timo, O.L., Rollet, A.: Remote testing of timed specifications. In: Yenigün, H., Yilmaz, C., Ulrich, A. (eds.) ICTSS 2013. LNCS, vol. 8254, pp. 65–81. Springer, Heidelberg (2013)
9. Vain, J., Kääramees, M., Markvardt, M.: Online testing of nondeterministic systems with reactive planning tester. In: Petre, L., Sere, K., Troubitsyna, E. (eds.) Dependability and Computer Engineering: Concepts for Software-Intensive Systems, pp. 113–150. IGI Global, Hershey (2012)
10. dTron - Extension of TRON for distributed testing. http://www.cs.ttu.ee/dtron
11. Behrmann, G., David, A., Larsen, K.G.: A tutorial on UPPAAL. In: Bernardo, M., Corradini, F. (eds.) SFM-RT 2004. LNCS, vol. 3185, pp. 200–236. Springer, Heidelberg (2004)
12. Anier, A., Vain, J.: Model based continual planning and control for assistive robots. In: Proceedings of International Conference on Health Informatics, pp. 382–385. SciTePress, Setúbal (2012)
13. Bengtsson, J.E., Yi, W.: Timed automata: semantics, algorithms and tools. In: Desel, J., Reisig, W., Rozenberg, G. (eds.) Lectures on Concurrency and Petri Nets: Advances in Petri Nets. LNCS, vol. 3098, pp. 87–124. Springer, Heidelberg (2004)
14. Tretmans, J.: Test generation with inputs, outputs and repetitive quiescence. Softw.-Concepts Tools **17**(3), 103–120 (1996). Springer
15. Segala, R.: Quiescence, fairness, testing, and the notion of implementation. In: Best, E. (ed.) CONCUR 1993. LNCS, vol. 715, pp. 324–338. Springer, Heidelberg (1993)

16. Hessel, A., Larsen, K.G., Mikucionis, M., Nielsen, B., Pettersson, P., Skou, A.: Testing real-time systems using UPPAAL. In: Hierons, R.M., Bowen, J.P., Harman, M. (eds.) FORTEST. LNCS, vol. 4949, pp. 77–117. Springer, Heidelberg (2008)
17. The Spread Toolkit. http://spread.org/
18. Kirilenko, A., Kyle, A., Samadi, M., Tuzun, T.: The flash crash: the impact of high frequency trading on an electronic market. In: Social Science Research Network (2015). http://www.cftc.gov/idc/groups/public/@economicanalysis/documents/file/oce_flashcrash0314.pdf

Linguistic Components of IS

Detection of Multiple Implicit Features per Sentence in Consumer Review Data

Nikoleta Dosoula, Roel Griep, Rick den Ridder, Rick Slangen,
Kim Schouten[✉], and Flavius Frasincar

Erasmus University Rotterdam, PO Box 1738, 3000 Rotterdam, DR, The Netherlands
{384964nd,416133rg,324065rr,362941rs}@student.eur.nl,
{schouten,frasincar}@ese.eur.nl

Abstract. With the rise of e-commerce, online consumer reviews have become crucial for consumers' purchasing decisions. Most of the existing research focuses on the detection of explicit features and sentiments in such reviews, thereby ignoring all that is reviewed *implicitly*. This study builds, in extension of an existing implicit feature algorithm that can only assign one implicit feature to each sentence, a classifier that predicts the presence of multiple implicit features in sentences. The classifier makes its prediction based on a score function and is trained by means of a threshold. Only if this score exceeds the threshold, we allow for the detection of multiple implicit feature. In this way, we increase the recall while limiting the decrease in precision. In the more realistic scenario, the classifier-based approach improves the F_1-score by 1.6 % points on a restaurant review data set.

Keywords: Feature detection · Aspect category detection · Sentiment analysis · Aspect level sentiment analysis

1 Introduction

In the last decade, a growing amount of retail activity is transferred from the street to the Web. Nowadays, people buy a wide range of consumer goods online using websites such as Amazon or Alibaba. These e-commerce companies often provide an easily accessible platform where consumers can share their experiences with and opinions about their purchases in the form of product reviews. As the required effort for writing these reviews becomes increasingly little, the number of product reviews on online retail shops sharply increased during the last decade. To illustrate this, in 2014 the number of reviews on Amazon exceeded the 10 million [3]. Furthermore, the number of online reviewing platforms, where consumers leave behind product or service reviews, continues to grow.

Using these product reviews for decision making has become increasingly popular [7]. Where some consumers might be looking for specific comments on their potential purchase, others might only be interested in the overall sentiment or in the sentiment per product aspect. However, the number of reviews can be

G. Arnicans et al. (Eds.): DB&IS 2016, CCIS 615, pp. 289–303, 2016.
DOI: 10.1007/978-3-319-40180-5_20

high for some (popular) products, which makes reading all those reviews very time consuming. In order to lower these information costs, one of three pillars in the classical transaction cost model [1], an automatic assessment of the overall sentiment within consumer reviews is asked for.

The main aim of this paper is to contribute to the existing research on the detection of implicit features within consumer reviews. In particular, we seek to extend the method proposed in [6] by adding a classifier that predicts the presence of multiple implicit features within a sentence. The evaluation of our method shows that we can significantly improve the F_1-measure by 1.6 % compared to [6], resulting in an F_1-measure equal to 64.5 %. Apart from increasing the F_1-measure, our method contributes to existing work by its suitability for a more realistic scenario in which sentences are allowed to have more than one implicit feature.

The remaining part of this paper is organized as follows. Section 2 reviews the relating literature and addresses the possible shortcomings of previously proposed methods. After presenting our method in Sect. 3, we discuss the data set used in our experiments in Sect. 4. Section 5 then discusses the implementation of our proposed method and we evaluate its performance in Sect. 6, also by comparing it to previous work in the literature. Section 7 concludes this paper and proposes possible avenues for future research.

2   Related Work

This section discusses the relevant literature in the field that is concerned with the automated assignment of implicit product features within consumer reviews. Our proposed method is motivated by the shortcomings of existing approaches.

The vast majority of approaches in the literature focuses on finding the explicit features in sentences. This limited approach is understandable because often in reviews most of the features are explicitly mentioned in a sentence. However, as addressed before, features that are implicitly mentioned in reviews are equally important. In feature-based sentiment analysis the detection of implicit feature therefore plays an essential role. However, in order to obtain reliable results, sophisticated methods that can infer implicit features from sentences are required. This section addresses some of the most relevant approaches.

A method of detecting implicit features is proposed in [5]. More specifically, the method refers to a two-phase co-occurrence association rule mining approach. In the first phase, [5] mines a set of association rules from co-occurrences between opinion words and explicit features. Therefore each opinion word is associated with a set of candidate features. In the second phase, the explicit features are clustered in order to obtain more powerful rules. If an opinion word is not linked with an explicit feature, the list of rules is checked in order to assign the most likely feature to this opinion word.

A similar approach to [5] is presented in [8]. Specifically, [8] mines as many association rules as possible between feature indicators and the corresponding features. Namely, the indicators are based on word segmentation, part-of-speech tagging, and feature clustering. As basic rules, the best rules in five different rule sets are chosen. In addition, three methods are proposed in [8] to find some set of rules: adding substring rules, adding dependency rules, and adding constrained topic model rules. In the final stage, the results of both approaches are compared where the latter one, using expanding methods, shows the best performance.

One pioneering method for the detection of implicit features is the one of [9], which originates from the following basic idea. A set of several selected opinion words is constructed and the reviews are scanned for so-called modification relationships between these opinions words and corresponding explicit feature words within the same sentences. In other sentences, these opinion words could appear without the presence of an explicit feature. Based on the modification frequencies, a set of candidate features is then determined for these sentences. Then, a co-occurrence matrix is built in which the numbers of co-occurrences between all notional words, i.e. also between non-opinion words and features, are calculated. Using this co-occurrence matrix, constrained by the set of candidates features, the algorithm in [9] selects features using information from all notional words within a sentence. The candidate features that are chosen have co-occurred with the corresponding opinion word before. For example, in the case of digital camera reviews, if the word 'good' appears within the same sentence as the explicitly mentioned features 'battery', 'lens' and 'material', these would be candidate features for an opinion word 'good'. From this set of candidate set of features, the implicit feature is inferred according to the associations between these candidate feature words and the rest of the notional words in the sentence, which are stored in the co-occurrence matrix.

It is important to keep the above described method in mind, since it forms an important building block of the method used by [6], on which this paper expands. The main difference in the approach by [6] is that it uses a supervised algorithm. Namely, consumers review data is used in which all implicit features are annotated. Therefore, co-occurrences can be calculated between these annotated implicit features and all words in the sentence. Based on the co-occurrences, scores are then assigned to potential implicit features, which in the case of [6] are *all* implicit features within the data set. Finally, the implicit feature with the highest score is assigned to the sentence. An advantage of [6], is that it can also be used to detect features that are not present explicitly within the data set. This is an improvement over the methods presented in [5,8,9], where implicit features can only be detected when they also appear explicitly in the data set. Nevertheless, it relies on the existence of training data that is annotated with implicit features.

Furthermore, [6] improves on [9] by introducing a trained threshold in the assignment of implicit features. Where in the method presented in [9] relative low co-occurrence scores could already lead to linking an opinion word to a feature, the algorithm in [6] only assigns an implicit feature to a sentence when its score exceeds the learned threshold. The idea behind this is that when the co-occurrence frequencies are low, it is questionable whether the sentence should be linked to any feature at all. Especially in the case when there are many sentences without any implicit feature, the improvements by using such a threshold show to be large [6].

However, one apparent disadvantage of the detection procedure by [6,9] is that it rules out the possibility that a sentence contains two or more implicit features. This seems an unrealistic constraint, especially in the field of product reviews, where people are explicitly asked for their opinion. In fact, sentences containing two or more implicit features appear quite frequently. For instance, [2] makes the following observation in tweets that were collected from Twitter for their sentiment analysis: even short sentences may contain multiple sentiment types, concerning possibly different topics, e.g. *#fun* and *#scary* in "*Oh My God* http://goo.gl/fb/K2N5z *#entertainment #fun #pictures #photography #scary #teaparty*". [10] sees the same tendency in product review data. From their Chinese restaurant review data, an intuitive example is extracted. In the sentence "the fish is great, but the food is very expensive", two obvious sentiment words can be noticed: 'great' and 'expensive'. Both these words implicitly refer to two different features which could be labeled respectively as 'quality' and 'price'.

3   Method

This section discusses our method that works as an extension on the algorithm developed by [6] in the sense that it allows for the extraction of multiple features per sentence. This more unrestrictive approach considers a more realistic scenario, in which sentences can be related to multiple implicit features.

We start with a short, formal description of the algorithm earlier presented in [6]. From the training data, the algorithm stores all unique annotated implicit features and all unique lemmas (which are the syntactic root form of a word) with their frequencies in list F and O. Furthermore, $|F| \times |O|$ matrix C stores the co-occurrences between all elements in F and O within sentences. Then, sentences in the test data are processed as follows. For each ith implicit feature $f_i \in F$, the sum of the ratios between the co-occurrence $c_{i,j} \in C$ of each jth word in the sentence and the frequency $o_j \in O$ of that word is calculated:

$$Score_{f_i} = \frac{1}{n} \sum_{j=1}^{n} \frac{c_{i,j}}{o_j}, \tag{1}$$

where n is the number of words in a sentence. Finally, the implicit feature with the highest score is assigned to the sentence when it exceeds a trained threshold. When there is no score that exceeds the threshold, no feature is assigned to the sentence. The training of the threshold is only based on the training data and is executed by simply finding the threshold value between 0 and 1 which yields the best performance.

One approach to extend the algorithm to a more realistic scenario is by selecting all implicit features that exceed the trained threshold (see Sect. 2). However, when only a small proportion of the data set consists of sentences that contain more than one implicit features, the precision of the algorithm would

---

Algorithm 1. Algorithm training using annotated data.

Construct list F of unique implicit features
Construct list O of unique lemmas with frequencies
Construct co-occurence matrix C
for all sentence $s \in$ training data **do**
  for all word $w \in s$ **do**
    if $\neg(w \in O)$ **then**
      add w to O
    end if
    $O(w) = O(w) + 1$
  end for
  for all implicit feature $f \in s$ **do**
    if $\neg(f \in F)$ **then**
      add f to F
    end if
    for all word $w \in s$ **do**
      if $\neg((w, f) \in C)$ **then**
        add (w, f) to C
      end if
      $C(w, f) = C(w, f) + 1$
    end for
  end for
end for
Train threshold for the classifier through linear search
Train threshold for the feature detection algorithm through linear search

---

suffer from such a crude selection mechanism. To understand this effect, one should realize that when specific words co-occur often with different implicit features, sentences in which these words are present consequently have a high score for more than one implicit feature. However, assigning more than one implicit feature to each of such sentences based on these scores might be naive when only few sentences are known to contain more than one implicit feature. Another approach to allow for multiple features is to use a classifier to determine the number of implicit features that is likely to be present within the sentence. Subsequently, the algorithm could assign features with top scores to a sentence, where now the number of assignments is based on the classifier's prediction. One should bear in mind however that this strategy now potentially suffers from the imperfect nature of both the classifier and the implicit feature extraction algorithm, which possibly leads to lower precision.

The method that we present works as a combination of the two above-mentioned methods such that we can utilize the advantages of both while minimizing their disadvantages. In particular, we use a classifier in order to detect for every sentence whether there it contains more than one implicit features. If the classifier predicts more than one implicit feature, all features with a score exceeding the threshold will be assigned to the sentence. Otherwise, only the feature with the highest score could be assigned to the sentence, that is, if it

Algorithm 2. Algorithm execution on new sentences in the test data.

Input: trained thresholds $kThreshold$ and $fThreshold$
Construct list NN with the number of nouns per sentence
Construct list JJ with the number of adjectives per sentence
Construct list CM with the number of commas per sentence
Construct list A with the number of 'and' words per sentence
Obtain $\hat{\beta}$'s from logistic regression using the full data set
for all sentence $s \in$ test data **do**
　　$kScore = \hat{\beta}_0 + \hat{\beta}_1 NN(s) + \hat{\beta}_2 JJ(s) + \hat{\beta}_3 CM(s) + \hat{\beta}_4 A(s)$
　　$currentBestFeature = empty$
　　$fScoreOfCurrentBestFeature = 0$
　　for all feature $f \in F$ **do**
　　　　$fScore = 0$
　　　　for all word $w \in s$ **do**
　　　　　　$fScore = fScore + C(w, f)/O(w)$
　　　　end for
　　　　if $kScore > kThreshold$ **then**
　　　　　　if $fScore > fThreshold$ **then**
　　　　　　　　Assign feature f to s
　　　　　　end if
　　　　else if $fScore > fScoreOfCurrentBestFeature$ **then**
　　　　　　$currentBestFeature = f$
　　　　　　$fScoreOfCurrentBestFeature = fScore$
　　　　end if
　　end for
　　if $\neg(kScore > kThreshold)$ **then**
　　　　if $fScoreOfCurrentBestFeature > fThreshold$ **then**
　　　　　　Assign $currentBestFeature$ to s
　　　　end if
　　end if
end for

---

exceeds the trained threshold. Hence, the classifier produces the binary result whether or not to allow for multiple features. The pseudocode describing the described method is shown in Algorithms 1 and 2.

The classifier calculates a score based on a number of sentence characteristics that are related with the number of implicit features k_s within a sentence s. When the score for a sentence exceeds another trained threshold, the classifier predicts multiple implicit features to be present. The score function uses the following variables: (i) number of nouns (#NN$_s$), (ii) number of adjectives (#JJ$_s$), (iii) number of commas (#Comma$_s$), and (iv) the number of 'and' words (#And$_s$). In order to determine the relation between these predictor variables and the number of implicit features, we estimate the following logistic regression equation by maximum-likelihood:

$$Score_{k_s} = \log\left(\frac{p_s}{1 - p_s}\right) = \beta_0 + \beta_1 \#NN_s + \beta_2 \#JJ_s + \beta_3 \#Comma_s + \beta_4 \#And_s, \quad (2)$$

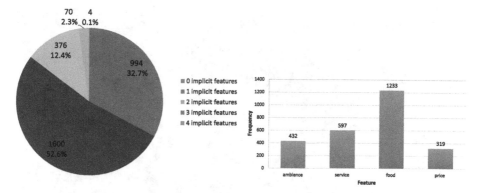

Fig. 1. (a) Distribution of the number of implicit features contained per sentence, in the restaurant review data set. (b) Frequencies of the four unique implicit features in our data set

where p_s is the probability that sentence s contains multiple implicit features. The coefficients are estimated using the full data set. The implementation of this regression approach is discussed in more detail in Sect. 5.

This extended algorithm is now trained in two steps, only using the training data. First, the threshold for the classifier is trained in terms of prediction performance. Second, the threshold for the feature detection algorithm, now using the prediction of the classifier, is trained (as described in the second paragraph of this section) to optimize the feature detection performance.

As a final remark, a limitation of this method is that it requires a sufficiently large data set in which the implicit features are annotated. The reason for this is that the training of the algorithm is executed on annotated implicit features. However, the benefit of this approach is that the algorithm is now able to detect all implicit features within the data set, and not only the features that are (also) *explicitly* present in the data set.

4   Data Analysis

The data set which is used to build up and validate the method proposed in the previous section consists of a collection of restaurants reviews [4]. Every review sentence is assigned to at least one of five so-called review aspect categories: 'food', 'service', 'ambience', 'price', and 'anecdotes/miscellaneous'. These aspect categories are generally not explicitly referred to in a sentence but can be inferred from each sentence. Therefore, these aspect categories operate as *implicit* features of the product, i.e., the restaurant. In the data set, both implicit and explicit restaurant features are labeled.

All 3,044 sentences in the restaurant data set contain at least one implicit feature. However, in order to obtain a better performance test of our classifier for the number of implicit features present in each sentence, the fifth category of 'anecdotes/miscellaneous' is removed from the data set. This particular category

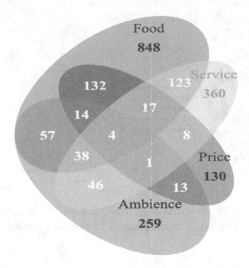

Fig. 2. Co-occurrence frequencies of the four unique implicit features in our data set.

seems most appropriate for removal, as it does not describe a unique implicit feature but refers to the general context 'miscellaneous'. In this way, the number of implicit features in our data set has a wider distribution because part of the set now consists of sentences without an implicit feature. As consumer review sentences generally do not always contain an implicit feature, the performance of our classifier on this more realistic scenario is interesting. Furthermore, in this setting the influence of the threshold parameter in the algorithm by [6] in combination with our classifier can be measured.

As clearly displayed in Fig. 1a, more than half of the sentences contain only one implicit feature. However, in a significant percentage of sentences, namely 12.4 %, two implicit features are mentioned. This motivates an approach that considers more than one implicit feature in a sentence.

Examining the frequency of the four implicit features in Fig. 1b, it is clear that all of them play an important role in customer's reviews. Interesting is the fact that each of them appears in more than 300 sentences which is because there is only a small set of features. More specifically, 'food' captures more than one third of the sentences in total and more than twice of any of the other categories. In terms of frequency, 'food' is followed by the feature 'service' appearing in nearly half as many sentences as 'food'. Feature 'ambience' is implicitly referred to in 432 sentences. Lastly, the least common feature is 'price', where the difference with 'food' is a factor of three.

As the main purpose of the method that we propose is to search for multiple implicit features in each sentence, it seems worthwhile to examine to what extent multiple features are present in one sentence. Figure 2 shows the frequency of all possible co-occurrences between the four unique implicit features. Clearly, most of the sentences in our data contain only one implicit feature, something that

Table 1. Coefficients of logistic regression (2) for the classifier.

Predictor variable	Coefficient	p-value
Constant	-3.019479	0.0000
#NN$_s$	0.116899	0.0002
#JJ$_s$	0.335530	0.0000
Comma$_s$	0.216417	0.0004
And$_s$	0.399415	0.0000

can also be seen in Fig. 1a. More than 4 % of the sentences implicitly refer to both 'food' and 'price', and almost the same percentage corresponds to the co-occurrence of 'food' and 'service'. The remaining combinations of two implicit features appear less frequently in the same sentence in our restaurant review data set.

5   Implementation

To predict the presence of multiple implicit features, we use the score function as given in Eq. 2. We think of this score function as a general rule for categorized review data such as our restaurant review data set. In order to specify the correct score function, however, sufficient amount of this type of consumer review data is required. Constrained by resources, however, only the same restaurant review data set is available to us. Therefore, the score function is not trained on a training part of the data set, and then tested on a test part. Instead, the full data set is used in order to maximize the information available to us.

We estimate the $Score_{k_s}$ function (2) using logistic regression. Table 1 displays the results. The p-values indicate that our variables are highly significant, i.e., for significance levels below 1 %. Apart from the variables that we include, we also test implementing the number of words in a sentence and the number of grammatical subjects in a sentence. Neither of these variables yield a significant improvement. Intuitively, this can be explained because the variables for the number of nouns and adjectives already capture the relevant information that lies within the number of words within a sentence. The number of subjects possibly does not perform better than the number of nouns because often the subject in a sentence is the product instead of the feature.

The regression is performed on the complete restaurant data set, as motivated above. However, one could argue that this could result in unfair performance, as the same data set is used to evaluate our algorithm. However, when the coefficients of the regression are robust for different subsamples, specifying a different score function based on an arbitrary train part of the data set will not alter the results heavily. Put differently, this would indicate that our approach of using the full data set does not provide an unfair edge. To check whether this is the case, we perform the logistic regression 1000 times on arbitrary subsamples

Table 2. Specifications of 1000 logistic regressions on 90 % subsamples.

Variable	Mean	Median	Std. dev.
#NN$_s$	0.117361	0.11768	0.011342
#JJ$_s$	0.335538	0.33536	0.014345
Comma$_s$	0.216409	0.21672	0.023185
And$_s$	0.399507	0.39892	0.023409

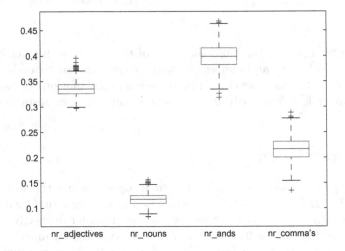

Fig. 3. Box-plot of the coefficients of the logistic regression.

containing 90 % of the data set. Figure 3 depicts the coefficients of the 1000 regressions in a box-plot and Table 2 provides descriptive statistics. The constant is excluded from the plot and table, because it does not influence the result with a trained threshold. We find that the values of the coefficients do not differ a lot for the different subsamples, so it is justified to use the complete data set when determining the coefficients.

The classifier predicts multiple implicit features for sentence s when Score$_{k_s}$ is larger than a certain threshold. We can therefore train the classifier by determining the optimal threshold. In order to do so, we isolate the performance of the classifier by assuming that the feature detection part of the algorithm is perfect. That is, if the classifier predicts the presence of multiple implicit features correctly, we assign all golden implicit features to that sentence; if the classifier predicts incorrectly, we assign either only one golden implicit feature (in case there are actually multiple implicit features), or one implicit feature too many (in case there are not actually multiple features) to the sentence. This way, the errors made by the classifier are isolated and can thus be minimized by means of altering the threshold.

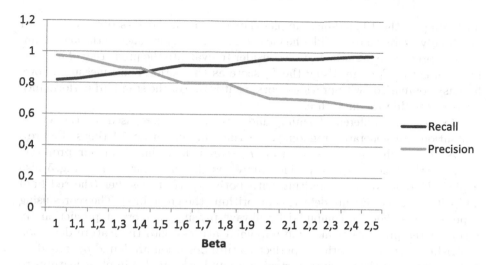

Fig. 4. Recall and precision of the classifier for different β's.

The classifier is optimized on F_β-score. Since the main goal of our classifier is to predict multiple implicit features when they are present, high recall is especially important. If it incorrectly predicts no multiple implicit features in the sentence, the recall of the final score will always decrease, because there can only be one implicit feature assigned to that sentence. However, when the classifier incorrectly predicts the presence of multiple implicit features, the precision of the final score does not necessarily decrease. The threshold in the feature detection part of the algorithm could prevent that multiple implicit features are assigned to a sentence. Figure 4 shows the precision and recall of the classifier with a trained threshold for different β's. It can be seen that with β larger than 1.8, the precision decreases relatively fast, while the recall only increases a little bit. Therefore, we use β equal to 1.8 in the F_β-score when training the classifier to emphasize recall.

Finally, the threshold is trained on an annotated training set containing 90 % of the data. To train the threshold, a range of threshold values needs to be defined. We use values between -3 and 3, with a step size of 0.1. With every threshold, the classifier is evaluated based on $F_{1.8}$. Hence, after linearly trying all possible thresholds, we use the threshold with the largest $F_{1.8}$-score.

6   Evaluation

Evaluation of the implemented method is based on 10-fold cross-evaluation. This means that the whole data set is split into two subsets: one part contains 90 % of the data, the other part 10 %. The algorithm is then trained on this 90 % of the data set. The trained algorithm then detects the implicit features in the remaining 10 % of the data. This procedure is repeated 10 times, where there is no overlap in the 10 hold-out samples. For each fold, the F_1-score is calculated and finally averaged to provide the measure for the performance of the algorithm.

The predictive performance we consider to evaluate the predicting of implicit features is the F_1-score. Using the F_1-score as the performance measure allows for easy comparison with previous work, as it is one of the standard performance measures within the literature.

Because the different training and test subsamples used in the cross-evaluation are generated randomly, we run our algorithm 10 times. Figure 5 shows the results, in terms of mean F_1-scores, following from our proposed method (the blue bars). To provide more insights into our results, Fig. 5 also depicts F_1-scores of the algorithm with both a perfect classifier (the red line) and with a perfect feature detection algorithm (the green line). The scores using a perfect classifier are computed by always passing the correct prediction (in terms of the presence of multiple implicit features) onto the feature detection algorithm. The scores with the perfect feature detection are found by, based on the prediction of the classifier, assigning a number of golden implicit features to the sentences.

Results are given for different part-of-speech filters, which are used to filter out possibly irrelevant words in the co-occurrence matrix that could be harmful to the performance of the algorithm. Figure 5 shows the scores for 16 different part-of-speech filters. The filters include only the words of types that are mentioned, where NN stands for nouns, VB for verbs, JJ, for adjectives and RB for adverbs. Examining the F_1-scores in Fig. 5, we find that the best results are obtained using the NN+JJ part-of-speech filter. That is, filtering for nouns and adjectives, we obtain an F_1-score equal to 64.5 %. We note that the F_1-score we find for using the NN filter is only marginally worse, namely 64.1 %.

Since our proposed method extends the one presented in [6], we start by evaluating the increase in performance as a result of our extension. In order to do so, we also evaluate the unextended algorithm as presented in [6] 10 times using the NN+JJ part-of-speech filter. Again, we note that each evaluation provides slightly different results due to the random nature of the cross-evaluation method. We find a mean F_1-score of 62.9 % for the algorithm without the classifier.[1] Hence, comparing this to our 64.5 %, we find an improvement of 1.6 % points. We test for significance by means of a two-sample t-test. This results in a t-test statistic equal to 12.0, which indicates a significant improvement at the significance level of 1 %.

At first sight, an improvement in mean F_1-score of 1.6 % points may not look large. However, in order to make a fair statement about the performance of our classifier, we first need to compare it to its potential, which is displayed by the red line in Fig. 5. We see that the potential is also at its largest for the NN+JJ filter, giving an F_1-score of 69.3 %. This means that our classifier captures 25 % of the maximum improvement that can be gained by adding a classifier that predicts the presence of multiple implicit features within a sentence, which is 6.4 %.

---

[1] We note that in [6], based on a number of runs, a maximum F_1-score of 63.3 % is reported.

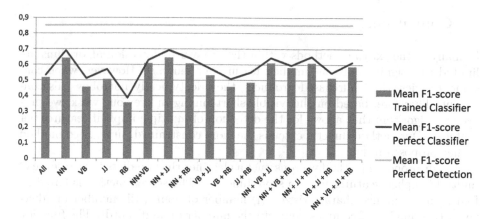

Fig. 5. Mean F_1-scores with different part-of-speech filters.

However, we note that the potential of such a classifier depends on the data set. In our restaurant review data set, 14.8 % of the sentences contain more than one implicit feature, where 12.4 % of the sentences contain two implicit features. Furthermore, calculations show that 20.4 % of the total possible implicit features remain to be detected when only one implicit feature per sentence is considered.[2] However, it is important to notice that the most apparent implicit feature in each sentence is already detected. As a result, the second implicit feature would be assigned with an already lower precision, tempering the improvement of the F_1-score due to a higher recall. Therefore, in light of these insights and considering the simplicity of our approach, we consider our gained improvement to be significant.

Lastly, we provide insights in our results by looking at the F_1-score that can be obtained by using our classifier in combination with a perfect feature detection algorithm, which is displayed by the green line in Fig. 5. Notice that our classifier is trained to maximize the $F_{1.8}$-score. For this reason, these 'potential' F_1-scores are hardly interpretable and a greater potential might be visible when the classifier is trained for the same measure by which it is now evaluated. However, training for $F_{1.8}$ yields best *overall* performance in our method, which is also motivated in Sect. 5. Nonetheless, these F_1-scores provide insight in what part in the loss of F_1-score can be attributed to the feature detection part of our algorithm. The F_1-scores with perfect detection, which do not rely on the part-of-speech filters, are 85.2 %. Comparing this result with the one in the previous paragraph, we conclude that improving the feature detection part of the algorithm shows greater potential than improving the prediction of the presence of multiple implicit features.

---

[2] Based on the distribution of the number of implicit features per sentence in our data set (see Fig. 1a), we have: $(12.4+2\cdot2.3+3\cdot0.1)/(52.6+2\cdot12.4+3\cdot2.3+4\cdot0.1) = 0.204$.

7   Conclusion

In many of the existing methods within the literature, detection algorithms are limited to assigning only one implicit feature per sentence. However, when consumers review their purchased products, they do typically not obey this constraint. Therefore, based on this visible shortcoming in previous work, we propose an algorithm that allows for the detection of multiple implicit features per sentence. Our method directly extends the more constrained, supervised method earlier proposed in [6].

In our proposed method we construct a classifier that predicts the presence of multiple implicit features using a score function. The score function is based on four simple sentence characteristics: (i) number of nouns, (ii) number of adjectives, (iii) number of commas, and (iv) the number of 'and' words. The function parameters are estimated by means of logistic regression and we train a threshold for better performance. Based on the prediction of the classifier for a given review, the feature detection part of our algorithm then looks for either one or multiple implicit features.

Considered on a restaurant review data set, our approach shows small but significant improvement with respect to the constrained method in [6]. That is, we improve the F_1-measure by 1.6% points. Based on analysis of the performance of our classifier we conclude that we capture a reasonable (considering its simplicity) part of the full potential of our approach. The performance and potential of the classifier is however dependent on the distribution of the number of implicit features per sentence within the data set. That is, when consumer reviews frequently cover multiple implicit features per sentence, our more realistic approach is desirable.

In our approach we determine a *general* relation between sentences written in consumer reviews and the number of implicit features. Nonetheless, it might be desirable to integrate the specification and estimation of this relation in the training part of the algorithm in order to make it specifically effective for a given data set. One promising path for future work is therefore to train a classifier for the number of implicit features by using more advanced machine-learning techniques, such as Support Vector Machines. Also, rule learning methods could be employed in order to determine more indicators for the presence of multiple implicit features.

Another interesting suggestion for future research may be to combine the classifier with sentiment analysis algorithms. Namely, when there are opposing sentiments within one sentence, it seems likely that the consumer is commenting on two different features of the product. To illustrate this idea, we provide the following example:

"The phone looks great, but the pictures it takes are of very low quality."

In this sentence, two features are implied: 'appearance' and 'camera'. Also, there are two sentiment polarities: the consumer is positive about the appearance, but negative about the camera.

Acknowledgments. The authors are partially supported by the Dutch national program COMMIT.

References

1. Dahlman, C.: The problem of externality. J. Law Econ. **22**(1), 141–162 (1979)
2. Davidov, D., Tsur, O., Rappoport, A.: Enhanced sentiment learning using Twitter hashtags and smileys. In: COLING Proceedings of the 23rd International Conference on Computational Linguistics, pp. 241–249 (2010)
3. Floyd, K., Freling, R., Alhoqail, S., Cho, H., Freling, T.: How online product reviews affect retail sales: a meta-analysis. J. Retail. **90**(2), 217–232 (2014)
4. Ganu, G., Elhadad, N., Marian, A.: Beyond the stars: improving rating predictions using review content. In: Proceedings of the 12th International Workshop on the Web and Databases (WebDB 2009), Rhode Island, USA (2009)
5. Hai, Z., Chang, K., Kim, J.: Implicit feature identification via co-occurrence association rule mining. In: Gelbukh, A.F. (ed.) CICLing 2011, Part I. LNCS, vol. 6608, pp. 393–404. Springer, Heidelberg (2011)
6. Schouten, K., Frasincar, F.: Finding implicit features in consumer reviews for sentiment analysis. In: Casteleyn, S., Rossi, G., Winckler, M. (eds.) ICWE 2014. LNCS, vol. 8541, pp. 130–144. Springer, Heidelberg (2014)
7. Senecal, S., Nantel, J.: The influence of online product recommendations on consumers' online choices. J. Retail. **80**, 159–169 (2004)
8. Wang, W., Xu, H., Wan, W.: Implicit feature identification via hybrid association rule mining. Expert Syst. Appl. **40**(9), 3518–3531 (2013)
9. Zhang, Y., Zhu, W.: Extracting implicit features in online customer reviews for opinion mining. In: Proceedings of the 22nd International Conference on World Wide Web Companion (WWW 2013 Companion), pp. 103–104. International World Wide Web Conferences Steering Committee (2013)
10. Zhu, J., Wang, H., Zhu, M., Tsou, B., Ma, M.: Aspect-based opinion polling from customer reviews. IEEE Trans. Affect. Comput. **2**, 37–49 (2011)

K-Translate - Interactive Multi-system Machine Translation

Matīss Rikters[(✉)]

University of Latvia, 19 Raina Blvd., Riga, Latvia
matiss@lielakeda.lv

Abstract. The tool described in this article has been designed to help machine translation (MT) researchers to combine and evaluate various MT engine outputs through a web-based graphical user interface using syntactic analysis and language modelling. The tool supports user provided translations as well as translations from popular online MT system application program interfaces (APIs). The selection of the best translation hypothesis is done by calculating the perplexity for each hypothesis. The evaluation panel provides sentence tree graphs and chunk statistics. The result is a syntax-based multi-system translation tool that shows an improvement of BLEU scores compared to the best individual baseline MT. We also present a demo server with data for combining English - Latvian translations.

Keywords: Machine translation · Hybrid machine translation · Syntactic parsing · Chunking · Natural language processing · Computational linguistics · Data services

1 Introduction

Multi-system machine translation (MSMT) is a subset of hybrid MT (HMT) where multiple MT systems are combined in a single system in order to boost the accuracy and fluency of the translations. It is also referred to as multi-engine MT, MT coupling or just MT system combination.

This paper presents an attempt to enrich an MSMT approach with language specific information and a clean, self-explanatory user interface. The experiments described use multiple combinations of outputs from two, three or four MT systems. Experiments described in this paper are performed for the English-Latvian language pair. Translating from English, French, and German to Latvian, English, French and German is currently supported, however the underlying framework developed within this work allows application of this strategy for other language pairs as well. The automatic evaluation results obtained with this hybrid system are analysed and compared with human evaluation. The code of the developed K-Translate system is freely available at GitHub[1]. A demo server[2] with data for combining English - Latvian translations is also available.

---

[1] K-Translate on GitHub - https://github.com/M4t1ss/K-Translate.
[2] K-Translate demo - http://k-translate.lielakeda.lv/.

© Springer International Publishing Switzerland 2016
G. Arnicans et al. (Eds.): DB&IS 2016, CCIS 615, pp. 304–318, 2016.
DOI: 10.1007/978-3-319-40180-5_21

The structure of this paper is as following: Sect. 2 compares the current tool with previous work. Section 3 describes the back-end and the evaluation mechanism. Section 4 outlines the main functionality of the graphical interface and Sect. 5 provides information about how the system performs under certain experiment. Finally, Sect. 6 includes a summary and Sect. 7 - aims for further improvements.

2 Related Work

Ahsan and Kolachina [1] describe a way of combining SMT and RBMT systems in multiple setups where each one had input from the SMT system added in a different phase of the RBMT system.

Barrault [2] describes a MT system combi-nation method where he combines multiple confusion networks of 1-best hypotheses from MT systems into one lattice and uses a language model for decoding the lattice to generate the best hypothesis.

Mellebeek et al. [2] introduced a hybrid MT system that utilised online MT engines for MSMT. Their system at first attempts to split sentences into smaller parts for easier translation by the means of syntactic analysis, then translate each part with each individual MT system while also providing some context, and finally recompose the output from the best scored translations of each part (they use three heuristics for selecting the best translation).

Heafield and Lavie [3] describe an open source MSMT system that consists of four components – hypothesis alignment (with METEOR aligner), definition of a search space on top of the alignments, definition of features for scoring hypotheses and a beam search decoder.

Freitag et al. [4] use a combination of a confusion network and a neural network model. A feedforward neural network is trained to improve upon the traditional binary voting model of the confusion network. This gives the confusion network the option to prefer other systems at different positions even in the same sentence.

3 Back-End System Workflow

The main workflow can be divided into three main constituents – (1) pre-processing of the source sentences, (2) the acquisition of a translations via online APIs (except for when translations are provided by the user) and (3) post-processing - the selection of the best translations of chunks and creation of MT output. A visualized workflow of the system is presented in Fig. 1.

For translation, four translation APIs are used. However, the system's architecture is flexible allowing to integrate more translation APIs easily. The system is set to be able to translate from English, German or French into Latvian, German, English or French. Nevertheless, the source and target languages can also be changed to other language pairs that are supported by the APIs, Berkeley Parser [4] parse grammars and KenLM [5] language models. Each new source language requires a grammar that is compliant with the Berkeley Parser. The parser is able to learn new grammars from treebanks. Each new target language requires a language model that is compliant with KenLM. New language models can be trained using the *lmplz* program included in KenLM.

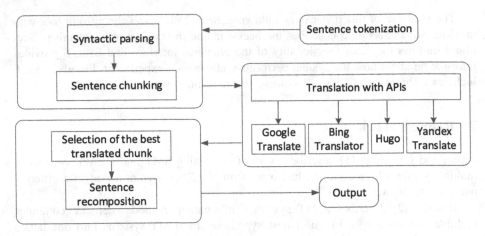

Fig. 1. General workflow of the translation process

3.1 Pre-processing

The first step is to tokenize the input. The tokenizer uses the whitespace and punctuation tokenizer from the NlpTools php library[3] that is included in the system. Tokenization is essential for proper functioning of all subsequent steps – the syntactic parser can misclassify a word or a phrase and the translation APIs can issue an incorrect translation. For example, the parser will not correctly understand a word that has a dot, comma or a colon as the ending symbol.

After tokenization it is necessary to divide sentences into linguistically motivated chunks that will be further given to the translation APIs. For this task the Berkeley Parser is used in conjunction with a chunk extractor (chunker). The parse tree of each sentence is processed by the chunker to obtain the parts of the sentence that will be individually translated and passed to the translation step.

Sentence Chunking. A need for splitting the possibly long sentences into smaller chunks is motivated by the hypothesis that modern MT systems can produce better results when given a reasonable length source sentence. To justify that this approach that uses the linguistically motivated chunks are much better as just cutting sentences into random chunks we performed three experiments. The sentence was split into 5-g in one experiment (+ one shorter n-g, if the last one is made up of less tokens), random 1-g to 4-g in the second experiment, random 1-g to 6-g in the third, and finally random 6-g to n-g of sentence length in the last experiment. We used the same 5-g JRC-Acquis language model for best translation selection. Results of these experiments (Table 1) fully confirmed the hypothesis of advantage of linguistically motivated chunks.

The chunker reads output of the Berkeley Parser and places it in a tree data structure. During this process each node of the tree is initialised with its phrase (NP, VP, ADVP, etc.), word (if it has one) and a chunk consisting of the chunks from its

---

[3] Natural language processing tools - http://php-nlp-tools.com/documentation/tokenizers.html.

Table 1. Influence of different chunk selection strategies on MT output

Chunks	BLEU
Linguistically motivated chunks	**18.33**
5-g	10.35
Random 1–4 g	7.33
Random 1–6 g	9.12
Random 6-max g	17.94

child nodes. To obtain the final chunks for translation the resulting tree is traversed bottom-up post-order and only the top-level subtrees are used as the resulting chunks. The chunking consists of steps shown in Fig. 2.

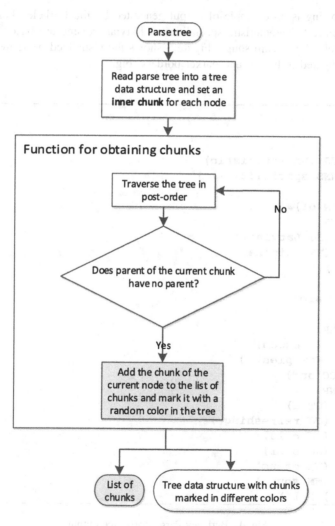

Fig. 2. Chunking process flowchart

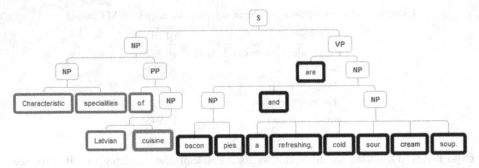

Fig. 3. Visualised tree with marked chunks

The following is an example of output generated by the Berkeley Parser for the English sentence "Characteristic specialities of Latvian cuisine are bacon pies and a refreshing, cold sour cream soup." Figure 3 shows the visualized parse tree with two chunks highlighted in lighter and darker borders (Fig. 4).

```
( (S
 (NP
   (NP
     (JJ Characteristic)
     (NNS specialities) )
   (PP
     (IN of)
     (NP
       (JJ Latvian)
       (NN cuisine)
     ) ) )
 (VP
   (VBP are)
   (NP
     (NP
       (NN bacon)
       (NNS pies) )
     (CC and)
     (NP
       (DT a)
       (JJ refreshing,)
       (JJ cold)
       (JJ sour)
       (NN cream)
       (NN soup.)
     ) ) ) ) )
```

Fig. 4. Berkeley Parser parse tree output

3.2   Translation with Online APIs

Currently four online translation APIs are included in the project – Google Translate[4], Bing Translator[5], *Yandex Translate*[6] and Hugo[7]. These specific APIs were selected because of their public availability and descriptive documentation as well as the range of languages that they support. One of the main criteria when searching for translation APIs was the ability to translate from English into Latvian.

Each translation API is defined with a function that has source and target language identifiers and the source chunk as input parameters and the target chunk as the only output. This makes adding new APIs very easy.

3.3   Selection of the Best Translated Chunk

The selection of the best translated chunk is done by calculating the perplexity for each hypothesis translation with KenLM. For reliable results a large target language corpus is necessary. For each ma-chine-translated chunk a perplexity score represents the probability of the specific sequence of words appearing in the training corpus used to create the language model (LM). Sentence perplexity has been proven to correlate with human judgments and BLEU scores, and it is a good evaluation method for MT without reference translations [6]. It has been also used in other previous attempts of MMT to score output from different MT engines as mentioned by [7, 8].

KenLM calculates probabilities based on the observed entry with longest matching history w_f^n:

$$p\left(w_n|w_1^{n-1}\right) = p\left(w_n|w_f^{n-1}\right) \prod_{i=1}^{f-1} b(w_i^{n-1})$$

where the probability $p\left(w_n|w_f^{n-1}\right)$ and backoff penalties $b(w_i^{n-1})$ are given by an already-estimated language model. Perplexity is then calculated using this probability:

$$b - \frac{1}{N} \sum_{i=1}^{N} \log_b q(x_i)$$

where given an unknown probability distribution p and a proposed probability model q, it is evaluated by determining how well it predicts a separate test sample x1, x2... xN drawn from p.

3.4   Sentence Recomposition

When the best translation for each chunk is selected, the translation of the full sentence is generated by concatenation of chunks. The chunks are recomposed in the same order as they were split up.

---

[4] Google Translate API - https://cloud.google.com/translate/.

[5] Bing Translator Control - http://www.bing.com/dev/en-us/translator.

[6] Yandex Translate API - https://tech.yandex.com/translate/.

[7] Latvian public administration machine translation service API - http://hugo.lv/TranslationAPI.

4 Translation Combination Panel

This section presents all the basic screens of the translation combination panel which is the graphical front-end of K-Translate. Figure 5 shows a schematic overview of the options available. Each of the two ways of combining translations consists of all or most of the steps covered in the previous section. An exception is when the user choses to input their own translations – this process skips translation with online APIs.

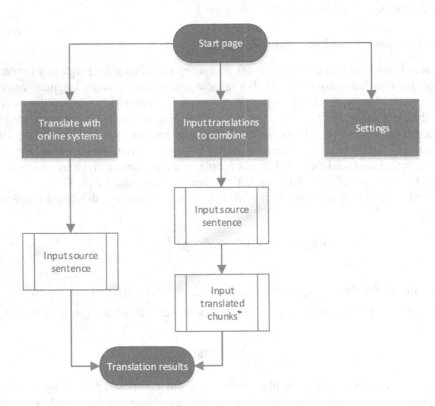

Fig. 5. Architectural visualization of the translation combination panel

4.1 Translating with Online Systems

The start-up screen of the translation combination panel allows to fully automatically get translations from several online MT systems that have APIs available, combine them and output the best fitting hybrid translation. The source sentence input screen is shown in Fig. 6 and the results look the same as when combining user provided translations (Fig. 11) with the exception of showing the name of the used online system as the source instead of MT1, MT2, etc.

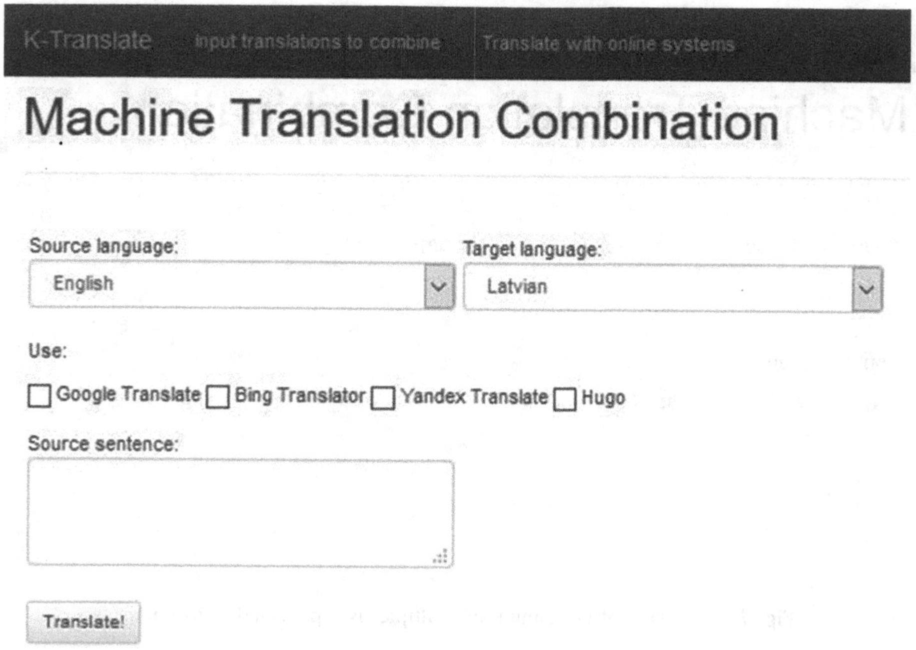

Fig. 6. Translating with online APIs

4.2 Combining Multiple User Provided Translations

The second option of the translation combination panel is intended for the more experienced MT professionals who already have several (two or more) translations of the input sentence from different MT systems and just want to obtain the combined result. At first the user must select source and target languages and input the sentence in a source language as shown in (Fig. 7).

In the next step K-Translate will have performed syntactic analysis on the input sentence and split it into smaller fragments or chunks as shown in Fig. 8[8]. The syntax tree with highlighted color-coded chunks will also be shown so that the user can better understand where and why the chunks have their boundaries (Fig. 9). These chunks will be given in a text box each in a new line for the user to translate with the chosen MT systems. Finally, the obtained translations must be pasted in the MT 1, MT 2, etc. text boxes (Fig. 10) below each chunk per line to move on to the last step.

In the last step (Fig. 11) K-Translate will provide the best fitting combined translation and highlight which chunks were used from which input. It also shows the source used for each chunk and the confidence level of each selection. The confidence is calculated by comparing chunk perplexities to each other.

---

[8] The process behind chunking is clarified in Sect. 3.1.

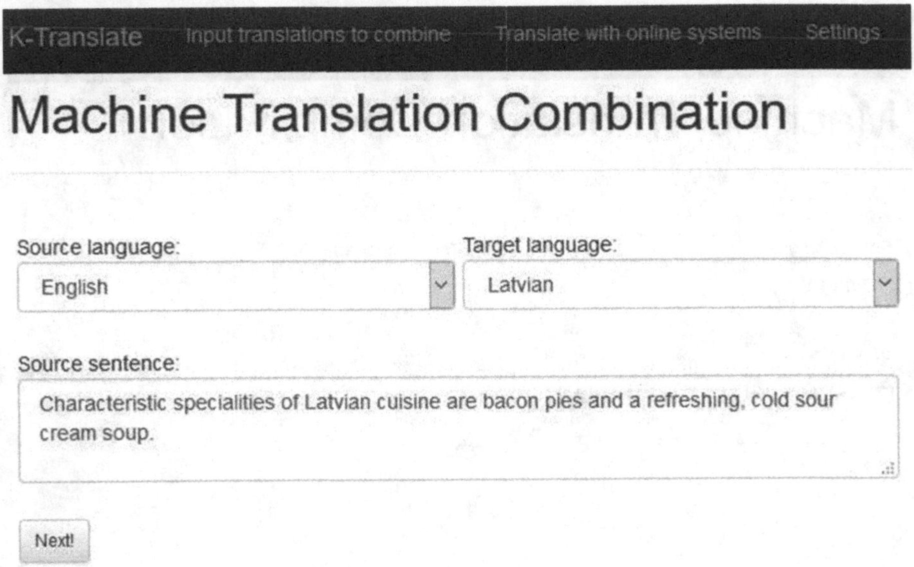

K-Translate   Input translations to combine   Translate with online systems   Settings

Machine Translation Combination

Source language:

| English | ⌄ |

Target language:

| Latvian | ⌄ |

Source sentence:

Characteristic specialities of Latvian cuisine are bacon pies and a refreshing, cold sour cream soup.

Next!

Fig. 7. First step of combining of multiple user provided translations

Source language:

| English | ⌄ |

Target language:

| Latvian | ⌄ |

Source sentence chunks:

Characteristic specialities of Latvian cuisine
are bacon pies and a refreshing, cold sour cream soup.

Chunks:

Characteristic specialities of Latvian cuisine   are bacon pies and a refreshing, cold sour cream soup.

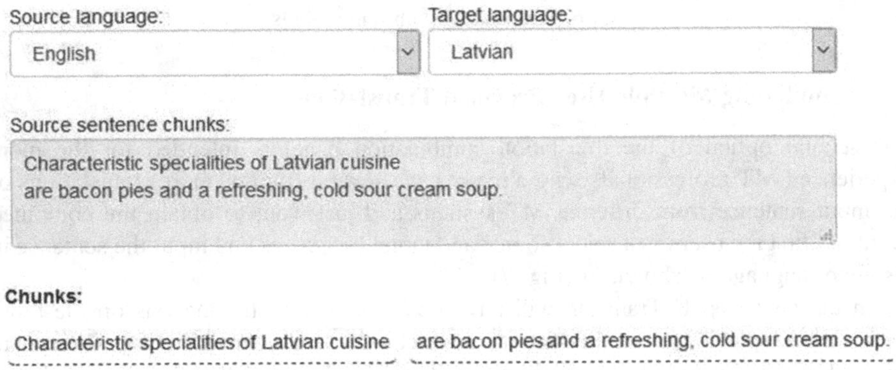

Fig. 8. Second step of combining of multiple user provided translations – part 1

4.3   Settings

Before any work with K-Translate can be done, one must first provide a Berkeley Parser compatible grammar file for each desired source language and a KenLM compatible language model file for each target language. Also, if usage of online APIs for translation is planned, the corresponding API settings are mandatory. The settings page allows for easy configuration of these values. The necessity of these requirements is explained in Sects. 3.1 and 3.2.

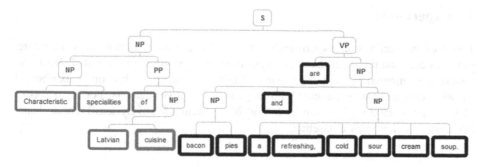

Fig. 9. Second step of combining of multiple user provided translations – part 2 – a syntax tree visualization

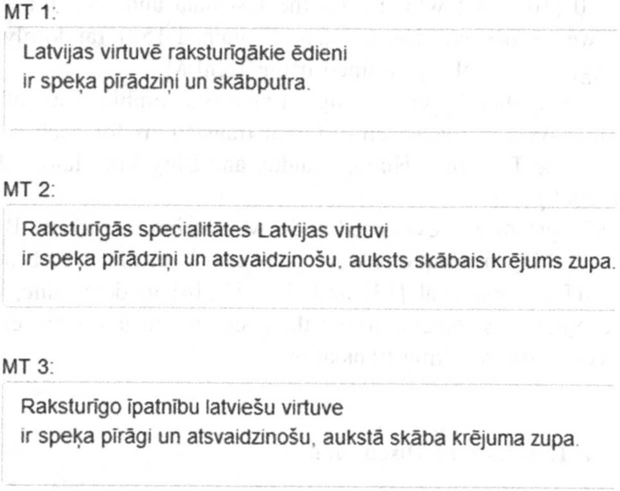

MT 1:

Latvijas virtuvē raksturīgākie ēdieni
ir speķa pīrādziņi un skābputra.

MT 2:

Raksturīgās specialitātes Latvijas virtuvi
ir speķa pīrādziņi un atsvaidzinošu, auksts skābais krējums zupa.

MT 3:

Raksturīgo īpatnību latviešu virtuve
ir speķa pīrāgi un atsvaidzinošu, aukstā skāba krējuma zupa.

Fig. 10. Second step of combining of multiple user provided translations – part 3 – input different translated chunks for source sentence chunks

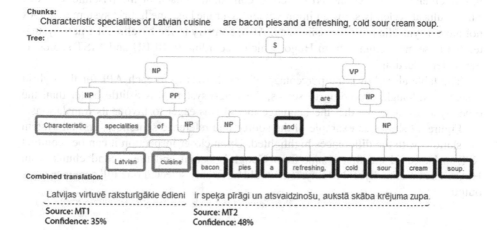

Fig. 11. Translation combination results page

5 Experiments

This section describes the experiments performed to test the workflow of K-Translate. Firstly, details on the input data and experiment methodology are provided. Next, the results are summarized and interpreted. Finally, a human evaluation is performed showing how the results coincide with judgement of native speakers. For the purposes of the experiment a slightly similar hybrid MT system - Multi-System Hybrid Translator [9] was chosen as a baseline.

5.1 Experiment Setup

The experiments were conducted on the English – Latvian part of the JRC Acquis corpus version 3.0 [10] from which both the test data and data for training of the language model were retrieved. The test data contained 1581 randomly selected sentences. A 5-g language model was trained using KenLM.

The method was applied by combining all possible combinations of two and then also all three APIs. As a result, seven different translations for each source sentence were obtained. Google Translate, Hugo, Yandex and Bing Translator APIs were used with the default configuration.

Output of each system was evaluated with two scoring methods – BLEU [11] and NIST [12]. The resulting translations were inspected with the Web-based MT evaluation platforms MT-ComparEval [13] and iBLEU [14] to determine, which system from the hybrid setups was selected to get the specific translation for each chunk and analyse differences in the resulting translations.

5.2 Experiment Results and Discussion

The results of the automatic evaluation are summarized in Table 2. Surprisingly all hybrid systems that include the Hugo API produce lower results than the baseline Hugo system. However, the combination of Google Translate and Bing Translator shows improvements in BLEU and NIST scores compared to each of the baseline systems. The results also clearly show an improvement over the baseline hybrid system that does not have a syntactic pre-processing step. Also, contrary to the baseline, the new system tends to use more chunks from Hugo, which, according to BLEU and NIST scores, is the better selection.

The table also shows the percentage of translations from each API for the hybrid systems. Although, according to scores, the Hugo system was a little better than the other systems, it seems that the language model was eager to favour its translations.

Figure 12 shows an example of the source and reference sentences, and all system translations with the differences highlighted. Upon closer inspection it can be seen that K-Translate used the first chunk from Google's output and the second chunk from Hugo. The baseline hybrid MT system would have only selected one full sentence as its output.

Table 2. Experiment results. B – Bing, G – Google, H – Hugo, L - LetsMT, Y – Yandex.

System	BLEU	NIST	Hybrid selection			
			G	B	H/L	Y
Google	16.19	8.37	100%	-	-	-
Bing	16.99	8.09	-	100%	-	-
Hugo	20.27	9.45	-	-	100%	-
LetsMT	20.55	9.48	-	-	100%	-
Yandex	19.75	9.30	-	-	-	100%
Baseline hybrid MT						
BG	**17.09**	8.41	56%	44%	-	-
GL	19.87	9.03	52%	-	48%	-
BGL	19.32	9.15	37%	29%	34%	-
K-Translate						
BG	**17.34**	8.54	74%	26%	-	-
GH	18.63	9.09	25%	-	74%	-
BH	18.98	8.97	-	24%	76%	-
HY	20.01	9.33	-	-	65%	35%
BGHY	18.33	8.67	17%	18%	35%	30%

Source	3 . the list referred to in paragraph 1 and all amendments thereto shall be published in the official journal of the european communities .
Reference	3 . sarakstu , kas minēts 1 . punktā , un visus tā grozījumus publicē eiropas kopienu oficiālajā vēstnesī .
Hugo	3 . punktā minēto sarakstu un visus grozījumus 1 ir publicēti Eiropas kopienu oficiālajā žurnālā .
Google	3 . sarakstu , kas minēts 1 . punktā un visus tā grozījumus publicē oficiālajā vēstnesī Eiropas Kopienu
Bing	3 . sarakstu , kas minētas punktā 1 un visi grozījumi tajos publicē Eiropas Kopienu Oficiālajā Vēstnesī .
Yandex	3 . sarakstu , kas minētas punktā 1 un visi grozījumi tajos publicē Eiropas Kopienu Oficiālajā žurnālā .
K-Translate	3 . sarakstu kas minēts 1 . punktā un visus tā grozījumus ir publicēti Eiropas kopienu oficiālajā žurnālā .

Fig. 12. Comparison of a sentence translations with the different systems with MT-ComparEval

5.3 Human Evaluation

A random 2 % (32 sentences) of the translations from the experiment were given to 10 native Latvian speakers with instructions to identify the most fluent and the most adequate translation for each source sentence. The results are summarized in Table 3. Comparing the evaluation results to the BLEU scores and the selections made by the syntax-based hybrid MT, a tendency towards the Hugo translation can be observed for the BLEU score and the selection of the hybrid method, that is not visible from the user ratings. The free-marginal kappa [15] for these annotations is 0.335 which indicates substantial agreement between the annotators.

Table 3. Human evaluation results

System	Fluency AVG	Accuracy AVG	K-Translate selection	BLEU
Google	35.29 %	34.93 %	16.83 %	16.19
Bing	23.53 %	23.97 %	17.94 %	16.99
Hugo	20.00 %	21.92 %	45.13 %	20.27
Yandex	25.93 %	27.07 %	20.10 %	19.75
K-Translate	21.18 %	19.18 %	–	18.33

The table shows that translations from the *Google Translate* system were recognized by annotators as most fluent and most adequate in 35 % of cases. This contradicts with the automatic evaluation results and the selections made by K-Translate where a tendency towards the *Hugo* translation is observed.

A broader analysis of this result was performed. The hypothesis is that *Hugo* was chosen less often by the annotators because of failure to translate dates or numbers in specific sentences while the rest of the sentence was very similar to the reference, hence scoring more BLEU points. Closer inspection revealed that three sentences from *Hugo* contained "βNUMβ" tag, which appears to be an error in the named entity processor during time of experiments. There were also five sentences that contained untranslated dates, e.g., "31 december 1992" or "february 1995." These errors account for *Hugo* not be selected by annotators in 25 % cases of the evaluation dataset, while in case of BLEU score, their influence was not so significant.

6 Conclusion

This paper described an interactive MT system combination approach that uses syntactic and statistical features and visualizes the intermediate steps. The main goals were to provide MT researchers with an intuitive and easy to use tool for combining translations and to improve translation quality [13] over the selected baseline.

All test cases showed an improvement in BLEU and NIST scores when compared to the baseline system. When used only with Google and Bing, the K-Translate scores 0.35 BLEU points higher than the best individual translation provided by the APIs.

In all hybrid systems that included the Hugo API a decrease in overall translation quality was observed. This can be explained by the scale of the engines - the Bing and Google systems are more general, designed for many language pairs, whereas the MT system in Hugo was specifically optimized for English – Latvian translations. This obstacle could potentially be resolved by creating a language model using a larger and cleaner training corpus and a higher order language model.

7 Future Work

The described system is in an early phase of its lifecycle and further enhancements are planned. There are several methods that could improve the current system combination approach. Improvements are intended for both – the front-end panel view and the back-end system. The visual side of K-Translate may become more comprehensible with additional graphs and charts that display how and why the system chose each individual chunk for the final translation. The back-end could benefit from improving the chunking step and the selection of the best translated chunk.

For now, the chunker splits sentences by all top-level chunks with no regard for sub-chunks or occasions when a chunk is only one word or symbol. The larger chunks should be split in smaller sub-chunks and the single-word chunks should be combined with one of the neighbouring longer chunks. It may be also more appropriate to divide certain phrases, e.g. noun phrases and verb phrases but not prepositional phrases, infinitive phrases, etc.

Adding alternative resources to select from in each step of the translation process could benefit the more advanced user base. For instance, the addition of more online translation APIs like Baidu Translate [16] or iTranslate.eu [17] would expand the variety of choices for translations. A configurable usage of different language modelling tools like RWTHLM [18] or character-aware neural language models [19] is likely to improve the chunk selection process.

There are also several possible areas of improvement for the selection of the best translation, for instance, to use a corpus cleaning tool on the corpora [20] before training a language model. Another would be to add a language model of morpho-syntactic tags.

References

1. Ahsan, A., Kolachina, P.: Coupling statistical machine translation with rule-based transfer and generation. In: AMTA-The Ninth Conference of the Association for Machine Translation in the Americas. Denver, Colorado (2010)
2. Barrault, L.: MANY: open source machine translation system combination. Prague Bull. Math. Linguist. **93**, 147–155 (2010)
3. Mellebeek, B., Owczarzak, K., Van Genabith, J., Way, A.: Multi-engine machine translation by recursive sentence decomposition. In: Proceedings of the 7th Conference of the Association for Machine Translation in the Americas, pp. 110–118 (2006)
4. Freitag, M., Peter, J., Peitz, S., Feng, M., Ney, H.: Local system voting feature for machine translation system combination. In: EMNLP 2015, Tenth Workshop on Statistical Machine Translation (WMT 2015), Lisbon, Portugal, pp. 467–476 (2015)
5. Petrov, S., Barrett, L., Thibaux, R., Klein, D.: Learning accurate, compact, and interpretable tree annotation. In: Proceedings of the 21st International Conference on Computational Linguistics and the 44th Annual Meeting of the Association for Computational Linguistics. Association for Computational Linguistics (2006)
6. Heafield, K.: KenLM: faster and smaller language model queries. In: Proceedings of the Sixth Workshop on Statistical Machine Translation. Association for Computational Linguistics (2011)
7. Gamon, M., Aue, A., Smets, M.: Sentence-level MT evaluation without reference translations: beyond language modeling. In: Proceedings of EAMT (2005)
8. Callison-Burch, C., Flournoy, R.S.: A program for automatically selecting the best output from multiple machine translation engines. In: Proceedings of the Machine Translation Summit VIII (2001)

9. Akiba, Y., Watanabe, T., Sumita, E.: Using language and translation models to select the best among outputs from multiple MT systems. In: Proceedings of the 19th International Conference on Computational Linguistics, vol. 1. Association for Computational Linguistics (2002)

10. Rikters, M.: Multi-system machine translation using online APIs for English-Latvian. In: ACL-IJCNLP 2015, p. 6 (2015)

11. Steinberger, R., Pouliquen, B., Widiger, A., Ignat, C., Erjavec, T., Tufis, D., Varga, D.: The JRC-Acquis: a multilingual aligned parallel corpus with 20 + languages. arXiv preprint cs/0609058 (2006)

12. Papineni, K., Roukos, S., Ward, T., Zhu, W.J.: BLEU: a method for automatic evaluation of machine translation. In: Proceedings of the 40th Annual Meeting on Association for Computational Linguistics. Association for Computational Linguistics (2002)

13. Doddington, G.: Automatic evaluation of machine translation quality using n-gram co-occurrence statistics. In: Proceedings of the Second International Conference on Human Language Technology Research. Morgan Kaufmann Publishers Inc. (2002)

14. Klejch, O., Avramidis, E., Burchardt, A., Popel, M.: MT-compareval: graphical evaluation interface for machine translation development. Prague Bull. Math. Linguist. **104**(1), 63–74 (2015)

15. Madnani, N.: iBLEU: interactively debugging and scoring statistical machine translation systems. In: 2011 Fifth IEEE International Conference on Semantic Computing (ICSC). IEEE (2011)

16. Zhongjun, H.E.: Baidu translate: research and products. In: ACL-IJCNLP 2015, p. 61 (2015)

17. Oravecz, C., Sass, B., Tihanyi, L.: 4.3 Evaluation campaign report (2012)

18. Sundermeyer, M., Schlüter, R., Ney, H.: rwthlm-the RWTH aachen university neural network language modeling toolkit. In: INTERSPEECH (2014)

19. Kim, Y., Jernite, Y., Sontag, D. Rush, A.M.: Character-aware neural language models. arXiv preprint arXiv:1508.06615 (2015)

20. Zariņa, I., Ņikiforovs, P., Skadiņš, R.: Word alignment based parallel corpora evaluation and cleaning using machine learning techniques. In: EAMT 2015 (2015)

Web News Sentence Searching
Using Linguistic Graph Similarity

Kim Schouten and Flavius Frasincar[✉]

Erasmus University Rotterdam, PO Box 1738, 3000 DR Rotterdam, The Netherlands
{schouten,frasincar}@ese.eur.nl

Abstract. As the amount of news publications increases each day, so does the need for effective search algorithms. Because simple word-based approaches are inherently limited, ignoring much of the information in natural language, in this paper we propose a linguistic approach called Destiny, which utilizes this information to improve search results. The major difference from approaches that represent text as a bag-of-words is that Destiny represents sentences as graphs, with words as nodes and the grammatical relations between words as edges. The proposed algorithm is evaluated using a custom corpus of user-rated sentences and compared to a TF-IDF baseline, performs significantly better in terms of Mean Average Precision, normalized Discounted Cumulative Gain, and Spearman's Rho.

Keywords: News search · Natural language processing · Graph similarity

1 Introduction

Nowadays, a significant portion of our mental capacity is devoted to the gathering, filtering, and consumption of information. With many things that are considered to be newsworthy, like updates from friends, twitter messages from people we follow, news messages on websites, and the more classical form of news like articles and news items, the amount of textual data (not to mention multimedia content) has become too large too handle. Even when considering only news items like articles, the number is overwhelming. And while some people can safely ignore lots of the news items, others are obliged to keep up with all the relevant news, for example because of their job.

While smart heuristics like skimming and scanning texts is of great benefit, it can only go so far. People, like investment portfolio managers, who have to monitor the stock of a certain group of companies, have to keep track of all news concerning these companies, including industry-wide news, but also that of competitors, suppliers, and customers. Therefore, being able to intelligently search news on the Web, for example to rank or filter news items, is paramount. Although text searching is very old, especially in computer science terms, the advance of new paradigms like the Semantic Web, has opened the way for new ways of searching.

© Springer International Publishing Switzerland 2016
G. Arnicans et al. (Eds.): DB&IS 2016, CCIS 615, pp. 319–333, 2016.
DOI: 10.1007/978-3-319-40180-5_22

This paper addresses one of these new search techniques, namely the search for news sentences. Searching for specific sentences enables the user to both search across and within documents, with the algorithm pointing the user to exactly the sentences that matches his or her query. With a previous publication [17] outlining the general concept of such a method, this paper aims to discuss the method in detail, providing additional analyses, and more insight into the actual workings of the algorithm.

2   Related Work

The over two decades worth of Web research has yielded several approaches to Web news searching. The most widely used approach is based on computing similarity by means of vector distances (e.g., cosine similarity). All documents are represented as a vector of word occurrences, with the latter recording either whether that word is in the document or not, or the actual number of times the word occurs in the document. Often only the stemmed words are used in these vector representations. The main characteristic of these methods is their bag-of-words character, with words being completely stripped of their context. However, that simplicity also allows for efficient and fast implementations, a useful trait when trying to provide a good Web experience. In spite of its simplicity, it has shown to perform well in many scenarios, for example in news personalization [1], but also in news recommendation [3]. Being the de facto default in text searching, TF-IDF [15], arguably the most well-known algorithm in this category, has been chosen to serve as the baseline for the evaluation of the proposed algorithm.

With the advance of the Semantic Web, a move towards a more semantic way of searching has been made. This includes the use of natural language processing to extract more information from text and storing the results in a formally defined knowledge base like an ontology. An example of such a setup can be found in the Hermes News Portal [8,16], where news items are annotated using an ontology that links lexical representations to concepts and instances. After processing the news items in this way, querying for news becomes a simple matter of selecting the ontology concepts of interest and all news items being annotated with these concepts are returned. Comparable to this is Aqualog [12], a question answering application which is similar in setup as Hermes, and SemNews [9], a news platform like Hermes using its own knowledge representation.

Unfortunately, because searching is performed in the ontology instead of the actual text, only concepts that are defined in the ontology and correctly found in the text can be returned to the user. A deeper problem however is caused by the fact that ontologies are formally specified, meaning that all information in the text first has to be translated to the logical language of the ontology. While translation always makes for a lossy transformation, in this case it is worse as the target language is known to be insufficient to represent certain natural language sentences. Barwise and Cooper [2] proved that first-order logic is inadequate for some types of sentences, and most ontologies are based on propositional or description logics which have even less expressive power.

3   Problem Definition

Using the linguistic principles [6] of homophonic meaning specification and com-
positionality, a natural way of representing text is a graph of interconnected
disambiguated words, with the edges representing the grammatical relations
between words. While this representation is not as rich in semantics as an ontology, it avoids the problems of ontology-based approaches while at the same time
providing more semantics than traditional word-based approaches.

With both the news items and the query represented by graphs, the problem of searching for the query now becomes related to graph isomorphism: the
algorithm needs to rank all sentence graphs in the news database according to
similarity (i.e., the measure of isomorphism) with the graph that describes the
user query. Since we need a measure of isomorphism instead of exact graph
isomorphism, we cannot simply implement Ullmann's algorithm [18].

This approximate graph isomorphism has a much larger search space than
regular graph isomorphism which already is an NP-complete problem [4]. There
are however some constraints that make the problem more tractable. Because
all graphs are both labeled and directed, they can be referred to as attributed
relational graphs, which are easier to deal with in this regard than unlabeled or
undirected graphs. Furthermore, missing edges in the query graphs are allowed
to be present in the news item graph (i.e., this is related to induced graph
isomorphism), a characteristic which also makes the problem easier to solve
since now the algorithm only has to check for the query's edges in the news
sentence graph and not the other way around.

We have chosen to use an augmented version of the backtracking algorithm
described in [13] to compute the graph similarities. The original algorithm iterates through the graph, checking with each step whether adding that node or
edge to the partial solution can still yield a valid final solution. Because of this
check, partial solutions that are known to be incorrect can be pruned, thus limiting the search space. Because parse graphs are labeled graphs, nodes can only be
matched to nodes when their labels are identical, again limiting the search space.
However, this will not work when considering measures of similarity or approximate matches. Then, its backtracking behavior is essentially lost as adding a
node never renders a solution invalid, only less relevant. Because of this we can
only speak of a recursive algorithm in the case of approximate matching. Such
a recursive algorithm would assign similarity scores to all nodes and edges in
the solution graph, and the sum of all these similarity scores would be the final
score for this solution.

4   The Destiny Framework

The Destiny framework is the implementation that follows from the above discussion. It has two main tasks: first, it transforms raw text into a graph, and
second, it ranks all graphs in a database based on similarity with a given user
graph. In the current use case, news items are transformed into graphs and stored

in a database. The user graph represents the user query which is executed on the database.

4.1 News Processing

A natural language processing pipeline has been developed that transforms raw text into a grammatical dependencies-based graph representation. The pipeline consists of a set of components with a specific natural language processing task that are consecutively ordered, each processing the result of the previous component, sending the outcome as input to the next component in the pipeline. The same pipeline is used to process both news items and user queries. An overview of the pipeline design is given in Fig. 1. The top half denotes the process of news transformation, whereas the bottom half denotes the process of searching the news.

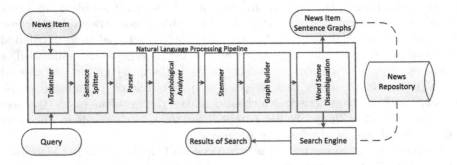

Fig. 1. Conceptual representation of framework

The pipeline is constructed on top of the GATE framework [5]. The same framework comes packaged with an extensive set of components, hence three out of the seven components are simply standard GATE components: the tokenizer to determine word boundaries, the sentence splitter to determine sentence boundaries, and the morphological analyzer to lemmatize words. While a default GATE component exists for the Stanford Parser [11], a slightly modified version is used to take advantage of a newer version of the parser itself. Porter's stemming algorithm [14] is used to determine the stem of each word.

The parser can be considered the main component of the pipeline, since it is responsible for finding the grammatical dependencies, thus directly influencing the graph output. Furthermore, it provides Part-of-Speech (POS) tags, essential information regarding the grammatical type of words (e.g., noun, verb, adjective, etc.). Based on the information extracted thus far, the graph builder component can construct a graph representation for all sentences. First, a node is generated for each word, encoding all known information about that word, like its POS, lemma, etc., in the node. Then, each syntactical dependency between words is used to generate an edge between the corresponding nodes, with the type

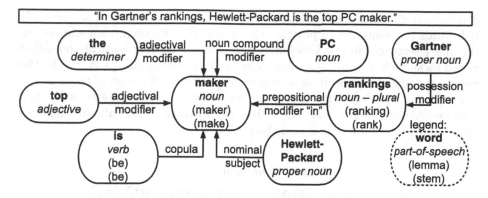

Fig. 2. Graph representation of the example sentence

of syntactical dependency encoded as an edge label. Even though a word can appear more than once in a sentence, each instantiation of that word has its own unique grammatical role in the sentence. As such it has its own dependencies, and is therefore represented as a unique node in the resulting graph as well.

An example of a graph dependencies representation of a sentence is shown in Fig. 2. As can be seen, some words are integrated into an edge label, in particular prepositions and conjunctions do not receive their own node. Integrating them in an edge label gives a tighter and cleaner graph representation.

The last step of this process is the disambiguation of words, where the correct sense of a word is determined and encoded in the corresponding node. Having the word senses allows the search process to compare words, not only lexically, which would not be very accurate in a number of situations, but also semantically. Even better, the search algorithm can effectively use this information to find relations of synonymy and hypernymy between words, something that would not be possible otherwise. Because the development of a word sense disambiguation algorithm is outside the scope of this paper, an existing, widely used, algorithm is implemented: the simplified Lesk algorithm [10].

4.2   News Searching

The news search algorithm is essentially a ranking algorithm, where all sentences in the database are ranked according to their similarity to the user query graph. As such, its core element is the part where the similarity between a sentence in the database and the user query is determined. This is the graph comparison, for which we decided to use a recursive algorithm.

However, an initial hurdle is the problem of finding a suitable starting point from where the graph comparison can commence. Since the structure of sentences can vary greatly, it would not suffice to simply start comparing at the root of both sentence graphs. On the other hand, comparing each node with every other node would be too computationally expensive. As a compromise, each noun and verb is indexed by stem and are used as starting location for the graph comparison, the intuition being that nouns and verbs are the most semantically rich words in a sentence. In practice, this means that for each noun and verb in the query sentence, an index lookup is performed, returning a list of nodes that would be suitable to start searching from for that node in the query graph. The recursive graph comparison algorithm is then executed for each of those nodes, however, each pair of (query sentence, news sentence) is associated with (and thus ranked according to) only the highest score over all runs. Suboptimal scores are discarded. This process is described in Eq. 1.

$$sentenceScore(query, sentence) =$$
$$\max_{startNode_k \in sentence} score(query, startNode_k) \tag{1}$$

where $startNode_k$ denotes the kth starting node for this query. The implementation of this formula is represented in the pseudocode of method SEARCH in Algorithm 1. This algorithm makes use of COMPARE, which is described in Algorithm 2, to compute the raw scores. Being a recursive function, Algorithm 2 also calls itself with the next set of parameters to be compared. The object holding the raw score is forwarded as a parameter as well, so that each recursive loop will add some points to the overall raw score when applicable. Algorithm 2 uses two methods: SIMILARITYEDGE and SIMILARITYNODE, which compute the similarity scores for the edges and nodes, respectively.

Algorithm 2 compares the two graphs by first comparing the two starting nodes in the query graph and a news sentence graph. Then, using the edges of both nodes, the most suitable set of two nodes is determined to continue the graph comparison. This is done by looking one node ahead: the algorithm compares each connected node of the 'query node' with each connected node of the 'news node' to find the best possible pair. By means of a threshold, any remaining pairs with a preliminary score that is below the threshold are discarded. An additional effect of this policy is that when the preliminary score of a node is too low to be visited, its children will be discarded as well. While discarding regions of the graph that are likely to be irrelevant saves time, errors can also be introduced. As such this is a design choice, trading off a possible increase in accuracy against a decrease in running time. In the pseudocode, the process of looking ahead and finding the best pair of nodes to continue the graph comparison, if any, is encoded as a call to GETBESTSCORINGEDGE, which can be found in Algorithm 2. The recursive process will thus continue until either all nodes have been visited or no suitable matching pair is available for the remaining unvisited nodes that are connected to at least one visited node.

The similarity score of a news sentence with respect to the query sentence is essentially the sum of similarity scores of all selected pairs of nodes and pairs of edges. As such the actual score is determined by the similarity function of two nodes and the corresponding one for edges. While edges only have one feature, nodes have many aspects that can be compared and proper weighting of all these features can substantially improve results. As such, all feature weights

Algorithm 1. Pseudocode for the search algorithm

```
1: function SEARCH(Document query, List of Documents processedDocuments) : SortedList
2:      Initialize finalResults as SortedList of Scores
3:      Initialize allScores as SortedList of Scores
4:      Initialize matchedSentences as List of Strings
5:      for all newsItem in processedDocuments do
6:          Intialize queryStartNodes as List of Nodes
7:          queryStartNodes = query.getStartNodes()
8:          for all qStartNode in queryStartNodes do
9:              Initialize newsStartNodes as List of Nodes
10:             newsStartNodes = newsItem.getNodes(queryStartNode.getStem())
11:             for all nStartNode in newsStartNodes do

        /* A new Score object is created. This object will be propagated through all recursive runs
        of the COMPARE method. */
12:                 Initialize score as Score

        /* The recursive method COMPARE as described in Algorithm 2 is started here. When it
        ends, the score collected over all recursive runs is saved. */
13:                 COMPARE(qStartNode,nStartNode,score)
14:                 allScores.add(score)
15:             end for
16:         end for
17:         finalResults.add(allScores.getHighestScore())
18:     end for
19:     return finalResults
20: end function
```

are optimized using a genetic algorithm, which was described in our previous paper [17].

The SIMILARITYEDGE function returns the similarity score for two edges. Since the only attribute edges have is their label, it returns a score only when the labels are identical. The exact score assigned to having identical edge labels is defined using a parameter which is optimized with the employed genetic algorithm. The SIMILARITYNODE function is slightly more complicated as nodes have more features that can be compared than edges. Each feature is again weighted using the genetic algorithm to arrive at a set of optimal weights for each of the features. The similarity score for nodes is computed as the sum of all matching feature scores that are applicable for the current comparison.

Nodes are compared using a stepwise comparison. First, a set of five basic features is used: stem, lemma, the full word (i.e., including affixes and suffixes), basic POS category, and detailed POS category. The basic POS category describes the grammatical word category (i.e., noun, verb, adjective, etc.), while the detailed POS category gives more information about inflections like verb tenses and nouns being singular or plural. For each feature, its weight is added to the score, if and only if the values for both nodes are identical.

If the basic POS category is the same, but the lemma's are not, there is the possibility for synonymy or hypernymy. Using the acquired word senses and WordNet [7], both nodes are first checked for synonymy and if so, the synonymy

---

Algorithm 2. Pseudocode for the raw score computation

1: **procedure** COMPARE(*currentQueryNode*, *currentNewsNode*, *score*)
2:　　Initialize *nodeScore* as double

　　/* The two nodes are compared, and their similarity score is added to the total score. Both nodes are now marked as being visited. */
3:　　*nodeScore* = SIMILARITYNODE(*currentQueryNode*,*currentNewsNode*)
4:　　*score*.addScore(*nodeScore*)
5:　　*currentQueryNode*.setVisited(true)
6:　　*currentNewsNode*.setVisited(true)

　　/* Now the parents and children of both nodes need to be compared. */
7:　　Initialize *queryEdges* as List of Edges
8:　　*queryEdges* = *currentQueryNode*.getEdges()
9:　　Initialize *newsEdges* as List of Edges
10:　　*newsEdges* = *currentNewsNode*.getEdges()

　　/* Using SIMILARITYEDGE and SIMILARITYNODE the best possible route for the recursion is determined by comparing *queryEdge* with each possible *newsEdge* in *newsEdges* in GETBESTSCORINGEDGE. This method also makes sure that parents are compared only with parents and children only with children. If an edge exist that is good enough, the recursion will continue through that node. */
11:　　**for all** Edge *queryEdge* in *queryEdges* **do**
12:　　　Initialize Edge *bestEdge*
13:　　　*bestEdge* = GETBESTSCORINGEDGE(*queryEdge*,*newsEdges*)
14:　　　**if** *bestEdge* ≠ ⊥ **then**
15:　　　　double *edgeScore* = SIMILARITYEDGE(*queryEdge*,*bestEdge*)
16:　　　　*score*.addScore(*edgeScore*)
17:　　　　*queryEdge*.setVisited(true)
18:　　　　*bestEdge*.setVisited(true)

　　　/* Recursion can only continue if there exists an unvisited node linked to *bestEdge* and one linked to *queryEdge*. */
19:　　　　*qNextNode* = GETNEXTNODE(*queryEdge*)
20:　　　　*nNextNode* = GETNEXTNODE(*bestEdge*)
21:　　　　**if** !*qNextNode*.isVisited() **then**
22:　　　　　**if** !*nNextNode*.isVisited() **then**
23:　　　　　　COMPARE(*qNextNode*,*nNextNode*,*score*)
24:　　　　　**end if**
25:　　　　**end if**
26:　　　**end if**
27:　　**end for**
28: **end procedure**

---

weight is added to the similarity score for this pair of nodes. If there is no synonymy, the hypernym tree of WordNet is used to find any hypernym relation between the two nodes. When such a relation is found, the weight for hypernymy, divided by the number of steps in the hypernym tree between the two nodes is added to the similarity score. In this way, very generic generalizations will not get a high score (e.g., while 'car' has a hypernym 'entity', this is so general it does not contribute much).

The last step in computing the similarity score of a node, is the adjustment with a significance factor based on the number of occurrences of the stem of that node in the full set of news items. For words which appear only once in the whole collection of news items, the significance value will be one, while the word that appears most often in the collection a significance value of zero will be assigned. Preliminary results showed that adding this significance factor, reminiscent of the inverse document frequency in TF-IDF, has a positive effect on the the obtained results. Equation 2 shows the formula used to compute the significance value for a sentence node.

$$significance_n = \frac{\log(\max \#stem) - \log(\#stem_n)}{\log(\max \#stem)} \tag{2}$$

where
n = a sentence node,
$\#stem_n$ = how often $stem_n$ was found in news,
$\max \#stem$ = the highest $\#stem$ found for any n.

Complexity Analysis. As with any action that would require a user to wait for the results to be returned, the speed of the search algorithm is important. The query execution speed is highly dependent on the size of the data set, as well as on the size of the query. Furthermore, the higher the similarity between the query and the data set, the more time it will take for the algorithm to determine how similar these two are, as the recursion will stop when encountering too much dissimilarity between the query and the current news item, as defined in the threshold parameter. To give some insight into the scalability of the algorithm with respect to the size of the data set and the size of the query, the complexity of the algorithm (in the worst case scenario) is represented in the big-O notation:

$$\mathbf{f}(n, o, p, q, r) = \mathbf{O}(no^2pqr) \tag{3}$$

where
n = the # of documents in the database,
o = the # of nodes in the query,
p = the average # of nodes in the documents in the database,
q = the # of edges in the query, and
r = the average # of edges in the documents in the database.
In order to attain this (simplified) complexity, it is assumed that the number of nodes in a query is equal to the average number of nodes in the documents in the database, and the number of nodes is roughly equal to the number of edges for each sentence. In practice, a query will usually be much smaller than the average size of the documents in the database. Furthermore, this complexity, as it is a worst-case scenario, assumes it will have to compare each node of the query to all other nodes from the database. Again, this is usually not the case because of the threshold value limiting the recursion.

Interestingly, when scaling this up, the o^5 will quickly be dwarfed by n, the number of documents in the data set. We can therefore conclude that the algorithm is linear in the number of documents in the database.

Implementation Notes. The system is developed in Java, using the Eclipse IDE (Helios). In order to have an easy and intuitive way of storing and retrieving the graph representations of text, we have chosen to use the object database provided by db4object (www.db4o.com). To access WordNet, the Java Word-Net Library (www.sourceforge.net/projects/jwordnet) is used. This also provides convenient methods for determining synonymy and hypernymy relations between synsets.

5   Evaluation

In this section, the performance of the Destiny algorithm will be measured and compared with the TF-IDF baseline. First, some insight is given into the used data set. Then the performance in terms of quality, including a discussion on the used metrics, and processing speed are given. Last, a section with advantages of using Destiny is included, as well as a failure analysis based on our experiments.

5.1   Setup

Since Destiny searches on a sentence level (i.e., not only among documents but also within documents), a corpus of sentences is needed where each sentence is rated against the set of query sentences. From 19 Web news items, the sentences were extracted and rated for similarity against all query sentences. The news items yielded a total of 1019 sentences that together form the data set on which Destiny will be evaluated. From this set, ten sentences were rewritten to function as queries. The rewritten sentences still convey roughly the same meaning, but are rephrased by changing word order and using synonyms or hypernyms instead of some original words. Each sentence-query pair is rated by at least three different persons on a scale of 0 to 3, resulting in a data set of over 30500 data points. For each sentence-query pair, the final user score is the average of the user ratings. From these scores, a ranking is constructed for each query of all sentences in the database.

The inter-annotator agreement, computed as the standard deviation in scores that were assigned to the same sentence-query pair, is only 0.17. However, this includes a lot of pairs with a score of zero. As the majority of the sentences in the data set is completely dissimilar, a fact easily recognized by most people, the standard deviation is severely impacted by these scores. When we exclude all scores of zero and recompute the standard deviation, we attain a standard deviation of 0.83, which is slightly worse.

As discussed in the previous section, the weights are optimized using a genetic algorithm. In order to have a proper evaluation, the data set is split into a training set and a test set. The split itself is made on the query level: the genetic algorithm is trained on 5 queries plus their (user-rated) results, and then tested on the remaining 5 queries. The results of the algorithm on those 5 queries are compared against the golden standard. This process is repeated 32 times, for 32 different splits of the data. All splits are balanced for the number of relevant query results, as some queries yielded a substantial amount of similar sentences, while others returned only a handful of good results.

5.2   Search Results Quality

The performance of Destiny is compared with a standard implementation of TF-IDF. As TF-IDF does not require training, the training set is not used and its scores are thus computed using the test set of each of the 32 splits only. The comparison is done based on three metrics: the Mean Average Precision (MAP), Spearman's Rho, and the normalized Discounted Cumulative Gain (nDCG) [12]. This gives a better view on the performance than when using only one metric, as each of these has its own peculiarities.

For this kind of data, the MAP is less suitable, as it assumes a Boolean similarity between query and candidates. A result is either similar, or it is not. This is in contrast with the graded similarity that is employed in this work. This means, that in order to compute the MAP, the user scores for all sentence-query pairs, ranging from 0 to 3, have to be mapped to either true or false. The cut-off value that will determine which user scores are mapped to dissimilar and which are mapped to similar is however rather arbitrary. We therefore made the choice to use a range of cut-off values, going from 0 to 3 with a stepsize of 0.1 and compute the MAP for each cut-off value. Hence, the MAP score reported in the next section is the average of these 30 computed MAP scores.

Both the nDCG and the Spearman's Rho do not suffer from the above problem and thus are more suitable metrics in this case. There are two concerns when computing Spearman's Rho. The first is that it computes the correlation between the ranked output of Destiny and the user scores over all sentence combinations in the list. This is not true in reality, as most users do not go through the whole list. Second, while the rankings are computed based on the degree of relevance, the latter is not directly used to compute the overall score. This means that it effectively assigns as much value to a top-ranking sentence being correct as to a lower- ranking sentence being correct.

In contrast, the nDCG only uses the first k number of results, and computes the added value of each of these k results for the total set by discounting for the position in the ranked results list. In this way the degree of relevance is also taken into account as results with a higher degree of relevance contribute more to the overall score than results with a lower degree of relevance.

To evaluate the performance of the Destiny algorithm, it is compared with the TF-IDF baseline on the ranking computed from the user scores. The results, shown in Table 1, clearly show that Destiny significantly outperforms the TFIDF baseline on the Spearman's Rho and nDCG ranking. The p-value is computed for the paired one-sided t-test on the two sets of scores consisting of the 32 split

Table 1. Evaluation results

	TF-IDF mean score	Destiny mean score	rel. improvement	t-test p-value
nDCG	0.238	0.253	11.2 %	<0.001
MAP	0.376	0.424	12.8 %	<0.001
Sp. Rho	0.215	0.282	31.6 %	<0.001

scores for both Destiny and TF-IDF, respectively. The reported scores are the average scores over all 32 splits.

5.3   Processing Speed

Query execution time is measured for the ten queries in our data set and compared with TF-IDF in Fig. 3. The average time needed to search with Destiny is about 1570 ms, while TF-IDF needs on average 800 ms to execute one query. As such, TF-IDF is on average approximately twice as fast as Destiny.

5.4   Advantages

Due to its focus on grammatical structure and word sense disambiguation, Destiny has some typical advantages compared to traditional search methods. The first is the focus on sentences rather than separate words. When searching is based on word occurrence in a document, the document can get a high score even though different words from the query are not related at all but simply occur somewhere in that document. By focusing on sentences, words from the

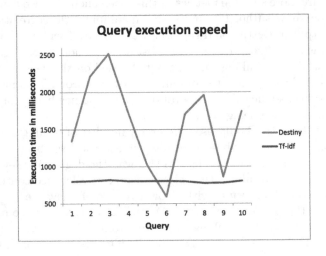

Fig. 3. Some query execution speed measures for Destiny and TF-IDF (Color figure online).

query are at least within the same sentence, making it much more likely that they are indeed semantically related.

Because grammatical relations are utilized when searching, users can actively use that to search for very specific information. While many different news items can be matched to the same bag-of-words, a group of words connected by a certain grammatical structure is much more specified. As such, it is more likely that a user will find his target when he can indeed specify his search goal by means of a sentence.

While grammar can be used to specify the query, the fact that the search algorithm employs synonyms and hypernyms improves the number of hits. Using synonyms and hypernyms, sentences can be found without explicit knowledge of the words in that sentence. This is obviously a great benefit compared to traditional word-based search algorithms which only take the literal word into account.

5.5 Failure Analysis

In order to analyze the errors made by Destiny and assess their origin, a failure analysis has been performed. This yielded a list of situations the algorithm is not able to handle well. These situations are summarized below.

With respect to dealing with named entities, Destiny is rather limited. Various versions of a name are for example not correctly identified as being the same, neither are different names belonging to the same concept. For example, "Apple" is not recognized to be the same as "Apple Inc." or "Apple Computers Inc.", nor is it matched properly to the ticker "AAPL". Another example of the same problem would be the mismatch of the algorithm between "United States of America" and "U.S.A." or just "USA". Also, co-reference resolution is missing, so pronouns are not matched to the entity they are referring to. A graph-based approach like [11] seems particularly well suited for this work.

Also problematic in terms of semantics are proverbs, irony, and basically all types of expressions that are not to be taken literally. This caused some specific errors in the evaluation as in the data set many equivalent expressions are used for "dying": "to pass away", "to leave a void", "his loss", etc. While word synonyms can be dealt with, synonymous expressions are not considered.

Another issue is related to the fact that the search algorithm, when comparing two graphs, cannot cope well with graphs of varying size. Especially the removal or addition of a node is something the algorithm is unable to detect. When comparing Destiny with an algorithm based on graph edit distance [8], it can only detect substitution of nodes in a certain grammatical structure. Additional or missing nodes can thus break the iterative comparison, resulting in a significantly lower score than expected. For example, in the sentence "Microsoft is expanding its online corporate offerings to include a full version of Office", it is Microsoft that is the one who will include the full version of Office, but instead of Microsoft being the grammatical subject of "include", it is the subject of "is expanding", which in turn is linked to "include". When searching for "Microsoft includes Office into its online corporate offering", a full match will therefore not be possible.

6 Concluding Remarks

We have shown the feasibility of searching Web news in a linguistic fashion by developing Destiny, a framework that uses natural language processing to transform both query and news items to a graph-based representation and then searches by computing the similarity between the graph representing the user query and the graphs in the database. In the graph representation, much of the original semantics are preserved in the grammatical relations between the words, encoded in graph as edges. Furthermore, the search engine can also utilize semantic information with respect to words because of the word sense disambiguation component: words can be compared on a lexical level, but also on a semantic level by checking whether two words are synonyms or hypernyms.

While Destiny is slower than the TF-IDF baseline because of all the natural language processing, it is, nevertheless, better in terms of search results quality. For all three used metrics (e.g., Mean Average Precision, Spearman's Rho, and normalized Discounted Gain), Destiny yielded a significantly higher score.

Based on the failure analysis in the previous section, it would be useful to improve the accuracy of the search results by adding a module to match named entities with different spelling or using abbreviations. Also co-reference resolution might be beneficial, as sentences later in a news item often use pronouns to refer to an entity previously introduced, while a query, being only one sentence, usually features the name of the entity. Last, as discussed in the previous section, some form of graph edit distance might be implemented to mitigate the problem of important nodes not being present in both graphs.

While not within range of real-time processing speed, the processing and query execution times of the prototype provide an acceptable basis for further development. Currently, the system is entirely single-threaded, so a multi-threaded or even distributed computing system (e.g., processing news items in parallel) is expected to improve the speed.

Acknowledgment. The authors are partially supported by the Dutch national program COMMIT.

References

1. Ahn, J., Brusilovsky, P., Grady, J., He, D., Syn, S.Y.: Open user profiles for adaptive news systems: help or harm? In: 16th International Conference on World Wide Web (WWW 2007), pp. 11–20. ACM (2007)
2. Barwise, J., Cooper, R.: Generalized quantifiers and natural language. Linguist. Philos. **4**, 159–219 (1981). http://dx.doi.org/10.1007/BF00350139
3. Billsus, D., Pazzani, M.J.: User modeling for adaptive news access. User Model. User-Adap. Inter. **10**(2–3), 147–180 (2000)

4. Cook, S.A.: The complexity of theorem-proving procedures. In: Third Annual ACM Symposium on Theory of Computing (STOC 1971), pp. 151–158. ACM (1971). http://doi.acm.org/10.1145/800157.805047

5. Cunningham, H., Maynard, D., Bontcheva, K., Tablan, V., Aswani, N., Roberts, I., Gorrell, G., Funk, A., Roberts, A., Damljanovic, D., Heitz, T., Greenwood, M.A., Saggion, H., Petrak, J., Li, Y., Peters, W.: Text Processing with GATE (Version 6), University of Sheffield Department of Computer Science (2011)

6. Devitt, M., Hanley, R. (eds.): The Blackwell Guide to the Philosophy of Language. Blackwell Publishing, Oxford (2006)

7. Fellbaum, C. (ed.): WordNet: An Electronic Lexical Database. MIT Press, Cambridge (1998)

8. Frasincar, F., Borsje, J., Levering, L.: A semantic web-based approach for building personalized news services. IJEBR **5**(3), 35–53 (2009)

9. Java, A., Finin, T., Nirenburg, S.: SemNews: a semantic news framework. In: The Twenty-First National Conference on Artificial Intelligence and the Eighteenth Innovative Applications of Artificial Intelligence Conference (AAAI 2006), pp. 1939–1940. AAAI Press (2006)

10. Kilgarriff, A., Rosenzweig, J.: English SENSEVAL: report and results. In: 2nd International Conference on Language Resources and Evaluation (LREC 2000), pp. 1239–1244. ELRA (2000)

11. Klein, D., Manning, C.: Accurate unlexicalized parsing. In: 41st Meeting of the Association for Computational Linguistics (ACL 2003), pp. 423–430. ACL (2003)

12. Lopez, V., Uren, V., Motta, E., Pasin, M.: AquaLog: an ontology-driven question answering system as an interface to the semantic web. J. Web Semant. **5**(2), 72–105 (2007)

13. McGregor, J.J.: Backtrack search algorithms and the maximal common subgraph problem. Softw. Pract. Experience **12**(1), 23–34 (1982)

14. Porter, M.F.: An algorithm for suffix stripping. In: Readings in Information Retrieval, pp. 313–316. Morgan Kaufmann Publishers Inc. (1997)

15. Salton, G., McGill, M.: Introduction to Modern Information Retrieval. McGraw-Hill, Maidenherd (1983)

16. Schouten, K., Ruijgrok, P., Borsje, J., Frasincar, F., Levering, L., Hogenboom, F.: A Semantic web-based approach for personalizing news. In: ACM Symposium on Applied Computing (SAC 2010), pp. 854–861. ACM (2010)

17. Schouten, K., Frasincar, F.: A linguistic graph-based approach for web news sentence searching. In: Decker, H., Lhotská, L., Link, S., Basl, J., Tjoa, A.M. (eds.) DEXA 2013, Part II. LNCS, vol. 8056, pp. 57–64. Springer, Heidelberg (2013)

18. Ullmann, J.R.: An algorithm for subgraph isomorphism. J. ACM **23**(1), 31–42 (1976)

Information Technology in Teaching and Learning

Heuristic Method to Improve Systematic Collection of Terminology

Vineta Arnicane(✉), Guntis Arnicans, and Juris Borzovs

Faculty of Computing, University of Latvia, 19 Raina Blvd., Riga LV 1586, Latvia
{vineta.arnicane,vineta.arnicane,juris.borzovs}@lu.lv

Abstract. In this paper, we propose an experimental tool for analysis and graphical representation of glossaries. The original heuristic algorithms and analysis methods incorporated into the tool appeared to be useful to improve the quality of the glossaries. The tool was used for analysis of ISTQB Standard Glossary of Terms Used in Software Testing. There are instances of problems found in ISTQB glossary related to its consistency, completeness, and correctness described in the paper.

Keywords: Glossary · Concept map · Quality · Graphs · Heuristic algorithms · Software testing

1 Introduction

Glossaries are alphabetical lists of the terms in a particular domain of knowledge with the definitions for those terms. Glossaries contain the explanations of numerous concepts of the certain field. It is important that the quality of glossaries is sufficiently high. The authors of this paper provide their view on the problem of glossaries' quality and propose some methods how to reveal the issues in glossaries and how to address them.

The authors use *Standard Glossary of Terms Used in Software Testing Version 3.01* (May 27, 2015) [7] (further in the text - Glossary). The Glossary is produced by the *Glossary* Working Group (GWG) of the International Software Testing Qualifications Board (ISTQB) [6]. The Glossary accumulates terms and their explanations from the most significant sources in the software testing field. The contribution was made from testing communities throughout the world. Co-author of this paper V. Arnicane is a member of this group. All authors of this paper are members of the Local Member Board of the Latvian Software Testing Qualification Board (LSTQB) and are doing a localization of the Glossary into the Latvian language together with the Terminology Commission of the Latvian Academy of Sciences.

GWG has worked almost ten years - the first good version V1.3 was issued on May 31, 2007. During next years, Glossary was substantially improved both in the range and in the quality. The Glossary contains 652 preferred terms that appear as an entry, with the total 170 synonyms indicated. GWG considers that glossary is almost complete and now concentrates on the improvement of its quality.

© Springer International Publishing Switzerland 2016
G. Arnicans et al. (Eds.): DB&IS 2016, CCIS 615, pp. 337–351, 2016.
DOI: 10.1007/978-3-319-40180-5_23

GWG has compiled the Glossary from the other related glossaries taking into account opinion of the industry, commerce and government bodies and organizations. The Glossary's scope is a little broader than software testing domain - "Some related non-testing terms are also included if they play a major role in testing, such as terms used in software quality assurance and software lifecycle models. However, most terms of other software engineering disciplines are not covered in this document, even if they are used in various ISTQB syllabi." [6].

We consider that Glossary is among the best the domain-oriented glossaries in the world. GWG did not follow principle stated in standard "ISO 704:2009 Terminology work – Principles and methods" – create concept system represented by graphical diagrams at first. Taking into account that the Glossary has very good quality, an attempt to create automatically lightweight ontology or concept map for software testing domain was made [2,3]. After analysis of results, it was concluded that the Glossary has inconsistency issues.

During the localization and translation of Glossary in Latvian, the authors realized that it is problematical to localize the terms of the whole domain of testing. There are difficulties to keep consistency both within the domain of testing and with the terms of related domains, for instance, software engineering, quality assurance, management and mathematics domains. It is hard to track down the consistency and mutual relationships between terms when there is such a plenitude of terms.

There are two major types of terminology work. The first one is an ad hoc work on terminology, which deals with a single term or a limited number of terms. The second one is a systematic collection of terminology, which deals with all the terms in a specific subject field or domain of activity, often by creating a structured ontology of the terms within that domain and their interrelationships. The contribution of this paper mostly is related to the second type of terminology work.

There is little research done in the field of the quality of glossaries, revealing the deficiencies and the elimination of them. Thus, the authors had to use the heuristic method - to learn, discover, and solve problems by experimental and especially by trial-and-error methods.

This paper is organized as follows. Section 2 describes the structure of glossaries, standards regarding terminology work and the quality evaluation of glossaries. Section 3 briefly explains GlossToolset developed by authors for analysis of glossaries. In Sect. 4 we analyze few problems revealed in ISTQB Software Testing Glossary. Section 5 summarizes paper and suggests issues for future research on glossary analysis.

2   Glossary Quality and Its Assessment

Glossary is a list of terms in a special subject, field, or area of usage, with accompanying definitions. From a perspective of automatic text analysis, the glossary is a semi-structured text document that contains descriptions of domain concepts and links among them. Some links are defined explicitly using keywords or text formatting means.

2.1   Glossary Structure

All records of Glossary that contain terms and definitions are arranged alpha-
betically. These records have various names: *entry, lemma, gloss.* We use term
entry in this paper. Figure 1 shows important structure elements of the glossary
such as *entry, term, definition, synonym, cross-reference, acronym,* and *source.*
The other elements also exist, for instance, a *variant of definition* in case the
term has many meanings, the context for which the definition is given.

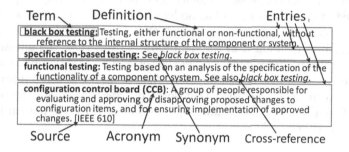

Fig. 1. The structure of the ISTQB Glossary v.2.2.

Each glossary may have different rules/symbols that determine how the entry
is composed, keywords that define links among terms, and formatting that can
express another links or properties (see Fig. 2). For instance, the glossary used
for Fig. 1 has an entry that contains term *specification-based testing* that is a
synonym for the preferred term *black-box testing,* and has no information about
synonyms for the preferred term. Otherwise, the glossary used for Fig. 2 has
terms that are preferred to other ones, in which case, the preferred term appears
as an entry, with the synonyms indicated, but synonyms do not have their individual entry at all.

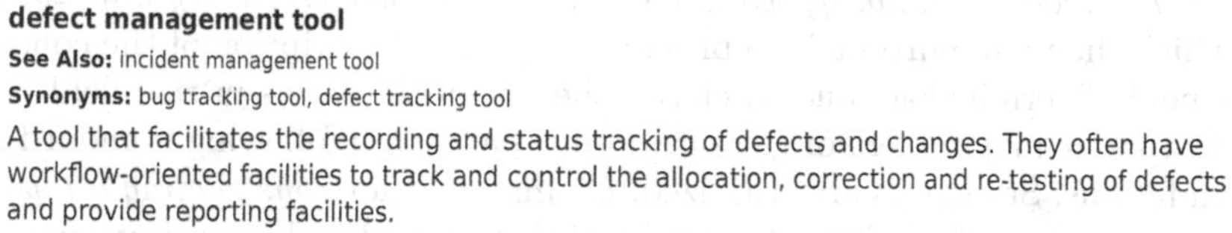

defect management tool
See Also: incident management tool
Synonyms: bug tracking tool, defect tracking tool
A tool that facilitates the recording and status tracking of defects and changes. They often have
workflow-oriented facilities to track and control the allocation, correction and re-testing of defects
and provide reporting facilities.

Fig. 2. An example of an entry from the ISTQB Glossary v.3.01.

For doing the automatic analysis of a glossary, a common approach is to
transform a glossary to a text format with convenient keywords or to a database
that stores glossary elements as separate units.

Most of the glossaries usually have terms that correspond to a *noun.* In such
case, entry is composited using pattern *X is a Y [that ...],* i.e., a term named
X (may be a phrase) is defined/explained by Y (may be a phrase). Usually,
a description, how X differs from Y and other information that is useful to

understand the meaning of X, is added. In linguistics, X is a *hyponym* and Y is a *hypernym*. A hyponym is a word or phrase whose semantic field is included within hypernym. In simpler terms, a hyponym shares a *type-of* relationship with its hypernym. In computer science, this relation is called *is-a* relationship. We prefer to use the later one. Few samples showed in the Fig. 1 are "black-box testing *is-a* testing", "functional testing *is-a* testing", "configuration control board *is-a* group of people" or "defect management tool *is-a* tool" showed in the Fig. 2.

Sometimes authors of glossary violate good practice of using the pattern *X is a Y*. For instance, entry in Fig. 3 has no hypernym given in an explicit way. The reader has to come to a conclusion himself that hypernym is *percentage* that is calculated as it is mentioned in the definition. In this case, the guessing is not difficult due to word *Percentage* used in the term. This observation is important – it is possible to establish *is-a* or *hyponym-hypernym* relationship from the term alone. We can make a conclusion that some entries use implicit pattern "X is an A and X is a B [and X is C, [and ...]]", i.e., term X as a hyponym has many hypernyms A, B[, C[, ...]].

Defect Detection Percentage (DDP)
See Also: escaped defects
Synonyms: Fault Detection Percentage (FDP)
The number of defects found by a test level, divided by the number found by that test level and any other means afterwards.

Fig. 3. A hypernym is not given in an explicit way.

2.2 Standards

Few standards are created regarding terminology work. For instance, *ISO 704:2009 Terminology work - Principles and methods* that is intended to standardize the essential elements of terminology work providing guiding principles; *ISO 1087-1:2000 Terminology work - Vocabulary - Part 1: Theory and application* which the main purpose is to provide a systemic description of the concepts in the field of terminology and to clarify the use of the terms in this field.

Creators or maintainers of glossaries only partly are following standard recommendations. Standards are criticized, for instance, for *constructing a typology of concept relations* for terminology work [14]. Unfortunately, there are not better standards, and we have to try to exploit or adapt standards at hand, i.e., to use the most appropriate ideas, principles, and recommendations.

The standard *ISO 704:2009 Terminology work – Principles and methods* states: "The goal of terminology work ... is ... a clarification and standardization of concepts and terminology for communication between humans".

The main activities of terminology work are:

1. Identifying concepts and concept relations.
2. Analysing and modelling concept systems on the basis of identified concepts and concept relations.

3. Establishing representations of concept systems through concept diagrams;
4. Defining concepts.
5. Attributing designations (predominantly terms) to each concept in one or more languages.
6. Recording and presenting terminological data, principally in print and electronic media.

By evaluating available public glossaries authors of this paper have made a conclusion that many of the terminologists do not follow these recommendations. Authors of glossaries base on models created in their mind. The main problem is in a fact that models with many hundreds of concepts have inconsistencies, and different people have different models in their minds. As a result, the created ontology suffers from various inconsistencies. Let us pay attention to the second activity of terminology work "Analysing and modelling concept systems on the basis of identified concepts and concept relations" and the third activity "Establishing representations of concept systems through concept diagrams".

2.3 Quality Evaluation

Redman formulated a definition of data quality. Data quality is the degree to which data meet the specific needs of specific customers. Note that one customer may find data to be of high quality for one usage of data, while another find the same data to be of low quality for another usage.

From the perspective of glossary user, the following quality attributes can be defined: structure of glossary, language correctness, glossary completeness regarding terms and relations, correctness of relations, unambiguity of definitions, understandability.

We focus on glossary completeness and correctness of relations, including consistency validation problem in this paper. Authors continue the further improvement of the tool that adopts the *term graph* building algorithm [2] and development of browsable concept map [3].

Since the Glossary is made without creating a concept system and its graphical representations before writing definitions, we propose to do reverse engineering – obtain the concept system for software testing domain from the Glossary. Thus, we can evaluate a quality of Glossary by analyzing the obtained concept system. The concept system can reveal the models that were in minds of terminologists.

3 GlossToolset

We have adopted our previously developed tools and improved them with better algorithms and functionality with a goal to use them for glossary analysis and quality evaluation. In order to refer to these tools easier, let us denote by *GlossToolset* a whole collection of our tools. The main goal of the GlossToolset is to generate various kinds of concept graphs or lightweight ontologies. Authors of

[5] had a similar goal - to create an ontology from the *"IEEE Standard Glossary of Software Engineering Terminology"*, but there are not shown any pieces of evidence in the paper that all process is automatic.

Automatic concept map or ontology construction from document collections is an actual problem that is not fully solved yet. The review [10] surveys what is possible now and what are current research directions. Authors of [12,15] and [13] solves similar problem as stated in this paper - how to extract semantic structure from glossaries automatically.

3.1   Parsing of Glossary

At the very beginning, one has to parse glossary and store into the database all terms, definitions, additional information and explicit relationships among terms. As a source of the glossary, one can take a text file obtained from initial glossary document or data (e.g., CSV files) received from glossary owner (for instance, the ISTQB Glossary is stored in the relational database).

The next step, according to works of Arnicans et al. [2,3], is to find *domain aspect names*. They introduce a new method for extracting the most significant words from a document in the form of a glossary. Then they assign a weight to each word in the glossary. The total weight of the word is a sum of the word weights in each entry. The weight of a word in a sentence depends on its position in the term and the definition. This new weighting method gives a better set of more significant words that characterizes the selected domain.

Authors of this paper improved weighting method by using natural language processing tool *Stanford CoreNLP* [9]. Authors of CoreNLP claim that the CoreNLP provides a set of natural language analysis tools. These tools can give the base forms of words, their parts of speech, whether they are names of companies, people, etc., normalize dates, times, and numeric quantities, and mark up the structure of sentences in terms of phrases and word dependencies, indicate which noun phrases refer to the same entities, indicate sentiment, extract open-class relations between mentions, etc.

Our improved weighting method has the following steps:

1. The GlossToolset transforms each entry into a sentence. By heuristic experiments we concluded that CoreNLP gives better results if (1) the term is enclosed in double quote marks; (2) an article "A" precedes the term; (3) an article "a" precedes the definition if it starts without any article. For instance, from entry *"branch testing: a white-box test design technique in which test cases are designed to execute branches"* we manually create *"A "branch testing" is a white-box test design technique in which test cases are designed to execute branches."* before CoreNLP processes it.
2. The CoreNLP processes each sentence, and all information is stored.
3. The GlossToolset analyze each obtained *Basic Dependencies tree* (see sample in Fig. 4) and extract hypernyms (it is a *white-box test design technique* in the given sample). Details of the extracting algorithm are out of the scope of this paper. After verifying results, we concluded that only a little bit over

80 % hypernyms are recognized correctly. Approximately a half of mistakes were produced by CoreNLP (some due to complex structure of definition). Other issues can be addressed by improving our algorithm or correcting the definition (see Sect. 2.1 that shows the definition without explicit hypernym).

4. The GlossToolset processes each entry and weights each word, that is some kind of noun. An instance of the word gets a weight calculated by formula 2^{-level} where the *level* is a level of word's instance in the Basic Dependencies tree (See Table 1). Total weight is calculated for each word as a sum of weights of each word's instance. For example, from the given entry a word *technique* receives 1, a word *test* receives $0.5 + 0.25 = 0.75$.

5. The GlossToolset creates a list of most important words/concepts. For instance, the Glossary has following most important 10 words called also significant words (weight rounded to whole integer is given in parentheses): *testing* (141), *test* (105), *technique* (51), *process* (43), *tool* (43), *software* (29), *capability* (29), *component* (27), *product* (26), *approach* (21).

Basic Dependencies:

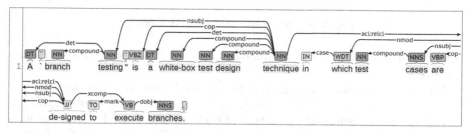

Fig. 4. Basic dependencies tree obtained by CoreNLP from sentence taken from entry's definition.

3.2   Annotated Definitions

Having at hand ordered list of the most significant words, we have created a collection of annotated definitions (Fig. 5). We set a threshold for significant words at 0.5. Each word from the list that satisfies the threshold is searched in the definitions. If the word is term it is enclosed in "{ }", otherwise, the word is enclosed in "| |". Annotated definitions help quicker evaluate whether most appropriate words define the term. Moreover, these annotations are very useful to organize localization work; annotations highlight words that are very important for consistent translation.

3.3   "explains" Graphs

The GlossToolset can create "explains" graphs and browsable concept map. The tool bases on principles described in [2,3].

First, the tool generates domain aspect graphs that can be considered as small concept maps. Authors of [4] also propose to create small concept maps to

Table 1. Weights of words from the tree shown in Fig. 4

Level	Word	Weight
0	Technique	$2^0 = 1$
1	White-box	$2^{-1} = 0.5$
	Test	
	Design	
2	Test	$2^{-2} = 0.25$
	Case	
3	Branch	$2^{-3} = 0.125$

anomaly: any **{condition}** that deviates from expectation based on {requirement}s **{specification}**s, |**design**| |**document**|s, |**user**| |**document**|s, **{standard}**s, etc. or from someone's perception or |**experience**|. Anomalies may be found during, but not limited to, **{review}**ing, **{testing}**, |**analysis**|, compilation, or use of **{software}** |**product**|s or applicable |**document**|ation.

Fig. 5. Sample of the annotated definition.

visualize domain. They take the important concept in focus and shows related concepts.

Second, all domain aspect graphs are merged into one large hypergraph. New concept maps are created on the fly by focusing any term; a subgraph from the hypergraph determine the structure of each concept map. The concept map helps to collect similar or related terms, immediately see definitions of terms and traverse through whole term graph (Fig. 6). The relation can be defined as "concept X *explains* concept Y". For instance, *peer review* explains *inspection*. ISTQB Syllabus level is shown using colors. Colors help to evaluate some quality aspects of syllabus too.

We have created graphs with different coloring principles, too. For instance, [8] used colors in order to classify terms in classes that follow Six Ws principle (Who, What, Where, When, Why, How).

3.4 "is-a" Graphs

"is-a" graphs show hierarchy among related concepts. Such kind of graphs is welcomed by ISO 704:2009 standard. We introduce two sorts of relations, one is term-hypernym_from_definition (black in a color image), and another one is term-hypernym_from_term (blue in the colored image) (Fig. 7). New terms (in ellipses) that are no part of the Glossary are generated using our algorithms. Terms in boxes are colored in order to expose their subdomain such as *pure testing, quality assurance, management, software engineering, mathematics*.

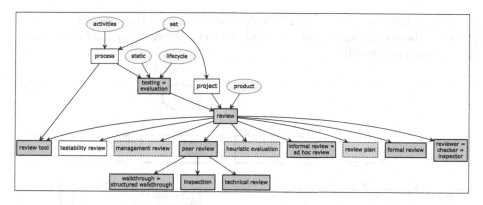

Fig. 6. Concept map with relationship *explains* for the focused node review as a part of the hypergraph (Color figure online).

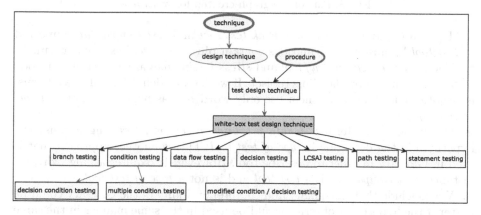

Fig. 7. Concept map with relationship *is-a* for the focused node *white-box test design technique* as a part of the hypergraph (Color figure online).

4   Analysis of ISTQB Software Testing Glossary

During our work in the localization of ISTQB Software Testing Glossary's terms in Latvian, we tried to preserve the level of Glossary's quality. In order to do that, we used GlossToolset, a set of our tools. As it later turned out, there are problems in the Glossary, too. In most cases there are inconsistencies, for instance, the same terms are used with different semantics or vice versa different terms are used for the same concept. Another problem is an inconsistent usage of hypernyms in definitions. Next subsections are devoted to a description of instances of problems found in Glossary using GlossToolset.

4.1   Inconsistent Usage of Hypernyms

Let us look at a group of terms related to concept *tool*. The graph is created by the GlossToolset, part of which is shown in Fig. 8. Definitions of all terms included in the figure are given in Table 2. The terms from the Glossary are represented by boxes and significant additional concepts revealed by the GlossToolset are represented by ellipses.

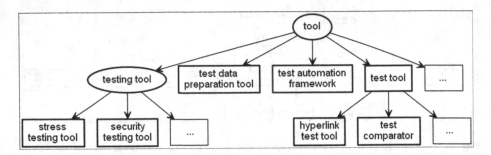

Fig. 8. Part of the graph created for word *tool*.

There are only two terms *hyperlink test tool* and *test comparator* connected to *test tool* because their definitions contain phrase *test tool* as a hypernym.

The terms *security testing tool* and *stress testing tool* are located under node *testing tool* offered by the GlossToolset. If we look at definitions of these terms, it is noticeable, that all of them use only word *tool* as hypernym, not *test tool* or *testing tool*.

Definition of the term *test data preparation tool* is interesting because the hypernym in the definition is *type of test tool*. This is another approach that is used in Glossary in order to describe tool. However, it requires asking, why, for instance, *test comparator* is a *test tool* and is not a *type of test tool*.

We conclude that the quality and comprehensibility of the Glossary could be better if the hypernyms of terms would be used in the same manner in the same context.

Sometimes definitions are formulated in such a way that the true hypernyms cannot be found automatically by the GlossToolset, but can be revealed by a domain expert. For instance, the definition of the term *test automation framework* explicitly shows semantics that the GlossToolset cannot recognize – this term belongs to the family of test tools.

GlossToolset generates graphs that demonstrate types of relationships among the terms by different colors and show the definitions of terms as tooltips for each node. This ability is very convenient for experts in order to notice consistency problems.

4.2   Different Meanings of the Same Term or Significant Word

Another case, when the automatic generation of concept map from glossary is very hard or impossible, is when terms have different meanings in different contexts. For instance, there are two different types of the significant word *framework* mentioned in the Glossary. In some definitions, it is used with the meaning as skeleton or outline of activities that should be done during some organizational

Table 2. Definitions of some Glossary terms related to tools

Term	Definition
Test tool	A software product that supports one or more test activities, such as planning and control, specification, building initial files and data, test execution and test analysis
Hyperlink test tool	A tool used to check that no broken hyperlinks are present on a web site
Test comparator	A test tool to perform automated test comparison of actual results with expected results
Security testing tool	A tool that provides support for testing security characteristics and vulnerabilities
Stress testing tool	A tool that supports stress testing
Test data preparation tool	A type of test tool that enables data to be selected from existing databases or created, generated, manipulated and edited for use in testing
Test automation framework	A tool that provides an environment for test automation. It usually includes a test harness and test libraries

process, for instance, *Capability Maturity Model Integration (CMMI)* or *TPI Next* or *SCRUM*, but sometimes *framework* means software tool, for instance, *test automation framework, unit test framework*. The definitions of *Capability Maturity Model Integration (CMMI)* and *unit test framework* as examples are shown in Table 3.

A *specified input* is a term used in detailed test cases which have described input values and specified expected output values or result. The term *specified input* does not mean variables as it can be supposed having the term *input* in the context. Such inconsistency leads to propose the term *specified input values* instead of *specified input*.

4.3   Usage Consistency of the Terms Test and Testing

According to the definitions provided by the Glossary shown in Table 5 the term *test* means a set of test cases while the term *testing* means process consisting of all activities that have to be done to test the software. At the same time, the definition of the term *test estimation* says that it is an "*approximation of a result related to various aspects of testing*". So, this term means the assessment not only of the set of test cases but also of process related aspects of the testing.

A similar situation is with terms *input, input value, specified input* whose definitions are shown in Table 4. The definition of *input* says, that *input* is a variable, and also *input value* has its definition as an instance of the *input*.

There are more cases of confusing and inconsistent usage of terms *test* and *testing* in the Glossary. For instance, the term *test level* by its definition is *a group*

Table 3. Definitions of the terms *Capability Maturity Model Integration (CMMI)* and *unit test framework*

Term	Definition
Capability Maturity Model Integration (CMMI)	A framework that describes the key elements of an effective product development and maintenance process. The Capability Maturity Model Integration covers best-practices for planning, engineering and managing product development and maintenance
Unit test framework	A tool that provides an environment for unit or component testing in which a component can be tested in isolation or with suitable stubs and drivers. It also provides other support for the developer, such as debugging capabilities

Table 4. Some definitions the Glossary terms related to concept *input*

Term	Definition
Input	A variable (whether stored within a component or outside) that is read by a component
Input value	An instance of an input
Specified input	An input for which the specification predicts a result

of activities. Consequently, this term has to be titled as *testing level* instead of *test level* according to the definitions of the terms *testing* and *test.*

If we look at the end part of the definition of the term *test level* there are named examples of those levels - *component test, integration test, system test,* and *acceptance test.* There are not such terms in the Glossary. Glossary contains terms *component testing* whose definition is shown in Table 5, *integration testing, system testing,* and *acceptance testing.*

4.4 Different Terms with the Same Meaning

Let us look at the definitions of the terms *test process* and *testing* shown in Table 5. In point of fact both of them have the same meaning. It means that one of them can be supposed as a redundant term in the glossary, for instance, *test process,* or maybe they have to be declared as synonyms.

4.5 The Loops in the Definitions of Terms

There are definitions of terms *component, software* and *system* shown in Fig. 9. The term *component* has term *software* in its definition, but the term *software* is explained by the term *system* because there is not the term *computer system* in the Glossary. The term *system* in its turn is explained using the term *component* in its definition. So, there is a loop where terms and their definitions explain each other.

Table 5. Usage of the terms *test* and *testing* in the Glossary

Term	Definition
Test	A set of one or more test cases
Testing	The process consisting of all lifecycle activities, both static and dynamic, concerned with planning, preparation and evaluation of software products and related work products to determine that they satisfy specified requirements, to demonstrate that they are fit for purpose and to detect defects
Test estimation	The calculated approximation of a result related to various aspects of testing (e.g., effort spent, completion date, costs involved, number of test cases, etc.) which is usable even if input data may be incomplete, uncertain, or noisy
Test process	The fundamental test process comprises test planning and control, test analysis and design, test implementation and execution, evaluating exit criteria and reporting, and test closure activities
Test level	A group of test activities that are organized and managed together. A test level is linked to the responsibilities in a project. Examples of test levels are component test, integration test, system test and acceptance test
Component testing	The testing of individual software components

component	A minimal {software} item that can be tested in isolation.
software	Computer programs, procedures, and possibly associated documentation and data pertaining to the operation of a computer {system}.
system	A collection of {component}s organized to accomplish a specific function or set of functions.

Fig. 9. A sample of the loop in the Glossary's definitions.

Experts have to analyze loops and make a decision whether some corrections in definitions are necessary. Navigli has solved similar problems offering an original algorithm for the automated disambiguation of glosses by finding cycles [11].

5 Conclusions and Future Work

The GlossToolset developed by authors of the paper emerged as useful and handy tool when it is necessary to analyze the glossaries. There can be different aims of analysis, for instance, improvement of glossary's quality or attempt to develop the ontology of the concepts held by a glossary. The GlossToolset can graphically show the different relationships between glossary's terms, for instance, relation hyponym-hypernym or which terms or significant words (concepts) are used to explain or define given term. The graph generated by the GlossToolset substantially facilitates recognition of different type inconsistencies that are in glossary thus helping to eliminate them and improve the quality of glossary.

There are also problems in glossaries that are not recognizable using the GlossToolset or another tool. In such case, a contribution of an expert is necessary. For instance, following problems require expert decision (1) true hypernym is verbally hidden in the definition of the term; (2) the same term of the concept is used in different meanings; (3) there are two or more terms for the same concept in the glossary.

Complimentary material to the paper, containing top domain aspects, browsable concept maps, etc., is available on our expanding site [1].

Acknowledgments. This paper partially has been supported by the European Regional Development Fund Project No. 2DP/2.1.1.3.1/11/APIA/VIAA/010.

References

1. Arnicane, V., Arnicans, G., Borzovs, J.: Improvement of systematic collection of terminology. http://science.df.lu.lv/aab16
2. Arnicans, G., Romans, D., Straujums, U.: Semi-automatic generation of a software testing lightweight ontology from a glossary based on the ONTO6 methodology. In: Caplinskas, A., et al. (eds.) Databases and Information Systems VII. Frontiers in Artificial Intelligence and Applications, vol. 249, pp. 263–276. IOS Press, Amsterdam (2013)
3. Arnicans, G., Straujums, U.: Transformation of the software testing glossary into a browsable concept map. In: Sobh, T., Elleithy, K. (eds.) Innovations and Advances in Computing, Informatics, Systems Sciences, Networking and Engineering. LNEE, vol. 313, pp. 349–356. Springer International Publishing, Switzerland (2015). doi:10.1007/978-3-319-06773-5_47
4. Deokattey, S., Bhanumurthy, K.: Domain visualisation using concept maps: a case study. DESIDOC J. Libr. Inf. Technol. **33**(4), 295–299 (2013)
5. Hilera, J.R., Pagés, C., Martínez, J.J., Gutiérrez, J.A., De-Marcos, L.: An evolutive process to convert glossaries into ontologies. Inf. Technol. Libr. **29**(4), 195–204 (2010)
6. ISTQB: Certifying software testers worldwide. http://www.istqb.org/
7. ISTQB: Standard glossary of terms used in software testing. http://www.istqb.org/downloads/category/20-istqb-glossary.html
8. Kuļešovs, I., Arnicane, V., Arnicans, G., Borzovs, J.: Inventory of testing ideas and structuring of testing terms. Baltic J. Mod. Comput. **1**(3–4), 210–227 (2013)
9. Manning, C.D., Surdeanu, M., Bauer, J., Finkel, J.R., Bethard, S., McClosky, D.: The Stanford CoreNLP natural language processing toolkit. In: ACL (System Demonstrations), pp. 55–60 (2014)

10. Medelyan, O., Witten, I.H., Divoli, A., Broekstra, J.: Automatic construction of lexicons, taxonomies, ontologies, and other knowledge structures. Wiley Interdisc. Rev. Data Min. Knowl. Disc. **3**(4), 257–279 (2013)
11. Navigli, R.: Using cycles and quasi-cycles to disambiguate dictionary glosses. In: Proceedings of the 12th Conference of the European Chapter of the Association for Computational Linguistics, pp. 594–602. Association for Computational Linguistics (2009)
12. Navigli, R., Velardi, P.: Ontology enrichment through automatic semantic annotation of on-line glossaries. In: Staab, S., Svátek, V. (eds.) EKAW 2006. LNCS (LNAI), vol. 4248, pp. 126–140. Springer, Heidelberg (2006)
13. Navigli, R., Velardi, P.: From glossaries to ontologies: extracting semantic structure from textual definitions. In: Ontology Learning and Population: Bridging the Gap Between Text and Knowledge, pp. 71–87 (2008)
14. Nuopponen, A.: Tangled web of concept relations. concept relations for ISO 1087-1 and ISO 704. In: Terminology and Knowledge Engineering 2014, pp. 10-p (2014)
15. Velardi, P., Faralli, S., Navigli, R.: Ontolearn reloaded: a graph-based algorithm for taxonomy induction. Comput. Linguist. **39**(3), 665–707 (2013)

The Application of Optimal Topic Sequence in Adaptive e-Learning Systems

Vija Vagale[(⊠)] and Laila Niedrite

University of Latvia, Raina boulv.19, Riga, Latvia
vija.vagale@gmail.com, laila.niedrite@lu.lv

Abstract. In an adaptive e-learning system an opportunity to choose a course topic sequence is given to ensure personalization. The topic sequence can be obtained from three sources: teacher-offered topic sequence that is based on teacher's pedagogical experience; learner's free choice that is based on indicated links between topics, and, finally, the optimal topic sequence acquisition method described in this article. The optimal topic sequence is based on previous learners' experience. With the help of the optimal topic sequence method, data about previous learners' course topic sequence and course results are obtained. After the data analysis the optimal topic sequence for the specific course is obtained based on the links between course topics. In this article the experimental test of this method is described.

Keywords: Optimal topic sequence · Adaptive e-learning system · Learner model

1 Introduction

Nowadays, the necessity for lifelong and asynchronous learning when learners and teachers are not tied to certain place or time is increasing [4]. It is not possible to use personalized approach for all learners in the process of synchronous learning in a big auditorium [12]. The aim of an adaptive e-learning system (AELS) is to overcome this problem. In the adaptive e-learning systems the diversity of content delivery is one of the most important factors to ensure high quality of an adaptive system [12]. The AELS adapts e-learning content (a) to satisfy learner's needs and desires; (b) to depend on learner's behavior; (c) to use obtained results to ensure further adaptation. The content adaptation for a learner is done based on: (a) learner's personal features (for example: learning style, basic knowledge in the course, the level of course acquisition difficulty); (b) course topics that are used based on suitable pedagogical strategies.

In this article, a discussion about AELS content organization focusing attention on course topics, their content, sequence, and choice opportunities is proposed. The course content for each learner in the experimental AELS is made dynamically based on a course learner group, where each learner is classified, and on the adaptation scenario that is suitable for appropriate group members [10]. The learner has opportunity to choose between one of three topic sequence versions: (a) a teacher's topic sequence, (b) a learner's topic sequence, or (c) an optimal topic sequence. The teacher-pointed topic sequence is based on teacher's pedagogical experience. In case when a learner

G. Arnicans et al. (Eds.): DB&IS 2016, CCIS 615, pp. 352–365, 2016.
DOI: 10.1007/978-3-319-40180-5_24

chooses the next topic on his own, it is obtained based on the indicated links between course topics. The optimal course topic sequence is made based on the optimal topic sequence method described in this article. This method is based on the data that are obtained from the previous course learner results and links between course topics.

The paper is organized as follows. Section 2 presents related work. Section 3 describes the background of the presented work. Section 4 describes the optimal topic sequence method. Section 5 describes opportunities to choose course topics. The experiment is described in Sect. 6. Finally, conclusions and plans for the future research are given in Sect. 7.

2 Related Work

Different approaches can be used for modeling learning content. For example, in the articles [2, 6, 7], description of relationships between unities of learning content is made using hierarchical structures. Their creation and maintenance takes a lot of time and system resources. Conditions that describe better learning strategy acquisition depending on learner's actions and results are used to organize sequence in all previous mentioned articles.

Madjarov and Betari [7] describe the network service-based learning infrastructure that uses SCORM content model. The learning objects are organized into a learning cluster tree and are ordered according to learner's actions and responses to the questions and exercises. The learning cluster contains text page with learning content, exercise, and questions. Modeling of the cluster tree is done using Petri Nets. In the offered system learning process is done in the following way. First of all, a learner receives an intro page and then a page with learning content. After familiarization with it, the system offers an exercise. A learner receives questions that are created especially for learner, depending on the exercise results. The goal of this pedagogical strategy is to evaluate the level of the learner's knowledge and his/her comprehension of the already acquired piece of content.

Jabari et al. [6] overviews personalization opportunities in e-learning and virtual classrooms. The problem of how to present learning object according to learner cognition style is solved in the article. Learning objects are arranged in hierarchical way. The delivery sequencing is ensured by IMS Simple Sequencing specification. Petri Nets are used for conditions. They are used to create system reaction to the results of the learning process.

In the article [1] Brusilovsky and Vassileva describe the system for dynamical course generation (DCG) and the concept-based courseware analysis system (CoCoA). CoCoA checks the consistency of the course and its quality in each moment of life. The DCG contains the domain authoring component and the adaptive course automatic generation component. It allows to generate an individual course according to learner's goals and previous knowledge and to adapt the course according to the learner's obtained knowledge.

In the article [5] Huang and Shiu describe a user-centric adaptive learning system (UALS), which creates learning material sequencing schemes based on users' collective intelligence and collaborative voting approach that use the item response theory

(IRT). The collective voting approach allows learners to cope with difficulties that occurred during learning process together. The system uses sequenced rules to personalize user-oriented learning ways. The IRT helps to evaluate students' skills and offers the most appropriate content for them.

In the article [2] Elouahbi et al. use pedagogical graph SMARTGraph for sequence modeling. Nodes of the graph are learning units and arcs – pedagogical restrictions between units. Creation of the sequence graph is done using Xlink (XML Linking) language.

The optimal topic sequence method that is offered in this article differs from the above-mentioned approaches by the fact that it uses advantages of costless course management system: learning course organization environment, tools for organizing learning process, processes, and a data base. It allows to save time and material resources to adapt the learning process to learner's needs. The offered OTS method is easy-to-use. Working object for these methods are topics that are used in any learning course. In the above-mentioned articles, sequences are organized for smaller content units. It makes the process of course content creation more complicated and time-consuming. In the related articles, hierarchical structures are used for data storage. It requires knowledge to manage the appropriate tools. In the offered OTS method, symbol sequences are used for topic sequence graph storage.

3   Background

A lot of research is made in the adaptive system development nowadays. Adaptive study environments are based on well-organized models and processes. The AELS consists of three main models: learner model (LM), content model (CM) and adaptation model (AM) [8]. Data about learner are described in the learner model. The content model describes the AELS content and its logical structure. The adaptation model offers appropriate learning content for a specific learner.

3.1   A Learner Model

The experimental AELS learner model is based on data life cycle in the model and includes eight data categories: Personal Data, Personality Data, Pedagogical Data, Preference Data, Device Data, System Experience, Current Moment's Knowledge, and learning process data (History Data) [8]. LM data can be divided into three basic classes by their life length or refreshing frequency: basic data (BasicData), additional data (AdditionalData), and learning process data (LearningProcessData). Wider description of LM data life cycle is given in the article [9]. The BasicData class includes learner data that are constant in system or are changing very rarely. These data include personal data about learner. They are gathered into PersonalData category. The additional data describe the learner individuality. This class contains five LM data categories: PersonalityData, PedagogicalData, PreferenceData, DeviceData, and SystemExperience. The examples of additional data are: learning styles (PersonalityData); data that organize learning process (learning course, course difficulty level, course pre-knowledge) (PedagogicalData); language of the course and course environment preferences (PreferenceData); experience in the course utilization (SystemExperience). Direct additional data are used to ensure adaptation in the system. Additional data are dynamic data. These data have a tendency to change in the long term. The LearningProcessData class contains data about learner knowledge at the current moment (CurrentMomentKnowledge) and data about learning process (HistoryData). The life length of these data in the learner model is the shortest, because during the learning process they are changed and supplemented constantly. The learning process data are accumulated and processed in the long term and new data that describe the learner are obtained. A more detailed description of the learner model is offered in articles [10, 11].

The source of data acquisition for the learner model are: (a) profile of the system user, (b) results of quizzes and tests, (c) individual choices of learner, (d) data of external system, (e) data from the learner group where a learner is classified, (f) data analysis results obtained in course of the learning process [11].

3.2 A Content Model

The content model used in the experimental system is based on the learning object (LO) and application of the different resource formats. A learning course described in the content model consists of one or multiple topics. Each topic is made of one or multiple learning objects. Each LO contains the description of the learning object, theoretical part, practical part, and evaluation part. More information on the content model is given in the article [10]. Each topic in this experiment consists of one learning object (Fig. 1). The description part explains the essence of the specific topic, its exercises, and place in the course structure. The theoretical part contains system-offered knowledge and ways of its representation. In case, when the learner has previous knowledge in the course, the basic content of the course is shown with system-offered knowledge represented by activities and resources. Otherwise, if a learner doesn't have pre-knowledge in the course, an expanded course topic content is shown. In this case, links to expanded definitions of the concepts are shown in the additional content. The practical part contains activities that are made to strengthen the acquired knowledge. In case of theoretical courses, practical part does not exist. The evaluation part contains activities that ensure course knowledge evaluation. The topic content also differs depending on the difficulty level of the chosen course. The course difficulty level (DL) gives a learner an opportunity to determine the maximal acceptable course grade for himself/herself. The offered exercise level, test difficulty, and also course final grade depends on the chosen DL.

The content model describes a wide range of resource types that are created based on learning styles. For instance, if a learner has "Aural" learning style, then in the theoretical part of the topic all resources that correspond to this learning style are shown. For instance, that includes audio recording and audio conferences. A wider range of resource types used in the system is given in articles [10, 11].

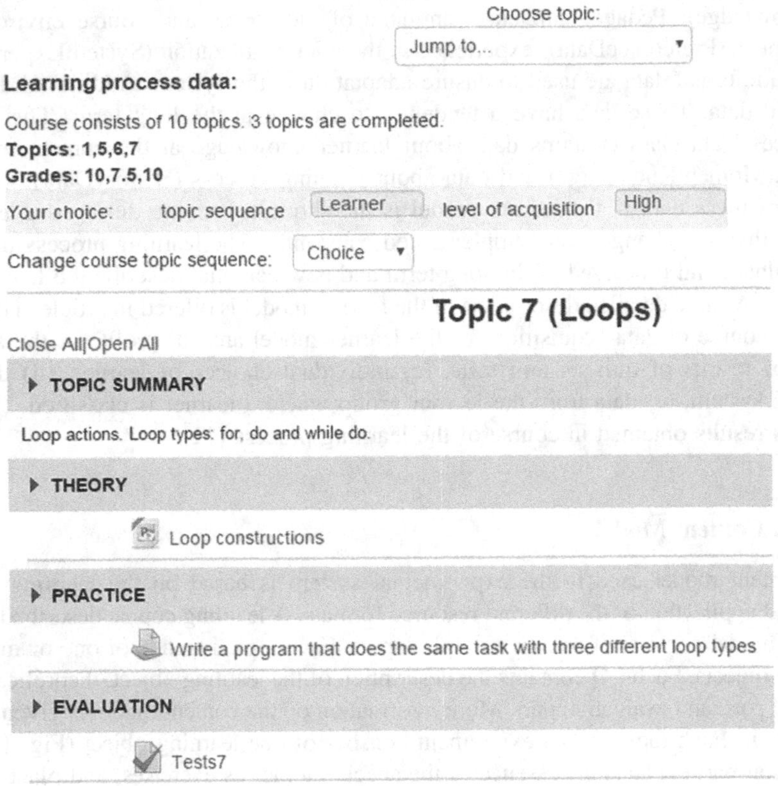

Choose topic:

Jump to...   ▼

Learning process data:

Course consists of 10 topics. 3 topics are completed.
Topics: 1,5,6,7
Grades: 10,7.5,10

Your choice: topic sequence   Learner   level of acquisition   High

Change course topic sequence: Choice ▼

Topic 7 (Loops)

Close All|Open All

▶ **TOPIC SUMMARY**

Loop actions. Loop types: for, do and while do.

▶ **THEORY**

[PS] Loop constructions

▶ **PRACTICE**

Write a program that does the same task with three different loop types

▶ **EVALUATION**

✓ Tests7

Fig. 1. The structure of the "Programming Foundations" course topic "Loops"

3.3   An Adaptive Model

A course is created by the course author or in some cases by the course teacher. The course author creates course structure, course content, selects features that will be used in course content adaptation and for learner group creation [11].

Before the course acquisition, the system checks whether it has all necessary data about the learner: (i) the learning style; (ii) the existence or absence of pre-knowledge in course; and (iii) the course difficulty level. Each course learner is classified into some of the created course learner groups based on data that the system has about the learner. In case when learner data are not enough for classification, the learner is classified into the "Group0". Members of this group receive the non-adapted course content. Learner classification into groups is described in the article [10].

4   The Method for Optimal Topic Sequence Creation

This chapter offers the method for optimal topic sequence creation for one course.

4.1 Definitions

All course topics can be considered as one set of topics. The subset of element set that contains elements with specific features is called a *selection*. The subset of such set that takes into consideration the sequence of elements is called an *ordered selection*. So, topic sequence is a course topic selection that is ordered by a specific feature.

In this article three type topic sequences are used: (i) the learner topic sequence, (ii) the teacher topic sequence, and (iii) the optimal topic sequence. *The learner topic sequence* is a topic sequence that is created from the learner-chosen topics during the course acquisition. *The teacher topic sequence* is a topic sequence that is offered by a teacher for acquisition of a specific course. This topic sequence is based on pedagogical experience of the teacher. *The optimal topic sequence* (OTS) is a topic sequence that ensures the best course acquisition results. The OTS is obtained using an optimal topic sequence method (OTSM), which is based on the learner-obtained topic sequences, acquired course results, and links between topics.

Each topic in the topic sequence has an order number or position that describes its position in the sequence. For example, if the course contains topics "1", "2", "3", "4", "5" and a learner has acquired these topics in the following sequence: "1", "2", "4", "5", "3", then the learner topic sequence set will be {1,2,4,5,3}. In this topic sequence, for instance, the number or position of the topic "4" is 3.

4.2 The Principle of the Optimal Topic Sequence Method Performance

Topic sequences (TS) obtained during the course acquisition are saved in the system database. The optimal topic sequence for one course can be obtained by selecting all topic sequences of one course and processing them with OTSM. The essence of the optimal topic sequence method is to take all topic sequences that are obtained from one course acquisition and to unite them in groups by similarities, where one group has equal sequences and for each sequence group the course acquisition grade is calculated. The group of sequences that has the highest average grade is the OTS that was searched for. In case when multiple topic sequence groups have the same grade, the optimal topic sequence is searched between the TS of these groups based on the highest repetition rate of the specific topic sequence position and existing links between topics that are indicated by the course author or teacher. The most frequently repeated topic is inserted into a specific position of the OTS, and this obtained topic has also a link to the last inserted OTS topic. If there is no link, then the next most frequent topic is taken.

The searching of OTS is based on the OTSM that is described by the following algorithm (Fig. 2):

1. All topic sequences and corresponding grades of the specific course are found.
2. The obtained data are grouped into groups with the same topic sequences.
3. A course average grade is calculated for each obtained group of the topic sequences.
4. In the course, the topic sequence group with the highest grade is found.

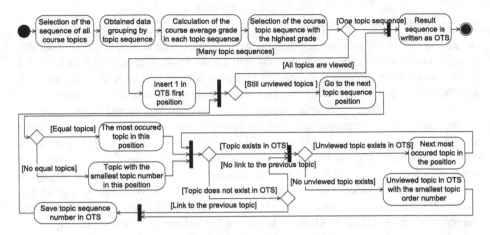

Fig. 2. An activity diagram of the optimal topic sequence creation

5. If there is only one such topic sequence, then it is assumed to be the optimal topic sequence, and the algorithm proceeds to the step 14. If multiple topic sequences are obtained with equal average grades, then a transition to the step 6 occurs in order to create a new OTS.

6. In the first position of the optimal topic sequence the first course topic is inserted.

7. Searching for unviewed topics. If all topics are viewed, the next step is 14. If still unviewed topics exist, the algorithm proceeds to the next step.

8. A transition to the position of the topic sequence.

9. If equal topics occur in the viewed position, the most frequent topic is taken. If all topics repeat only once, then topic with the lowest order number is taken.

10. If the next topic is already in the OTS, then the next step is 12, in case if it is not in the OTS, then the algorithm goes to the next step.

11. If the last topic of the OTS has a link to the chosen topic then the next step is 13, otherwise – step 12.

12. The next most frequent topic is taken, transition to the step 10. If there are no more topics in this position, then an unviewed topic with the lowest order number is taken and the algorithm proceeds to the step 13.

13. The order number of the topic is inserted into the OTS and the next step is 7.

14. The OTS creation is finished.

In the step 12 of OTS algorithm, when none of viewed TS topics are suitable for positions, the topic with the lowest order number that does not appear in the created sequence is inserted into in OTS. In this case, there is no testing whether a link from the last OTS topic to the chosen topic exists. It is not necessary, because OTS always begins with the first topic with the order number 1. It means that the way to the topic with the lowest order number always exists in the graph of course topics (for example, Fig. 3).

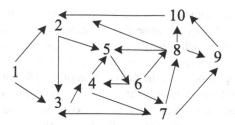

Fig. 3. A graph of course topics

4.3 An Example of the Optimal Topic Sequence Method Activity

The optimal topic sequence algorithm activity is described with the help of a specific example. Let's assume that a course consists of ten compulsory topics. Connections between topics are represented as an oriented graph in Fig. 3. An arrow indicates a link between two topics. Links between topics define which one will be offered as next. For instance, there are two arrows going out of topic "1" to topics "2" and "3". It means that in case when the version of the learner's topic choice is used, the system offers topics "2" and "3" as next topics after the first topic is acquired successfully. A learner must choose the next topic from the topic set offered by the system.

Let's assume that learners have acquired the previously-described course, and three different learner topic sequences with equal course grades are obtained as shown in the top three rows: LTS1, LTS2, and LTS3 of the Fig. 4. The lowest row shows the topic sequence that is obtained from these three topic sequences. Let's take a closer look at how OTS is obtained using the OTS method described in this chapter. Multiple topic sequences have an equal course grade that is why the OTS algorithm starts from the step 6. While TS still has unviewed positions, we will keep reviewing them. Only the OTS steps with different methods applied will be reviewed:

Position=	1.	2.	3.	4.	5.	6.	7.	8.	9.	10.
LTS1=	1	2	5	6	8	9	10	4	7	3
LTS2=	1	3	5	7	9	10	6	4	8	2
LTS3=	1	2	3	8	9	10	6	4	7	5
OTS=	1	2	5	6	8	10	3	4	7	9

Fig. 4. An example of the optimal topic sequence creation

Step 1. Acquisition of the course starts with the topic 1, therefore, the topic "1" is inserted into the 1st position of OTS, OTS = {1}.

Step 2. There are two topics "2" in the 2nd position of TS that is why this topic is taken. The topic "2" still does not appear in the created OTS, and a link between the topics "1" and "2" exists, which is why the topic "2" is inserted into the 2nd position of the OTS, OTS = {1,2}. Similar case is in the step 2, OTS = {1,2,5}.

Step 4. No equal topics occur in the 4th position of the TS that is why topic with the lowest order number is taken, which is the topic "6". This topic still does not appear in OTS and a link between the topics "5" and "6" exists, therefore, OTS = {1,2,5,6}.

Step 5. Two equal topics "9" appear in the 5th position of TS, but a link from the topic "6" to the topic "9" does not exist that is why the topic "8" is taken. The topic "8" still can't be found in the OTS, and a link between topics "6" and "8" exists that is why this topic is added to the OTS, OTS = {1,2,5,6,8}. The next step is similar to the step 2, where OTS = {1,2,3,6,8,10}.

Step 7. Two topics "6" appear in the 7th position of the TS. Since the topic "6" is already added to the OTS, the topic "10" is taken as the next one. The topic "10" is also already inserted into the OTS. Then, the topic with the lowest order number that still is not added to the OTS is taken. In our case, it is the topic "3", OTS = {1,2,5,6,8,10,3}. Next two steps are similar to the step 2, OTS = {1,2,5,6,8,10, 3,4,7}.

Step 10. All topics in the 10th position of TS are single. So, the topic with the lowest order number is taken, which is the topic "2". This topic already appears in OTS. The same case is for the topics "3" and "5". Since no more topics for this position exist, a still uninserted topic which has the lowest order number is taken as the next one in the OTS. Such topic is the topic "9", OTS = {1,2,5,6,8,10,3,4,7,9}.

Now all topic sequence positions are viewed, it means that the OTS creation is finished.

4.4   The Topic Sequence Module

The topic sequence module (TSM) was made and implemented in Moodle system for the optimal topic sequence acquisition. Figure 5 shows the interaction of the TSM with the system components. Colored figures show TSM that is added to the system and its new tables. Arrows show directions of the data flow. TSM interacts with the system offered course page, the system content model, the learner model, and the adaptation model that are described in Sect. 3. For TS purposes the following model activity tables were created (Fig. 6): course_teaching_parametrs, user_learning_data, and course_teaching_parametrs. Course acquisition parameters such as links between course topics (topic_link), a teacher-indicated topic sequence (teacher_topic_sequence), and an optimal course topic sequence (optimal_topic_sequence) are stored in course_teaching_parametrs table. Data that are necessary for the learning process and data obtained from learning process, e.g., a learner topic sequence (topic_sequence), ways to choose topic sequence (user_ts_choice), topic grades (topic_grade), an optimal topic sequence granted to a learner (user_ots) are stored in user_learning_data table.

The main functions and data flow of the TSM are shown in the picture Fig. 6. The main TSM actions are:

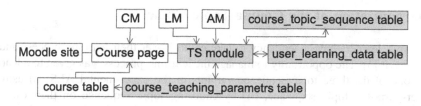

Fig. 5. Interaction of the OTS module and system components

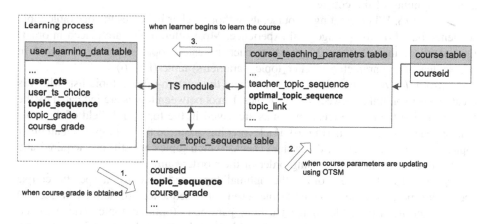

Fig. 6. A sequence of the data flow for OTS acquisition

- When a learner starts the course acquisition, TSM checks whether a specific course has the optimal topic acquisition sequence in the table course_teaching_parametrs. In case if it has, it is written into user_learning_data table. The newly-created course has no data about learner topic sequences that is why this course has no optimal topic sequence. At the beginning, teacher topic sequence can be assumed as the optimal topic sequence;

- The next step in TSM is collection of the learner study process data into user_learning_data table;

- After the successful course acquisition, the topic sequence created for a specific course is saved in course_topic_sequence table;

- To refresh course parameters, a teacher creates a new OTS. TSM the topic sequence created for a specific course and course grades from course_topic_sequence table. A new course OTS is created and saved in course_teaching_parametrs table with the help of OTS method.

During the course acquisition the OTS does not change for a specific learner. Course OTS can change after the course parameters are refreshed, and for other learners it will be different.

5   Organization of the Course Topic Choice

Nowadays, learners have a tendency to take control over their learning process and to determine it [3]. The opportunity to manage the learning process was given to a learner using one of the three topic sequence variants in the experimental ALS: (i) using a teacher-purposed topic sequence, (ii) by choosing topic sequence on their own, or (iii) using a system-offered optimal topic sequence that is described in previous section. Regardless of the chosen topic sequence variant, the first course topic is always offered at the beginning of the course.

Teacher TS. When creating a course, its author/teacher indicates the desirable topic sequence based on the pedagogical experience. Most often topics are placed in order, first of all 1, then 2, then 3, etc. A teacher-indicated topic sequence is stored in course_teaching_parametrs (teacher_topic_parametrs) table (Fig. 6).

Learner TS. A learner chooses a topic from the system-offered topic list, which is created based on links between course topics. Links between topics are indicated by the teacher during the course creation. Links are saved in the topic_link field of the table course_teaching_parametrs (Fig. 6). If a learner has acquired the last topic, but all other topics are still not acquired, the system checks which topics are still not acquired and offers these topics in an increasing order of their order numbers.

Optimal TS. A system offers the optimal topic sequence for a specific course acquisition based on the previous learner-used topic sequences (described in Sect. 4) that is taken from the table course_teaching_parametrs (optimal_topic_sequence) and saved as learner study process data in the table user_learning_data (user_ots) (Fig. 6).

The OTS is based on the previous learner data, which is why in case of a new course it is not offered or might be similar to the teacher topic sequence. It is useful to create the OTS for a course only when the amount of the learner topic sequences obtained is large enough (for example, when one learner group has already finished the course).

6   An Experiment

The optimal topic sequence method used in a topic sequence module was created for testing purposes (Sect. 4.4). It was implemented in AELS of the standard learning course management system Moodle that is described in Sect. 3. The structure of the learning course "Programming Foundations" was created according to the described AELS content model. The structure of the course consists of 10 topics: (1) "C++ program structure. Data output", (2) "Data types. Data input", (3) "Mathematical functions", (4) "Conditional constructions", (5) "User-defined functions", (6) "Parametric functions", (7) "Cyclic constructions", (8) "One-dimensional numeric arrays", (9) "Multi-dimensional numeric arrays", and (10) "Symbolic arrays". An order number of each topic is given in brackets. Each course topic is divided into four parts: topic summary, theory, practice, and evaluation (Fig. 1). Evaluation part of each topic consists of the created tests depending on the course difficulty level chosen by a learner. A course teacher indicated the links between topics and a teacher-suggested topic acquisition sequence. The following links between topics were indicated:

{1:2,3,5;2:3,4,5;3:4,5;4:5,7;5:6,7;6:7;7:5,8,10;8:9,10;9:10}. The order number of a topic is separated from links with ":" sign, and links between themselves are denoted with "," sign. One portion of "topic:link" is divided from next one with ";" sign. For example, a topic with order number 1 has links to topics number 2, 3, and 5. The order of the teacher-advised topic acquisition sequence is the following: {1,2,3,4,5, 6,7,8,9,10}.

38 first-year students of the professional higher-education bachelor study program "Information technologies" of the University of Daugavpils took part in the experiment in the academic year 2015/2016. The number of experiment participants is based on real number of students in course "Programming Foundations" that was used in experiment. Data about learners that would provide adaptation of the course content were obtained. Tests and quizzes, created in Moodle, helped to acquire data about learning styles of each learner (visual, aural, kinesthetic, reading, visual-aural), pre-knowledge in the course (yes, no), and the desirable course acquisition level (low, average, high) (Sect. 3.3). There were 30 learner groups made in the course based on the possible values of the learning style, course pre-knowledge, and course acquisition level [11]. Learner groups created in the course defined a scenario that was used for adaptation (Sect. 3.3).

25 students were classified into 13 groups created in course based on the obtained data about learners. Other students with lack of enough data were classified into "Group0". In this group participants do not get any adaptation scenario. Learner groups in this experiment were used to make analysis of the learner model based on the course acquisition results. LM analysis is out of scope of this article. Learner group utilization does not affect the results of the OTS method application.

The experiment lasted for one semester. As the result of acquisition of the course "Programming Foundations", 38 learner-created topic sequences were obtained. These topic sequences were divided into 11 groups where each group had equal topic sequences. A course average grade was calculated for each obtained topic sequence group. Obtained data are shown below: topic sequence group number in square brackets, followed by topic sequence and calculated average course grade in parenthesis: [1] 1,2,3,4,5,6,7,10,8,9 (6.944); [2] 1,2,3,4,5,6,7,8,9,10 (7.487); [3] 1,2,3,4,5,7,6,8,9,10 (7.917); [4] 1,2,3,4,7,10,5,6,8,9 (8.167), [5] 1,2,3,4,7,5,6,8,9,10 (7.917), [6] 1,2,3,6,4,5,7,9,8,10 (9.375), [7] 1,2,5,3,4,6,7,8,9,10 (7.542), [8] 1,2,5,6,7,8,9,10,3,4 (6.167), [9] 1,3,2,4,5,7,8,10,6,9 (8.292), [10] 1,3,5,7,8,9,10,2,4,6 (8.917), [11] 1,5,6,7,10,2,3,4,9,8 (8.583). The highest grade 9.375 was for the topic sequence {1,2,3,6,4,5,7,9,8,10}. This topic sequence is also taken as optimal topic sequence described in the experiment.

7   Conclusions

This article presents a method for optimal topic sequence (OTS) acquisition that is based on previous learner results obtained during the learning process. The offered optimal topic sequence method gives multiple benefits:

- It gives learner an opportunity to manage his/her own learning process based on his/her wishes and interests;
- It gives an opportunity to obtain data that can be useful for learner course evaluation;
- The offered method can also be implemented in a standard course management system to ensure a dynamic change of the topic sequence;
- The OTS method is easier than one of the approaches described in the related work (Sect. 2). The OTS method takes advantage of the content units (i.e. topics) of any study course. A system should ensure the ability to save learner-used topic sequences and grades received in the course. Then, having analized those topic sequences with the OTS method, a new optimal topic sequence for this course is acquired, which can be further used by succeeding learners for better results;
- The OTS method works with simple data structures such as symbol sequences.

The experiment was performed by allowing students to manage their learning process on their own. For experiment purposes, "Programming Foundations" course with 10 topics was prepared. Students had an opportunity to use the teacher topic sequence or to use a choice option of the learner topic sequence depending on links between course topics. 37 students used topic sequences, and according to them course acquisition results were obtained in the experiment. The optimal topic acquisition sequence for a described experimental course (Sect. 6) was created using the method for optimal topic sequence acquisition presented in Sect. 4. The optimal topic sequence acquired is OTS = {1,2,3,6,4,5,7,9,8,10}.

Future work will be connected with effectivity tests of the obtained OTS, detailed topic sequence choice development, and also a search for a solution to change the OTS in cases when the course structure is being changed.

A lot of study courses have been developed for a certain accreditation period. The target of OTS application was a course with a stable structure developed during several years. Even when changes in a course appear, they mostly affect the content of the topic rather than topics as such. For courses with still unstable structure, two cases must be forseen: (i) essential, and (ii) not essential content changes. In case of essential changes, a teacher-advised sequence can be taken as an OTS. In case of not essential changes, when the basic structure of a course remains the same, it is important to save the obtained OTS. In this case, new topics can be added at the end of OTS in ascending order.

References

1. Brusilovsky, P., Vassileva, J.: Course sequencing techniques for large-scale web-based education. Int. J. Continuing Eng. Educ. Life Long Learn. **13**(1/2), 75–94 (2003)
2. Elouahbi, R., Abghour, N., Bouzidi, D., Nassir, M.A.: A flexible approach to modelling adaptive course sequencing based on graphs implemented using XLink. Int. J. Adv. Comput. Sci. Appl. **3**(2), 7–14 (2012)
3. Fung, A.C.W., Yeung, J.C.F.: An object model for a web-based adaptive educational system. In: 16th World Computer Congress 2000, Proceedings of Conference on Educational Uses of Information and Communication Technologies, pp. 420–426. (2000)

4. Hrastinski, S.: Asynchronous and synchronous e-learning. Educause Q. **31**(4), 51–55 (2008)
5. Huang, S.L., Shiu, J.H.: A User-centric adaptive learning system for e-learning 2.0. J. Edu. Technol. Soc. **15**(3), 214–225 (2012)
6. Jabari, N.A., Hariadi, M., Purnomo, M.H.: Intelligent adaptive presentation and e-testing system based on user modeling and course sequencing in virtual classroom. Int. J. Comput. Appl. **50**(9), 23–31 (2012)
7. Madjarov, I., Betari, A.: Adaptive learning sequencing for course customization: a web service approach. In: Asia-Pacific Services Computing Conference, APSCC 2008, pp. 530–535. IEEE (2008)
8. Vagale, V., Niedrite, L.: Learner model's utilization in the e-learning environments. In: Čaplinskas, A., Dzemyda, G., Lupeikiene, A., Vasilecas, O. (eds.) Local Proceedings 10th International Baltic Conference on Databases and Information Systems, Materials of Doctoral Consortium, pp. 162–174. Žara, Vilnius (2012)
9. Vagale, V.: Eportfolio data utilization in LMS learner model. In: Hammoudi, S., Maciaszek, L., Cordeiro, J., Dietz, J. (eds.) 15th International Conference on Enterprise Information Systems, ICEIS 2013, vol. 2, pp. 489–496. SCITEPRESS, Portugal (2013)
10. Vagale V., Niedrite, L.: Learner classification for providing adaptability of e-learning systems. In: Haav, H.-M., Kalja, A., Robal, T. (eds.) 11th International Baltic Conference, Databases and Information Systems, Baltic DB&IS 2014, pp. 181–192. TUT Press, Tallin (2014)
11. Vagale, V., Niedrite, L.: Learner group creation and utilization in adaptive e-learning systems. In: Haav, H.-M., Kalja, A., Robal, T. (eds.) 11th International Baltic Conference on Databases and Information Systems, Baltic DB&IS 2014. Frontiers in Artificial Intelligence and Applications, vol. 270: Databases and Information Systems VIII, pp. 189–202. IOS Press, Amsterdam (2014)
12. Vassileva, D.: Adaptive e-learning content design and delivery based on learning styles and knowledge level. Serdica J. Comput. **6**(2), 207–252 (2012)

Initial Steps Towards the Development of Formal Method for Evaluation of Concept Map Complexity from the Systems Viewpoint

Janis Grundspenkis[✉]

Riga Technical University, Riga, Latvia
Janis.Grundspenkis@rtu.lv

Abstract. An advantage of intelligent tutoring systems (ITS) is their ability to adapt to the current knowledge level of each learner by offering most suitable tasks for him/her at the current phase of learning process. The main problem is how to evaluate the degree of task difficulty, which, as a rule, is done subjectively. The paper presents the first results of ongoing research, the final goal of which is to develop the formal method for evaluation of the degree of concept map-based task difficulty. The basic idea is to interpret and use for evaluation of concept map (CM) complexity the four aspects applied for estimation of systems complexity – the number of system's elements and relationships between them, attributes of systems, their elements, and relationships, and the organizational degree of systems. The proposed approach is described using as an example relatively simple hierarchical CMs which complexity as systems is estimated.

Keywords: Intelligent tutoring system · Concept map · Degree of task difficulty · Complexity of concept map

1 Introduction

At present practitioners of teaching and learning intensively are developing new approaches and tools for adaptation of educational process to requirements of individual learners (for example, student-centered learning, group learning, problem-oriented learning, etc.) and usage of advanced information and communication technologies (ICT). These developments enable more effective turning of information into knowledge and increase effectiveness of knowledge acquisition. Intelligent tutoring systems (ITS) are one of the most advanced tools for achieving the abovementioned objectives. The advantages of ITSs are at least twofold. First, to the certain extent they achieve the same operation flexibility and adaptivity that can be manifested by human teachers. Second, they provide flexible knowledge self-assessment and assessment, which is a real problem concerning teachers' workload in case of large number of students registered in one course [1]. A concept map (CM) based ITS offers an acceptable balance of workload comparing with objective tests, which can be assessed quickly but

© Springer International Publishing Switzerland 2016
G. Arnicans et al. (Eds.): DB&IS 2016, CCIS 615, pp. 366–380, 2016.
DOI: 10.1007/978-3-319-40180-5_25

provide too simplified knowledge assessment, while evaluation of free-text essays demands processing of natural language [2].

CMs introduced as pedagogical tools by Novak and Gowin [3] nowadays are used both with and without technological support; however, their main advantage is ability to externalize the internal mental structures of learners' knowledge. For instance, Ruiz-Primo claims that well-structured knowledge is an aspect of competence in a particular field [4]. Looking at CMs from the technological point of view, their undoubted advantage is that they can be constructed (generated), visualized, and assessed using ITSs.

Regardless the fact that at the moment there is a wide variety of tools supporting different activities with CMs, the intelligent knowledge assessment system IKAS, developed at the Department of Artificial Intelligence and Systems Engineering of Riga Technical University, has several advantages [2].

Firstly, the IKAS provides assessment of students' CMs in contrast with greatest part of other known systems, which support only such functions as CM construction, navigation, and sharing. Secondly, instead of rather simple scoring systems, the IKAS uses a mathematical model for scoring students' CMs. Thirdly, the IKAS supports the learning process using adaptation mechanisms to students individual characteristics (learning style, preferences, etc.) and their current knowledge level.

The latter is supported by monitoring the learning process and, depending on the conclusion about the current level of knowledge of particular student, the IKAS offers CM-based tasks with different degree of difficulty. The current version of IKAS differentiates between only six classes of tasks, namely, four "fill-in-the-map" tasks when the structure of CM is given and two "construct-the-map" tasks when a learner must construct his/her CM from scratch. The increase of the degree of task difficulty is represented in Fig. 1.

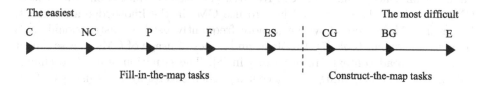

C – a structure of CM is given, all linking phrases and some concepts are already put in correct places

NC – a structure of CM is given, linking phrases are already put in correct places

P – a structure of CM is given, a list of concepts is given, linking phrases are not required

F – a structure of CM is given, lists of concepts and linking phrases are given

ES – only an empty structure of CM is given (not used in the IKAS)

CG – a list of concepts is given, linking phrases are not required

BG – lists of concepts and linking phrases are given

E – learners must contruct a CM from scratch (not used in the IKAS)

Fig. 1. The increase of the degree of task difficulty

Such rough and imperfect division of all CM-based tasks in accordance with the degree of their difficulty is one of the drawbacks of IKAS and serves as motivation to start the research, the final goal of which is the development of formal method for evaluation of CM complexity.

The paper is organized as follows. Section 2 is devoted to the short description of research directions in the field of CMs. In Sect. 3, the used terminology of CMs and corresponding graphs is introduced. Section 4 is devoted to the proposed framework for estimation of complexity of CMs. Conclusions, open questions, and some directions of future work are given at the end of the paper.

2   Related Work on Concept Mapping

During the three decades since J.D. Novak and D.B. Gowin published their book "Learning how to learn" [3], CMs have become valuable tools for teaching, assessment, and learning [5]. A lot of research on CMs has been done, methods and tools developed, and various experiments carried out. The results of this work may be found in numerous publications, including proceedings of biyearly organized international conferences on concept mapping. Back in 2008, a group of authors published a review on CM research based on materials of the first two conferences on concept mapping [6]. The analysis of presented papers clearly confirms that the mainstream of research on concept mapping is done in applications of CMs as teaching and learning tools. Besides, innovative ways of using CMs also were studied. The use of CMs to organize instructions and to gather student feedback on learning, research on teaching and learning related to a variety of content areas, study of different groups of learners, collaborative concept mapping, and curriculum planning and design are only few examples of works in abovementioned direction. Scoring and assessment is the second largest research direction on concept mapping. CM research has been conducted on how to score CMs of individual learners and how to use CMs in the knowledge assessment. However, CMs till now have been more frequently used as instructional tools than as assessment tools [7]. Problems and issues of usage of CMs as assessment tools in a broad context are described in [8]. The situation started to change only at the very beginning of 21st century when several systems using CMs as assessment tools appeared (see [9–11]), including ITSs [12,13], which provide automated assessment, thus reducing the workload of teachers.

It is worth to point that CMs are computable in the sense that the assessment of learners' answers given in the form of CMs is based on some scoring system [8]. In general, the assessment using CMs is based on comparison of teacher's (expert's) and learners' CMs. It is a serious drawback because such mere comparison does not conform to cognitive principles by forcing learners to construct their knowledge in a way that mimics the knowledge constructed by the teacher [11]. Another drawback is caused by the fact that there is a wide variety of CMs and CM-based tasks. This causes the variety of possible scoring criteria and methods and, as a consequence, difficulties in choosing the most appropriate one. The detailed analysis of this issue is given in [7].

The survey of large number of papers devoted to concept mapping shows that only the developers of IKAS have focused on the problem of CM-based task difficulty and estimation of its degree, while the issue of possibility of formal evaluation of the degree of task difficulty and its interconnectedness with CM complexity has not been studied at all.

3 Corresponding Terminology of Concept Maps and Graphs

CMs are used to support different activities where knowledge needs to be organized and represented, predominantly in educational settings [14]. CMs are semi-formal tools for representing semantic knowledge and its conceptual organization (structure). They specify the concepts of a knowledge domain and relationships among them. Mathematically defined, a CM is undirected or directed graph consisting of a finite, non-empty set of nodes $V = \{v_1, \ldots, v_n\}$, which represent concepts of a knowledge domain, and a finite, non-empty set of arcs (undirected or directed) $Q = \{q_1, \ldots, q_m\}$, which represent the relationships between pairs of concepts. A CM may be represented by an attribute graph. In this case, the set of arcs contains attributes (labels) or so called *linking phrases* used to specify the relationships between concepts [11]. Besides, the corresponding graph may be homogeneous, if all its arcs have the same weight, or heterogeneous, if weights of arcs are different. The latter represents a teacher's (expert's) point of view that some relationships are more important than others.

The 3-tuple $\langle$concept, relationship, concept$\rangle$ is called a *proposition*, which is a semantic unit of CM or, in other words, a representation of declarative knowledge that is a meaningful statement about some object or event in the problem domain [15]. In terms of graphs, a proposition corresponds to a 3-tuple $\langle v_i, q_r, v_j \rangle$.

Table 1 summarizes the correspondence between basic elements of CMs and graphs. The variety of CMs represented by graphs is shown in Fig. 2.

Table 1. Correspondence between basic elements of CMs and graphs

Concept map	Graph
A concept	A node
A relationship	An arc (undirected or directed)
A CM with linking phrases	An attribute graph
The importance of a relationship	The weight of an arc
A proposition	A 3-tuple $\langle v_i, q_r, v_j \rangle$

Consequently, CMs are viable, computable, and theoretically sound solution to the problem of expressing and assessing learners' current knowledge level. The undoubted advantage of CMs is their ability to represent part of a learner's cognitive structure, revealing his/her particular understanding of a specific knowledge

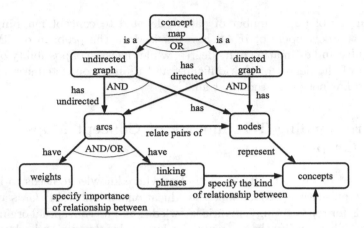

Fig. 2. Variety of concept maps (Source: [16])

area. The representation of knowledge structure is the topmost quality of CMs, which justifies their usage as alternative knowledge assessment tools concurrently with different tests and full-text responses.

A wide variety of CM representations as graphs entails an enormous variety of CM tasks, which makes CM-based knowledge assessment adaptable to learners' knowledge levels and preferences. According to [8], tasks vary with regard to task demands, task constraints, and the content structure of tasks. Task demands are related to two commonly used classes: fill-in-the-map and construct-the-map tasks, while task constraints refer to constraints that are defined for the concept set and/or linking phrase set. Definitely, both abovementioned factors allow to a certain extent to adapt CM-based tasks to learners' knowledge levels, learning style, and preferences. At the same time, such, let's call it, meta-adaptation is not effectively applicable if the degree of task difficulty is not at least estimated. This is the reason more sophisticated methods for evaluation of the degree of task difficulty should be needed for constructing flexible and adaptable ITSs.

The proposal is to correlate the degree of CM-based task difficulty with the complexity of the CM as a whole, that is, as a system. Following this point of view, focus is on two types of graphs which represent CMs, namely, hierarchical CMs and network CMs [8]. Both result from the roots of CMs, namely, Ausubel's assimilation theory and Deese's associationist memory theory, respectively. Based on Ausubel's theory, Novak and Gowin argued that CMs should be: (1) hierarchical with superordinate concepts at the top level; (2) labeled with appropriate linking phrases; (3) crosslinked such that relations between subbranches of the hierarchy are identified [8]. Their arguments are justified by expansion of hierarchy according to the principle of progressive differentiation: new concepts and new relationships are added to the hierarchy either by creating new branches or by further differentiating existing ones. The crosslinks are links between segments of the CM and represent the integrative connection among different domains of structure. Fig. 3 presents a hierarchical CM.

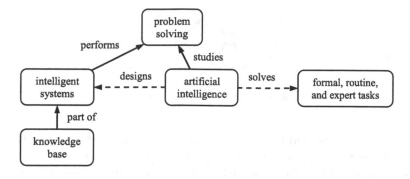

Fig. 3. A fragment of a hierarchical CM (dashed arcs represent crosslinks)

A network CM characterizes cognitive structure as a set of concepts and their interrelations and, in fact, is a semantic network with concept nodes linked by directed arcs with labels to produce propositions [17]. Associationist memory theory places similar requirements on concept mapping with the exception that CMs are not hierarchical. An example of a network CM is presented in Fig. 4.

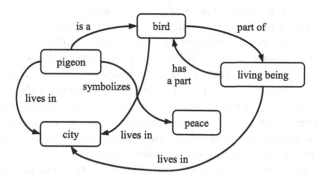

Fig. 4. A fragment of a network CM

It is necessary to stress that in graph theory, the corresponding types of graphs are trees (usually representing hierarchies) and networks [18]. In this paper, as a starting point and a test-bed for the development of formal method for evaluation of complexity of CMs, only one subtype of trees is chosen, namely, an incoming tree, an example of which is presented in Fig. 5.

Such tree has the simplest structure and corresponds to all definitions given in graph theory for undirected trees: (1) the number of directed arcs $m = n - 1$, where n is the number of nodes; (2) all paths starting in apex nodes always end in the root node (corresponding to the superordinate concept); (3) the underlying undirected graph (directions of links ignored) has no cycles. Besides, the chosen type of incoming tree represents so called proper hierarchy where arcs of a node at l-th level are only between $(l \pm 1)$-th levels, and nodes at the same level are not adjacent (a proper hierarchy has no crosslinks).

Fig. 5. Example of incoming tree

The motivation behind the decision to choose the simplest possible structure of graph for representing a system is straightforward – if the proposed approach will be applicable then more complex systems' structures should be introduced and studied in the future.

4   The Framework for Evaluation of Complexity of Concept Maps

The central idea of the approach is based on interpretation of CMs as systems (as a whole) and application of criteria used in Systems Theory for estimation of complexity of a system to CMs. In Systems Theory, as a rule, there are used two quantitative parameters – the number of system's elements and the number of implemented relationships. Logically, it is declared that simple systems have a small number of elements and relationships, while complex systems consist of a large number of elements and relationships [19]. These parameters are relative and only shallowly evaluate the complexity of systems. Some improvements are known which suggest using expert's knowledge who evaluates the complexity of each element and then summing up these evaluations to get a conjunctive parameter of complexity. An awkward attempt to ask experts to evaluate complexity of relationships comparing them with complexity of elements is also proposed despite the fact that such approach is useless in practice. Authors in [20] have shown that complexity also depends on other aspects, such as the knowledge about organization of system and attributes of its specific elements, which may substantially change the evaluation of system's complexity, so that a very complex system at first sight, in fact, is simple for an expert.

Taking abovementioned into account, the following criteria are proposed [20]:

- The number of elements
- The number of relationships
- The attributes of specific elements of the system
- The organizational degree of the system

The one-to-one correspondence between these criteria, which are supplemented with introduced additional ones, namely, attributes of the system and attributes of a relationship, in case of systems and in case of CMs is defined in Table 2.

As it was already mentioned, this paper is focused only on graphs with a specific topology – incoming trees, including chains as the trivial case. All aspects

Table 2. Correspondence of complexity criteria

No	System	Concept map
1	The number of elements	The number of concepts
2	The number of relationships	The number of arcs
3	Attributes of the system	Linking phrases (their number and variety of categories and/or the number of synonyms of concepts)
4	Attributes of an element	The structural importance of a concept
5	Attributes of a relationship	The weight of an arc
6	The organizational degree of the system	The topological features of the corresponding graph

and solutions will be shown using incoming trees $T(V, Q)$ with $|V| = 5$ and $|Q| = 4$, which have been chosen as a trade-off between trivial cases ($|V| = 2, 3, 4$) and more complicated ones ($|V| = 6, 7, \ldots$), which have much more categories of different structures. Table 3 represents all categories of different structures (topologies) of such trees, including T_1 – a chain, and T_2 – a bipartite graph.

It is obvious that similarly with the general case, the first two criteria help nothing because all trees shown in Table 3 have the same complexity. As a consequence, one can get a more complex CM by adding new concepts and increasing in such a way the number of concepts and arcs.

The situation changes if the third criterion, namely, system's attributes is introduced, despite that at first sight nothing changes because the number of arcs (relationships) and linking phrases is the same. Interpretation of system's attributes in case of CMs is tightly connected with semantics of concepts and linking phrases. For example, if CMs can be constructed with free vocabulary, different learners can use different words or linking phrases for the same concept and arc, respectively. The CM is more complex comparing with a practically identical CM with the only difference being that all concepts and linking phrases are predefined unambiguously. This conclusion refers to the both cases – construction of a CM and its assessment by comparison with an expert's CM. The latter task leads to the graph matching problem [21].

Thus, speaking about concepts, the conclusion is that the complexity of a particular CM increases if the number of synonyms is growing. Linking phrases also are expressed in a natural language which is not unambiguous. If linking phrases are not given to the CM creator then he/she may use any expression that seems appropriate according to his/her understanding of how concepts are related in a particular domain. Moreover, the semantics of relationships of the same two concepts can vary depending on the context in which they are used [22], as well as there can even be cases when it is meaningful to represent more than one relationship between two concepts [22,23]. Such situation is not inspiring

Table 3. Categories of topologies of incoming trees ($|V| = 5$, $|Q| = 4$)

Symbol	Visualization of the tree

because the variance of linking phrases theoretically is indefinite. For example, research completed by Strautmane [24] shows that for inheritance relationship alone, there are more than 50 ways how to label it. That is the reason why researchers of semantic networks and CMs have defined typical linking phrases, such as "is a", "part of", "kind of", "is an example of", "has value", "characterizes", etc. along with others. In any case, the variety of linking phrases of CMs exists, and it is possible to consider how to use this variety.

The working hypothesis is that the complexity of CM (more precisely, the complexity of CM-based task) increases if the variety of linking phrases (the variety of semantics) increases and vice versa.

As the first approximation, for incoming trees, only six categories of linking phrases will be taken into account, namely, "is a", "is part of", "is a kind of", "is an attribute of", "is a value of", and linking phrases with unrestricted semantics, like, "made of", "has property", "is opposite of", etc. It is obvious that the number of arcs may restrict the maximum number of linking phrase categories which may be used. For incoming trees discussed in this paper, the maximum number of linking phrase categories is four out of six because $|Q| = 4$.

To evaluate the complexity of CM using the third criterion, the results are as follows:

- If only one category of linking phrases is used, the complexity of the CM is minimal (if only the third criterion is taken into consideration).
- If the number of used linking phrase categories is equal to the number of relationships then the complexity of the CM is maximal (if only the third criterion is taken into consideration).
- Intermediate values of complexity depend on the chosen scale, development of which is one of the future tasks.

For example, in Fig. 6, the CM (a) has the minimal complexity, while the CM (b) has the maximal complexity.

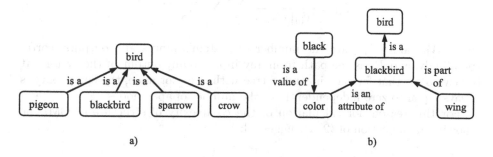

a)   b)

Fig. 6. Examples of CMs with different complexity

First approximation in terms of quantity can use the following formula:

$$C = N \cdot m \,, \tag{1}$$

where C is the complexity of CM if only the third criterion is taken into account, N is the number of categories of linking phrases, and m is the number of arcs. Thus, CMs in Fig. 6 have the following complexity: (a) 4 and (b) 16.

As it was already pointed out above, in this paper, only the first steps towards the development of complete method for evaluation of complexity of CMs are worked out. There are many open questions for future research. For example, is the complexity different if both CMs have the same two categories of linking phrases? Suppose that the first CM has only one linking phrase of one category (for example, "is a"), but the other three linking phrases are from another category (for example, "part of"), while the second CM has two linking phrases from each of these categories. If such open questions are ignored at the moment, it is clear that the full set of CMs with the same number of concepts and relationships are divided into subsets in accordance with the number of used categories of linking phrases.

Now, let's consider how to evaluate the complexity of CMs according to the topological features of the corresponding graphs. In this paper, it is proposed to use one criterion which is borrowed from scoring systems used for CMs – the number of valid levels of hierarchy N_H, which shows where on the general–specific continuum each concept lays in respect to the domain being represented. The number of levels of hierarchy is related to the extent to which the learner subsumes more specific knowledge under more general knowledge [25]. In graph theory, the number of hierarchy levels is equal to the diameter of the tree. Such criteria as the complexity of structure, the relative weight of each hierarchy level, and the degree of centralization of structure are borrowed from [26]. Other graph theory criteria which already are used in structural modeling [27] will be applied in future work.

First, it is worth to stress that the following parameter for evaluation of complexity of systems, which is based on the consideration of complexity of structural analysis [26], is not applicable:

$$C_S = \frac{1}{|V_{in}| \cdot |V_{out}|} \sum_{i=1}^{V_{in}} \sum_{j=1}^{V_{out}} P_{ij} - 1 \, , \tag{2}$$

where $|V_{in}|$ and $|V_{out}|$ are the number of system's inputs and outputs, correspondingly, and P_{ij} is the path from any input to any output of the system. It is easy to see that for each incoming tree with one root node, the complexity is always equal to zero independently of the number of apex nodes.

For this reason, for calculation of the complexity of structure of CM, the following modification of (2) is suggested:

$$C_S = \frac{1}{|V_{apex}|} \sum_{i=1}^{V_{root}} P_{i,root}^W - 1 \, , \tag{3}$$

where $|V_{apex}|$ is the number of apex nodes in the incoming tree, V_{root} is the root node, and $P_{i,root}^W$ is the weighted path from any apex node to the root. The $P_{i,root}^W$ is found as follows:

$$P_{i,\text{root}}^{\text{W}} = d_{i,\text{root}} + \sum_{j=1}^{S_i} d_{j,\text{root}} , \tag{4}$$

where $d_{i,\text{root}}$ is the distance from the apex node to the root, S_i is the number of descendants of apex node i, and $\sum_{j=1}^{S_i} d_{j,\text{root}}$ is the sum of distances from all descendants of apex node i to the root. For example, the complexity of structure of the incoming tree T_5 (see Table 3) is

$$C_{\text{S}}^{T_5} = \frac{1}{2} \cdot (1 + (3 + 2 + 1)) - 1 = 2.5 .$$

The T_5 has four valid hierarchy levels (the root node is always placed at 0-level), and the relative weights of hierarchy levels are the following: 0-level – 0.2, 1st level – 0.4, 2nd level – 0.2, 3rd level – 0.2.

The degree of centralization of structure is calculated as follows [26]:

$$D_{\text{C}} = \frac{1}{(n-1)(\rho_{\max}-1)} \sum_{i=1}^{n} (\rho_{\max} - \rho_\Sigma(V_i)) , \tag{5}$$

where n is the number of nodes, $\rho_\Sigma(V_i) = \rho^+(V_i) + \rho^-(V_i)$, where $\rho^+(V_i)$ and $\rho^-(V_i)$ denote outdegree and indegree [18] of the node V_i, and $\rho_{\max}$ is the maximum value of ρ_Σ for the given structure. For example, the degree of centralization of T_5 is:

$$D_{\text{C}}^{T_5} = \frac{1}{(5-1)(2-1)}((2-2)+(2-1)+(2-2)+(2-2)+(2-1)) = \frac{2}{4} = 0.5 .$$

The structural modeling approach [27] offers also other parameters of evaluation of topological characteristics of structure, for instance, redundancy of arcs (not applicable for trees, which have minimum number of arcs), compactness of structure, structural importance of node, and the dispersion of ranks of nodes, which are useful for network CMs and are not discussed in this paper.

For comparison of values of chosen parameters, these values for the example – incoming tree $T(V, Q)$ with $|V| = 5$ and $|Q| = 4$ are collected in Table 4. A shallow analysis of parameter values in Table 4 shows that the least complexity of structure has the bipartite graph T_2. Taking into account that this graph also has the least number of hierarchy levels, it is justifiable to conclude that the corresponding CM has the least complexity comparing with other CMs whose corresponding graphs belong to the same class of incoming trees. In addition, if all linking phrases belong to only one category then one gets the easiest CM-based task where 5 concepts must be related with 4 relationships. At first, it may look strange that T_1 (the chain), which is the simplest structure of graphs, in case of CMs has the maximum value of C_S in this class of incoming trees. But it precisely corresponds to the statement given above that the number of levels of hierarchy is related to the ability of learner to subsume more specific knowledge under more general knowledge.

Due to the limited scope of the paper, more detailed analysis of the results is not presented.

Table 4. Parameter values of the example

Category	N_H	The relative weight of the level	D_C	C_S
T_1	5	0.2 each	0.2	9
T_2	2	0.2 – 0-level	1	0
		0.8 – 1st level		
T_3	3	0.2 – 0-level	0.875	0.67
		0.6 – 1st level		
		0.2 – 2nd level		
T_4	3	0.2 – 0-level	0.5	2
		0.4 – 1st level		
		0.4 – 2nd level		
T_5	4	0.2 – 0-level	0.5	2.5
		0.4 – 1st level		
		0.2 – 2nd level		
		0.2 – 3rd level		
T_6	3	0.2 – 0-level	0.875	1.33
		0.4 – 1st level		
		0.4 – 2nd level		
T_7	3	0.2 – 0-level	1	2
		0.2 – 1st level		
		0.6 – 2nd level		
T_8	4	0.2 – 0-level	0.875	3.5
		0.2 – 1st level		
		0.4 – 2nd level		
		0.2 – 3rd level		
T_9	4	0.2 – 0-level	0.875	5
		0.2 – 1st level		
		0.2 – 2nd level		
		0.4 – 3rd level		

5   Conclusions

The paper presents an attempt to interpret CMs as systems and to apply the criteria used in Systems Theory for estimation of systems' complexity. As it represents a very early phase of research, the paper contains fewer results, while putting forward a lot of open questions. For example:

- How to use the results of topological feature analysis for evaluation of organizational degree of a CM?
- Moreover, how to aggregate the values of each criterion in a joint parameter which characterizes the complexity of a CM?

- How and why changes of values of such parameters as D_C and C_S within different classes of incoming trees or, in more general cases – in different hierarchies and networks, take place?
- If the expression for evaluation of CM complexity will be found, how will it correlate with the degree of CM-based task difficulty?
- How the degree of CM-based task difficulty found from CM complexity should be mapped on the scale for learners' knowledge assessment?

The listed are only few of open questions which clearly manifest that a lot of work should be done in future before answers will be found. At the same time, the first results are encouraging enough for motivation of future work and reaching the final goal – the development of formal method for evaluation of CM complexity from the systems viewpoint.

Acknowledgments. This work was supported by the Latvian National research program SOPHIS under grant agreement Nr. 10-4/VPP-4/11.

References

1. Grundspenkis, J.: Concept map based intelligent knowledge assessment system: experience of development and practical use. In: Ifenthaler, D., Spector, M.J., Kinshuk, Isaias, P., Sampson, D.G. (eds.) Multiple Perspectives on Problem Solving and Learning in the Digital Age, pp. 179–198. Springer, New York (2011)
2. Anohina-Naumeca, A., Grundspenkis, J., Strautmane, M.: The concept map based assessment system: functional capabilities, evolution and experimental results. Int. J. Continuing Eng. Edu. Life-Long Learn. **21**(4), 308–327 (2011)
3. Novak, J.D., Gowin, D.B.: Learning How to Learn. Cornell University Press, New York (1984)
4. Ruiz-Primo, M.A.: On the use of concept maps as an assessment tool in science: what we have learned so far. Rev. Electrónica de Investigación Educativa **2**(1), 29–52 (2000)
5. Moon, B.M., Hoffman, R.R., Novak, J.D., Cañas, A.J.: Applied Concept Mapping. CRC Press, Boca Raton (2011)
6. Daley, B.J., Conceição, S., Mina, L., Altman, B.A., Baldor, M., Brown, J.: Advancing concept map research: a review of 2004 and 2006 CMC research. In: 3rd International Conference on Concept Mapping, pp. 159–166 (2008)
7. Strautmane, M., Grundspenkis, J.: Determination of the set of concept map scoring criteria. In: International Conference on E-Learning and the Knowledge Society, pp. 137–142. ASE Publishing House, Bucharest (2011)
8. Ruiz-Primo, M.A., Shavelson, R.J.: Problems and issues in the use of concept maps in science assessment. J. Res. Sci. Teach. **33**(6), 569–600 (1996)
9. Cimolino, L., Kay, J., Miller, A.: Incremental student modelling and reflection by verified concept-mapping. In: 11th International Conference on Artificial Intelligence in Education, Supplementary Proceedings, pp. 219–227 (2003)
10. Gouli, E., Gogoulou, A., Papanikolaou, K., Grigoriadou, M.: COMPASS: an adaptive web-based concept map assessment tool. In: 1st International Conference on Concept Mapping, pp. 128–135 (2004)

11. da Rocha, F.E.L., da Costa, J.V.Jr., Favero, E.L.: An approach to computer-aided learning assessment. In: 3rd International Conference on Concept Mapping, pp. 100–107 (2008)
12. Anohina, A., Grundspenkis, J.: Prototype of multiagent knowledge assessment system for support of process oriented learning. In: 7th International Baltic Conference on Databases and Information Systems, pp. 211–219. IEEE, Piscataway (2006)
13. Grundspenkis, J.: Usage experience and student feedback driven extension of functionality of concept map based intelligent knowledge assessment system. Commun. Cogn. **43**(1–2), 13–32 (2010)
14. Gava, T.B.S., Menezes, C.S., Cury, D.: Applying concept maps in education as a metacognitive tool. In: 3rd International Conference on Engineering and Computer Education, pp. 127–135 (2003)
15. Cañas, A.J.: A Summary of literature pertaining to the use of concept mapping techniques and technologies for education and performance support. Technical report, The Institute for Human and Machine Cognition (2003)
16. Grundspenkis, J.: Concept maps as knowledge assessment tool: results of practical use of intelligent knowledge assessment system. In: IADIS International Conference on Cognition and Exploratory Learning in Digital Age 2009, pp. 258–266 (2009)
17. Luger, G.F.: Artificial Intelligence: Structures and Strategies for Complex Problem Solving. Pearson Education, Harlow (2005)
18. Agnarsson, G., Greenlaw, R.: Graph Theory: Modeling, Applications, and Algorithms. Pearson Education, Upper Saddle River (2007)
19. Skyttner, L.: General Systems Theory: Problems, Perspectives, Practice. World Scientific, Singapore (2005)
20. Schoderbek, P.P., Schoderbek, C.G., Kefalas, A.G.: Management Systems: Conceptual Considerations. Richard D. Irwin, Burr Ridge (1990)
21. Souza, F.S.L., Boeres, M.C.S., Cury, D., Menezes, C.S., Carlesso, G.: An approach to comparison of concept maps represented by graphs. In: 3rd International Conference on Concept Mapping, pp. 205–212 (2008)
22. Jonassen, D.H.: Computers in the Classroom: Mindtools for Critical Thinking. Prentice Hall, Upper Saddle River (1996)
23. Shavelson, R.J., Lang, H., Lewin, B.: On concept maps as potential "authentic" assessments in science. Technical report, National Center for Research on Evaluation, Standards, and Student Testing (1994)
24. Strautmane, M.: Usage of semantics of links as the basis for learner's knowledge structure assessment: the pros and cons. In: Joint International Conference on Engineering Education and International Conference on Information Technology, pp. 294–303 (2014)
25. Borda, E.J., Burgess, D.J., Plog, C.J., DeKalb, N.C., Luce, M.M.: Concept maps as tools for assessing students' epistemologies of science. Electron. J. Sci. Edu. **13**(2), 160–185 (2009)
26. Nikolayev, V.I., Bruk, V.M.: Sistemotehnika: Metodi i Prilozeniya (in Russian). Masinostroyeniye, Leningrad (1985)
27. Grundspenkis, J.: Reasoning supported by structural modelling. In: Wang, K., Pranevicius, H. (eds.) Lecture Notes of the Nordic-Baltic Summer School "Intelligent Design, Intelligent Manufacturing and Intelligent Management", pp. 57–100. Kaunas University of Technology, Kaunas (1999)

Author Index

Printed in the United States
By Bookmasters